NATO ASI Series

Advanced Science Institutes Series

A series presenting the results of activities sponsored by the NATO Science Committee, which aims at the dissemination of advanced scientific and technological knowledge, with a view to strengthening links between scientific communities.

The Series is published by an international board of publishers in conjunction with the NATO Scientific Affairs Division.

A Life Sciences B Physics	Plenum Publishing Corporation London and NewYork
C Mathematical and Physical Sciences D Behavioural and Social Sciences E Applied Sciences	Kluwer Academic Publishers Dordrecht, Boston and London
F Computer and Systems Sciences G Ecological Sciences H Cell Biology I Global Environmental Change	Springer-Verlag Berlin Heidelberg NewYork Barcelona Budapest Hong Kong London Milan Paris Santa Clara Singapore Tokyo

Partnership Sub-Series

1. Disarmament Technologies	Kluwer Academic Publishers
2. Environment	Springer-Verlag / Kluwer Academic Publishers
3. High Technology	Kluwer Academic Publishers
4. Science and Technology Policy	Kluwer Academic Publishers
5. Computer Networking	Kluwer Academic Publishers

The Partnership Sub-Series incorporates activities undertaken in collaboration with NATO's Cooperation Partners, the countries of the CIS and Central and Eastern Europe, in Priority Areas of concern to those countries.

NATO-PCO Database

The electronic index to the NATO ASI Series provides full bibliographical references (with keywords and/or abstracts) to about 50 000 contributions from international scientists published in all sections of the NATO ASI Series. Access to the NATO-PCO Database compiled by the NATO Publication Coordination Office is possible in two ways:

- via online FILE 128 (NATO-PCO DATABASE) hosted by ESRIN, Via Galileo Galilei, I-00044 Frascati, Italy.
- via CD-ROM "NATO Science & Technology Disk" with user-friendly retrieval software in English, French and German (© WTV GmbH and DATAWARE Technologies Inc. 1992).

The CD-ROM can be ordered through any member of the Board of Publishers or through NATO-PCO, B-3090 Overijse, Belgium.

Series F: Computer and Systems Sciences, Vol. 155

Springer
Berlin
Heidelberg
New York
Barcelona
Budapest
Hong Kong
London
Milan
Paris
Santa Clara
Singapore
Tokyo

Cooperation: Game-Theoretic Approaches

Edited by

Sergiu Hart

Department of Economics and Department of Mathematics
The Hebrew University of Jerusalem
Givat Ram, 91904 Jerusalem, Israel

Andreu Mas-Colell

Facultat de Ciències Econòmiques i Empresarials
Universitat Pompeu Fabra
c/ Ramon Trias Fargas, 25–27, E-08005 Barcelona, Spain

Springer
Published in cooperation with NATO Scientific Affairs Division

Proceedings of the NATO Advanced Study Institute on Cooperation: Game-Theoretic Approaches, held at SUNY, Stony Brook, New York, July 18–29, 1994

Library of Congress Cataloging-in-Publication Data

Cooperation : game-theoretic approaches / edited by Sergiu Hart, Andreu Mas-Colell.
p. cm. -- (NATO ASI series. Series F, Computer and systems sciences ; no. 155)
"Proceedings of the Nato Advanced Study Institute on Cooperation: Game-Theoretic Approaches, held at SUNY, Stony Brook, New York, July 18-29, 1994"--T.p. verso.
Includes bibliographical references.
Additional material to this book can be downloaded from http://extra.springer.com.

ISBN-13:978-3-642-64413-9 e-ISBN-13:978-3-642-60454-6
DOI: 10.1007/978-3-642-60454-6

1. Game theory--Congresses. I. Hart, Sergiu. II. Mas-Colell, Andreu. III. Nato Advanced Study Institute on Cooperation: Game -Theoretic Approaches (1994 : Stony Brook, N.Y.) IV. Series: NATO ASI series. Series F, Computer and systems sciences ; no. 155.
QA269.C66 1997
519.3--DC20 96-27102
CIP

CR Subject Classification (1991): H.1.1, J.4

ISBN-13:978-3-642-64413-9

Softcover reprint of the hardcover 1st edition 1997

Typesetting: Camera-ready by editor
Printed on acid-free paper
SPIN: 10537423 45/3142 - 5 4 3 2 1 0

Preface

This volume includes the proceedings of the NATO Advance Study Institute on "Cooperation: Game Theoretic Approaches", that took place at Stony Brook, NY, USA, from July 18 to July 29, 1994.

The Institute was a success and it is already part of a well established biannual Stony Brook tradition to which many researchers around the world look forward to.

We thank the institute for Decision Sciences of the State University of New York at Stony Brook for hosting this event. It is a particular pleasure to thank Colleen Wallahora and Eileen Zapia, for the very successful organization of this ASI.

June 1996 Sergiu Hart and Andreu Mas-Colell

List of Authors

Prof. Beth Allen, University of Minnesota, Minneapolis

Prof. Robert J. Aumann, The Hebrew University of Jerusalem

Prof. Joseph Greenberg, McGill University, Montreal

Prof. Sergiu Hart, The Hebrew University of Jerusalem

Prof. Vijay Krishna, Pennsylvania State University at University Park

Prof. Andreu Mas-Colell, Harvard University,
and Universitat Pompeu Fabra, Barcelona

Prof. Abraham Neyman, The Hebrew University of Jerusalem,
and State University of New York at Stony Brook

Prof. Philip J. Reny, University of Pittsburgh

Prof. Reinhard Selten, Universität Bonn

Prof. Tomas Sjöström, Harvard University

Prof. Sylvain Sorin, Université de Paris X, Nanterre,
and École Normale Supérieure, Paris

Prof. William Thomson, University of Rochester

Prof. Rajiv Vohra, Brown University, Providence

Table of Contents

Introduction

Sergiu Hart [a] **and Andreu Mas-Colell** [b]

[a] The Hebrew University of Jerusalem
[b] Harvard University and Universitat Pompeu Fabra, Barcelona

This book constitutes a systematic exposition of the various game theoretic approaches to the issue of cooperation.

Game Theory is the study of decision making in multi-person situations, where the outcome depends on everyone's choice. The goal of each participant is to maximize his own utility, while taking into account that the other participants are doing the same. In such interactive situations, *cooperation* between the agents may lead to results that are better, for everyone, that the non-cooperative outcomes. A simple - but extensively studied - example is the so-called "Prisoners' Dilemma": Assume each on of the two players can ask a generous donor either to give him 1 million dollars, or to give 4 million dollars to the other player, the donor will carry out the instructions of both players (thus, for example, if player 1 asks for $1M to himself and player 2 asks for $4M to the other, then player 1 gets $5M and payer 2 gets nothing). Clearly, whatever the other player does, it is strictly better for each player to ask for $1M to himself (more precisely, it will always lead to an additional " 1M). This yields $1M for each; cooperation, whereby each one asks for $4M to the other, would have yielded each $4M instead! The Prisoner's Dilemma is by no means an artificial example. The economic competition between firms exhibits similar phenomena: keeping a commodity in short supply may be to the advantage of all producers; at the same time, it may be better for any single producer to unilaterally increase his own production.

The problems that need to be addressed are, first, whether cooperation can be reached at all; second, by what procedures are agreements reached; and third, which ones will be indeed attained. This volume will survey some of the contributions of game theory to these questions, from its early traditional theories to its current approaches.

Game theoretical approaches are usually classified as either "cooperative" or "non-cooperative". This should not be viewed as an exclusive division; these are two ways of looking at the same problem. The Introductory Remarks of R. J. Aumann that follow this Introduction address this point in some detail.

Part A, which opens this volume, surveys the classical cooperative approach. This starts by assuming that binding agreements are possible, and it abstracts away from the detailed bargaining procedures. The selection of the appropriate cooperative outcome is usually based on a set of desired postulates or axioms, which, when applied to a class of problems, characterize one or another solution concept. Chapters 1 and 2 by W. Thomson cover the pure bargaining problems, where only the grand coalition of all players can reach a beneficial agreement; Chapter 1 deals with the classical approaches that originate with Nash's 1950 seminal paper, and Chapter 2 deals with recent axiomatizations based on internal consistency properties (the " reduced game property"). Chapters 3 and 4, by S. Hart, survey the general n-person problems where subcoalitions of players can reach agreements as well, and this of course influences the final outcome. The classical cooperative solution concepts that arise are grouped into " core-like" notions and "value-like" notions. The former include the core, the stable sets of von Neumann and Morgenstern, the bargaining set, the kernel and the nucleolus; the latter include the Nash bargaining solution, the Shapley value and their many extensions and generalizations. Chapter 5 by B. Allen deals with games of incomplete information, i.e. , games where some of the participants may possess private information not known to the others. Here, the questions of cooperation are further complicated by the need to address the informational issues; how to ensure that the players have incentive to reveal the appropriate information.

Part B is devoted to non-cooperative approaches, namely, non-cooperative models that lead to cooperative solutions. One may start from a non-cooperative bargaining model, like the Ståhl-Rubinstein " alternating offers" procedure, characterize its strategic equilibria, and relate the resulting outcomes to various cooperative solutions. Or, one may start from a cooperative solution, and construct games whose equilibria yield precisely this given solution. Either way, one establishes connections between non-cooperative and cooperative setups, that further strengthen and reinforce one another. In the literature, all this is usually referred as " bargaining procedures", " non-cooperative foundations", or " implementation" . The distinctions are not always clear , in particular since some of the recent implementation literature is concerned with " natural" and " simple" games. Chapter 6 by A. Mas-Colell covers bargaining procedures that lead to value-like cooperative solutions, and the second part of Chapter 7 by P. Reny and Chapter 8 by B. Allen, for the case of complete information and incomplete information, respectively. Chapter 9 by R. Vohra discusses coalitional non-cooperative approaches -i.e., models where not only individuals, but also coalitions may act strategically. Chapter 10 by J. Greenberg surveys the theory of " social situations" , which looks for a stable standards of behavior in general coalitional interactions.

Part C deals with dynamic models, that is models of long-term interactions between the participants. Returning, for example, to the Prisoners' Dilemma, it seems clear that if the same participants play it again and again , then cooperation may indeed be attained. However, this is by no means always so; for instance, in a fixed finite-horizon repetition, it is very difficult to escape the non-cooperative outcome of $1M each. There is by now a large and deep literature on "repeated games" -starting with the so-called "Folk Theorem" - that shows the extent to which cooperation may arise. The complete information case is covered in Chapter 11 by S. Sorin[1]. Chapter 12, also by S. Sorin, then goes on to survey models of communication; namely, one examines the effect of the players being able to communicate among themselves before the game is played, and also, in the case of a multi-stage game, during the play. This leads to correlation and cooperation. Another important issue in multi-stage interactions is that they require, by their very nature, extremely complex strategic considerations. This suggests considering models where the assumption that players are fully rational - i.e., that they are restricted in one way or another in their choices. Chapter 13 by R.J. Aumann discusses some of the underlying ideas and approaches of this kind. The case where strategies are implemented by automata of bounded complexity is then studied in Chapter 14 by A. Neyman. Chapter 15 by V. Krishna and T. Sjöström is devoted to a simple but interesting learning model, known as the "fictitious play": players assume that the past behavior of their opponents is, in a certain sense, and appropriate predictor of their future behavior. Chapter 16, also by V. Krishna and T. Sjöström, studies another type of bounded rationality models: the "evolutionary models ". These are based on the biological paradigm of natural selection and evolution, where there is no conscious optimization at all; instead, it is the dynamics of the evolution of the population that leads ultimately to equilibria and stable outcomes.

Part D, that concludes this volume, is concerned with "descriptive" results. One looks at the actual behavior of participants in various interactive situations. The question is not "what should rational players do", but rather "what do they do" in specific experiments. Chapter 17 by R. Selten surveys some of the large literature on experimental game theory, in particular relating to issues of cooperation. Since the outcomes are at times at odds with those predicted by the various theories of rational behavior, there is much need to understand what exactly are the principles leading to the different behaviors.

---

[1] The incomplete information case was also covered in the lectures. The reader is referred to the *Handbook of Game Theory with Economic Applications* (edited by R. J. Aumann and S. Hart, North-Holland, volume I: 1992, volumeii: 1994, volume III: forthcoming), for surveys of this topic (see Chapters 5 and 6 volume I), as well as of many other related topics.

Introductory Remarks

Robert J. Aumann

The Hebrew University of Jerusalem

There is a broad division of game theory into two approaches: the cooperative approach and the noncooperative approach. These approaches should not be considered as analyzing different kinds of games; rather, they are different ways of looking at the same game. As Joachim Rosenmüller has said, the game is one "ideal" of which the cooperative and the noncooperative approaches are two "shadows".

The noncooperative theory is strategy oriented. It studies what we expect the players to *do* in the game. The cooperative theory, on the other hand, studies the *outcomes* we expect. In the cooperative approach we look directly at the space of outcomes, not the nitty-gritty of how one gets there. The noncooperative theory is a kind of micro theory; it involves precise descriptions of what happens. In the cooperative theory we are interested in what the players can *achieve*; thus we ask how coalitions can form, what coalitions will form and how the coalitions that do form divide what they achieve.

Why do we call that shadow of the game "cooperative"? "Cooperation" seems to indicate more that that. Indeed, though this term is somewhat misleading, it does have a basis in the theory. In the cooperative theory we are interested in *feasible* outcomes. Thus anything that the players *could* get is taken into consideration, even if it is not incentive compatible for them. For example, in the prisoner's dilemma we are interested also in the cooperative outcome. This is done by assuming that the players have enforceable contracts available to them; i.e., they can make commitments. The players can get into a coalition and agree on a joint course of action, and hence on outcomes; and it is assumed that the players must honor their commitments. We assume that there is some mechanism, like a court, that enforces these contracts, so that all feasible outcomes should be considered. This idea of writing a contract is at least reminiscent of cooperative action.

The distinction between cooperative and noncooperative goes back to the dawning of game theory. It appears already in the works of Nash and others in the early fifties, and I remember a conference in 1955 (attended by von Neumann and Morgenstern) where the issue of cooperative vis-à-vis noncooperative was discussed. However it was only in the 60's that

Harsanyi had the insight of distinguishing commitment as differentiating the cooperative from the noncooperative model.

Summing up, the cooperative (or coalitional) approach studies games from a macro point of view, focusing on the feasible outcomes that can be obtained by enforceable commitments.

PART A

CLASSICAL COOPERATIVE THEORY

Cooperative Theory of Bargaining I: Classical

William Thomson

Department of Economics, University of Rochester
Rochester, NY 14627, USA

In these notes we deal with the so called "bargaining problem" (Nash, 1950). Our approach is axiomatic. We search for solutions satisfying some desirable properties (axioms).

1 Domain

Let $N = \{1, 2, \ldots, n\}$ be the set of agents. A **bargaining problem** is a pair (S, d) interpreted as follows: The group of agents N can get any point in the **feasible set** $S \subseteq \mathbb{R}^N$ if they agree on it, and they get $d \in S$, the **disagreement point**, if they fail to agree on any point. We assume that

- S is a compact and convex set,
- There is at least one point in S that strictly dominates d. (See Figure 1).

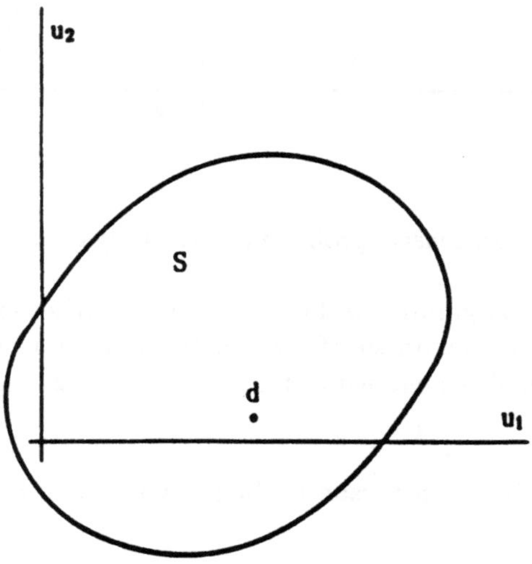

Figure 1: Bargaining problem

For convenience we also assume that

- $d = 0$,
- $S \in \mathbb{R}^N_+$,
- S is comprehensive: If $x \in S$ and $x \geq y \geq 0$, then $y \in S$.[1] (See Figure 2a).

Comprehensiveness of the feasible set is an implication of the assumption of free disposal of utility. We sometimes limit our attention to strictly comprehensive problems; that is problems such that the Pareto efficient boundary of the feasible set does not contain a segment parallel to an axis. For such a problem utility transfers are always possible along the boundary. (See Figure 2b). As we fixed the disagreement point at $d = 0$ it suffices to specify a feasible set to define a bargaining problem.

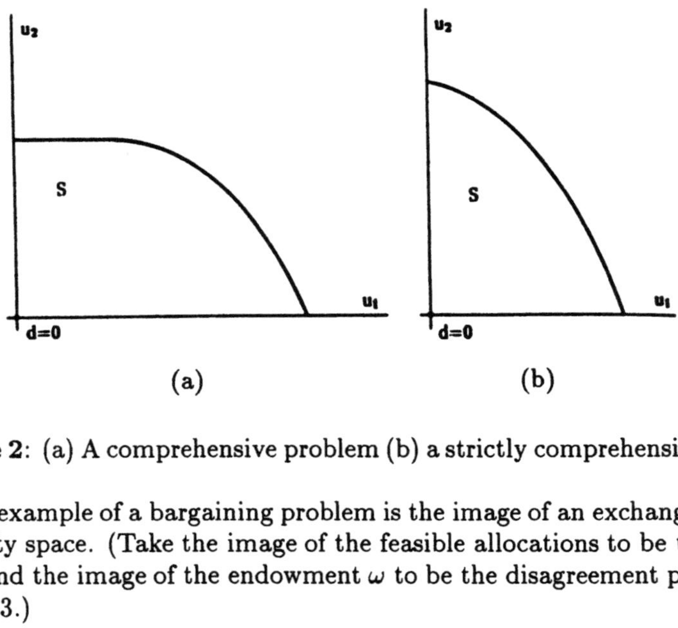

Figure 2: (a) A comprehensive problem (b) a strictly comprehensive problem

An example of a bargaining problem is the image of an exchange economy in utility space. (Take the image of the feasible allocations to be the feasible set S and the image of the endowment ω to be the disagreement point d. See Figure 3.)

We denote the class of n-person problems with $d = 0$ by Σ^n_0.

[1]Given $x, x' \in \mathbb{R}^n, x \geq x'$ means $x_i \geq x'_i$ for all i.

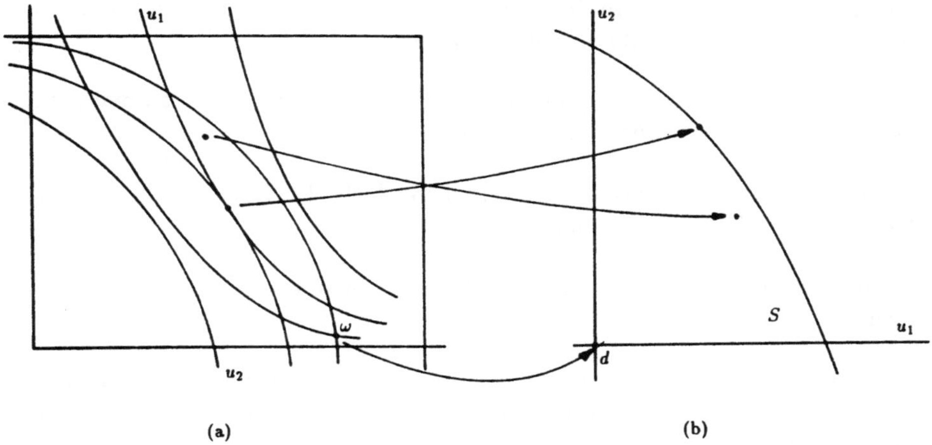

Figure 3: Construction of a bargaining problem from an exchange economy

2 Solutions

A **solution** F is a method to choose a feasible point for each problem. Formally, a solution is a function F from Σ_0^n to $\mathbb{R}^n$ such that $F(S) \in S$ for all $S \in \Sigma_0^n$.

Usually on economic domains most solutions are multi-valued, but in bargaining theory there are many interesting single-valued solutions and the theory has been mainly developed under the assumption of single-valuedness. We first define a number of the solutions that have been considered, starting with the most natural ones.

The idea of equal gains is central to economic reasoning in a variety of contexts. This idea is the motivation for the following solution:

Egalitarian solution, E (Figure 4a): $E(S)$ is the maximal point of S of equal coordinates.[2]

The next solution can be seen as a normalized version of the egalitarian solution. Define the **ideal point** of S, $a(S)$ as follows: $a_i(S) = \max\{x_i \mid x \in S\}$

[2]When the bargaining problem is presented in classroom to students that have never been exposed to the theory, the solution that they come up most often is the egalitarian solution. The second most often proposal is the Kalai-Smorodinsky solution.

for all i. The Kalai-Smorodinsky solution sets utility gains from the origin proportional to the agents' maximal utilities.

Kalai-Smorodinsky solution, K (Figure 4b): $K(S)$ is the maximal point of S on the segment connecting the origin to $a(S)$.

The best known solution, introduced by Nash (1950), select the point at which the product of utility gains is maximized.

Nash solution, N (Figure 4c): $N(S)$ is the maximizer of the product $\prod x_i$ over S.

The next two solutions are extreme cases of solutions favoring one agent at the expense of others. Solutions in the same spirit often appear in social choice theory.

Dictatorial solutions, D^i and D^{*i} (Figure 4d): $D^i(S)$ is the maximal point x of S with $x_j = 0$ for all $j \neq i$. $D^{*i}(S)$ is the Pareto optimal point with maximal i^{th} coordinate.

The next two solutions are representatives of a family of solutions based on processes of balanced concessions: imagine agents working their way from their preferred alternatives to a final position by moving from compromise to compromise. For the solution defined first, the initial compromise is obtained by choosing the halfway point between agents' most preferred alternatives. This point is not in general Pareto efficient. The next step is choosing the halfway point between the agents' most preferred alternatives among these alternatives that both prefer to the previous compromise. The procedure is repeated until a Pareto efficient allocation is reached.

The discrete Raiffa solution, R^d (Figure 4e): $R^d(S)$ is the limit point of the sequence $\{z^t\}$ defined by: $x^{i0} = D^i(S)$ for all i; for all $t \in \mathbb{N}, z^t = (\sum x^{i(t-1)})/n$, and x^{it} is the weakly Pareto optimal point with $x^{it}_j = z^t_j$ for all $j \neq i$.

The definition of the next solution might not be very transparent but this solution has a very appealing characterization based on an additivity condition.

The Perles-Maschler solution, PM (Figure 4f): For $n = 2$. If ∂S (the weakly Pareto optimal boundary of S) is polygonal, $PM(S)$ is the common limit point of the sequences $\{x^t\}, \{y^t\}$, defined by: $x^0 = D^{*1}(S), y^0 = D^{*2}(S)$; for each $t \in \mathbb{N}, x^t, y^t$ are Pareto optimal with $x^t_1 \geq y^t_1$, the segments $[x^{t-1}, x^t], [y^{t-1}, y^t]$ are Pareto optimal and the products $|(x^{t-1}_1 - x^t_1)(x^{t-1}_2 -$

$x_2^t)|$ and $|(y_1^{t-1}-y_1^t)(y_2^{t-1}-y_2^t)|$ are equal and maximal. If ∂S is not polygonal, $PM(S)$ is defined by approximating S by a sequence of polygonal problems and taking the limit of associated solution outcomes.

The next solution exemplifies a family of solutions for which compromises are evaluated globally:

The Equal Area solution, A (Figure 4g): For $n = 2$. $A(S)$ is the Pareto optimal point x such that the area of S to the right of the vertical line through x is equal to the area of S above the horizontal line through x.

The next solution plays an important role in welfare economics:

Utilitarian solution, U (Figure 4h): $U(S)$ is a maximizer in S of $\sum x_i$. Note that the utilitarian solution is not single-valued in general. To obtain a well-defined solution one needs to make a selection from the set of maximizers.

Agents cannot simultaneously obtain their preferred outcomes. An intuitively appealing idea is to try to get as close as possible to satisfying everyone. The following one-parameter family of solutions reflects the flexibility that exists in measuring how close two points are from each other:

Yu solutions, Y^p (Figure 4i): Given $p \in (1, \infty)$, $Y^p(S)$ is the point of S for which the p-distance to the ideal point of S, $(\sum |a_i(S) - x_i|^p)^{1/p}$, is minimal.

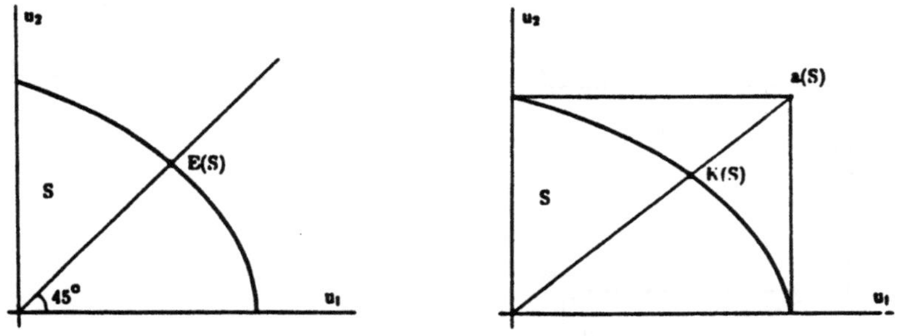

(a) (b)

Figure 4a-b: Solutions (a) the Egalitarian solution, (b) the Kalai-Smorodinsky solution

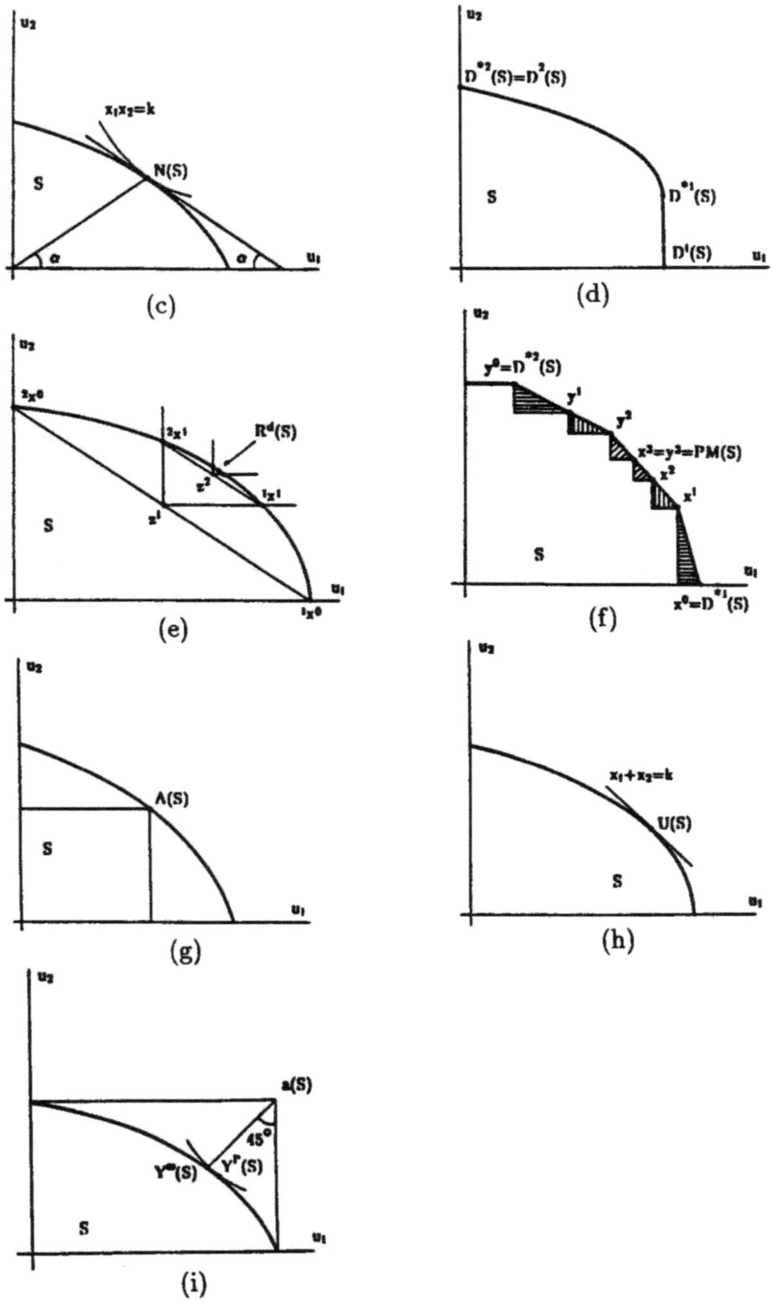

Figure 4c-i: Solutions (c) the Nash solution, (d) the Dictatorial solutions, (e) the Discrete Raiffa solution, (f) the Perles-Maschler solution, (g) the Equal Area solution, (h) the Utilitarian solution, (i) the Yu solutions

3 Axioms and Main Characterizations

In this section we introduce some desirable properties of solutions (the **axioms**) and study their implications. The first axiom we study requires that if opportunities expand, then all agents should (weakly) gain.

Strong monotonicity: For all $S, S' \in \Sigma_0^2$ if $S' \supseteq S$, then $F(S') \geq F(S)$. (See Figure 5a).

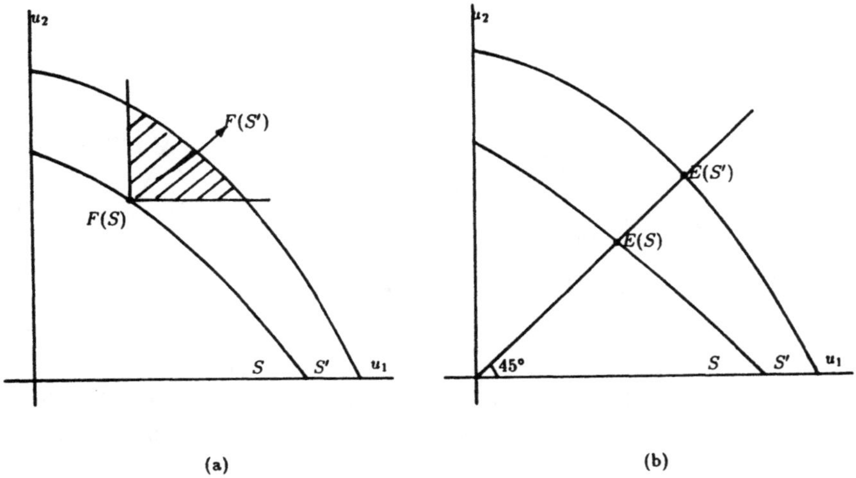

Figure 5: (a) *Strong monotonicity:* if opportunities expand, all agents should (weakly) gain (b) the egalitarian solution is *strongly monotonic*

The egalitarian solution is *strongly monotonic* (See Figure 5b), but neither the Kalai-Smorodinsky solution, nor the Nash solution is. One wonders if there are other solutions than the egalitarian solution that are *strongly monotonic*. Of course the solution that selects the disagreement point for all problems is *strongly monotonic*. However it is not *Pareto optimal*, and not even *weakly Pareto optimal*:

Pareto optimality: For all $S \in \Sigma_0^2$,

$$F(S) \in PO(S) = \{x \in S \mid \nexists x' \in S \text{ with } x' \geq x\}$$

Weak Pareto optimality: For all $S \in \Sigma_0^2$,

$$F(S) \in WPO(S) = \{x \in S \mid \nexists x' \in S \text{ with } x' > x\}$$

Note that the egalitarian solution is only *weakly Pareto optimal*. It is *Pareto optimal* on the class of strictly comprehensive problems.

What about the solution that selects the maximal point on the 30° line for all problems? (See Figure 6a). This solution is *strongly monotonic*, as well as *weakly Pareto optimal.* More generally, consider the following class of solutions: Fix a monotone path in $\mathbb{R}^2_+$ emanating from the origin, and for each problem choose the maximal feasible point on this path. We call any such solution a **monotone path solution**. (See Figure 6b).

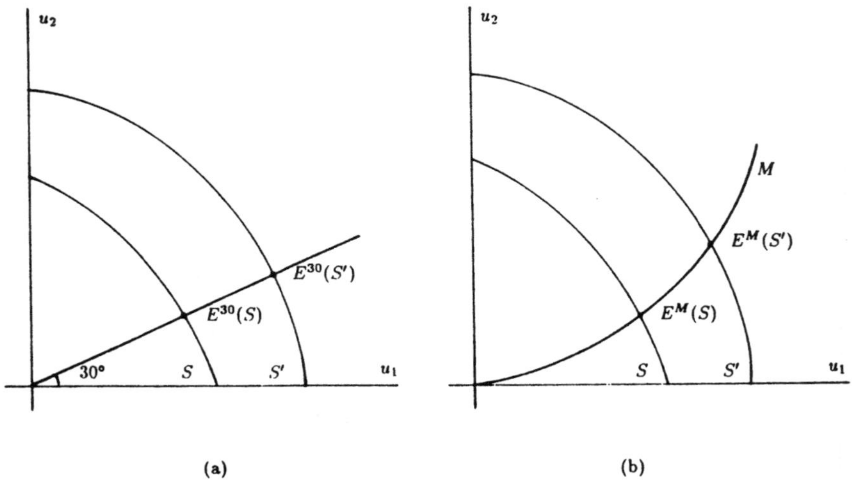

Figure 6: (a) The solution that selects the maximal point on the 30° line is *strongly monotonic* (b) monotone path solutions are *strongly monotonic*

Monotone path solutions are *strongly monotonic* and *weakly Pareto optimal.* But none of them (except the egalitarian solution) is *symmetric*:

Symmetry: If S is invariant under all exchanges of agents, $F_i(S) = F_j(S)$ for all i, j.

Next we have the first theorem:

Theorem 1 (Kalai, 1977): The egalitarian solution is the only solution satisfying *weak Pareto optimality, symmetry,* and *strong monotonicity.*

Proof: It is easy to show that the egalitarian solution satisfies *weak Pareto optimality, symmetry,* and *strong monotonicity.* Conversely let F satisfy the three axioms. We first consider the class of strictly comprehensive problems. Let $S \in \Sigma_0^2$ be strictly comprehensive. We need to show that $F(S) = E(S)$. Let $E(S) = x$ and $S' = cch\{x\}$.[3] $F(S') = x$ by *weak Pareto optimality*

[3]Given $A \subset \mathbb{R}^n_+$, $cch\{A\}$ denotes the "convex and comprehensive hull" of A: it is the smallest convex and comprehensive subset of $\mathbb{R}^n_+$ containing A. If $x, y \in \mathbb{R}^n_+$ we write $cch\{x, y\}$ instead of $cch\{\{x, y\}\}$.

and *symmetry*. Furthermore $S \supseteq S'$. Therefore $F(S) \geq F(S') = x$ by *strong monotonicity*. Yet x is the only feasible point to satisfy this inequality. Therefore $F(S) = x = E(S)$. (See Figure 7).

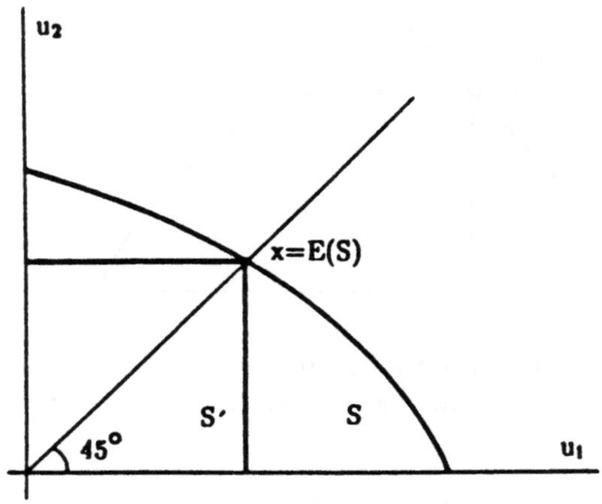

Figure 7: Characterization of the egalitarian solution on the basis of *strong monotonicity*: the case of strictly comprehensive problems

Next we consider the class of comprehensive problems. Let $S \in \Sigma_0^2$. If $E(S)$ is *Pareto efficient* then we can conclude as before. Suppose then that $E(S)$ is not *Pareto efficient*. (Figure 8). We need to show that $F(S) = E(S) = x$. By the previous argument applied to S and S', we have $x \leq F(S) \leq y$. Suppose $F(S) = z \neq x$. Then we construct a strictly comprehensive problem $T \supset S$ such that $E(T)$ is to the northwest of z. We have $F(T) = E(T)$ as T is strictly comprehensive. Hence $F(T) = E(T)$ and $F(T)$ is to the northwest of $F(S) = z$, contradicting *strong monotonicity*. Therefore $F(S) = E(S) = x$. Q.E.D.

Note that the axioms of Theorem 1 are independent. If we drop *weak Pareto optimality*, the solution which selects the disagreement point for all problems satisfies *symmetry* and *strong monotonicity*. If we drop *symmetry*, any monotone path solution satisfies *weak Pareto optimality* and *strong monotonicity*. If we drop *strong monotonicity*, the Nash solution (and most of the solutions mentioned above) satisfies *weak Pareto optimality* and *symmetry*.

What if we drop the comprehensiveness assumption? Then the egalitarian solution is not necessarily *weakly Pareto optimal*. In fact it may suggest the worst possible feasible point.

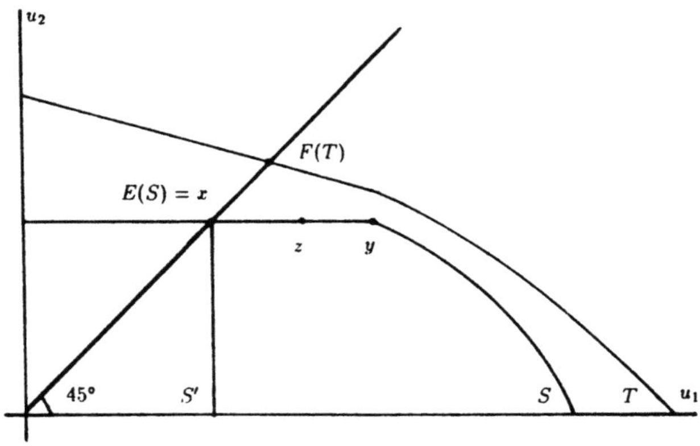

Figure 8: Characterization of the egalitarian solution on the basis of *strong monotonicity*: the case of comprehensive but not strictly comprehensive problems

Can we recover *weak Pareto optimality* in this domain? Not as long as the solution is *strongly monotonic*. On this domain there is no solution satisfying *weak Pareto optimality* and *strong monotonicity*. To see this, consider the example of Figure 9. Suppose F satisfies *weak Pareto optimality* and *strong monotonicity*. Then $F(S) = x$ by *weak Pareto optimality* and then $F(S'') = F(S) = x$ by *strong monotonicity*. By the same reasoning $F(S'') = F(S') = y \neq x$, a contradiction.

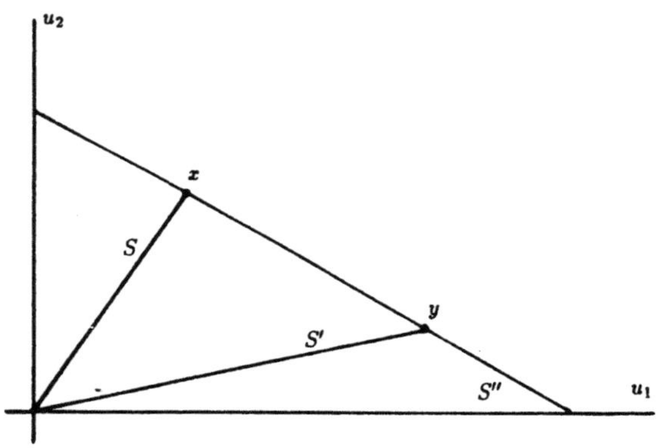

Figure 9: An impossibility: there is no solution satisfying *weak Pareto optimality* and *strong monotonicity* if non-comprehensive problems are permitted

One of the weaknesses of the egalitarian solution is that it is not fully *Pareto optimal*. However there is a very natural way to adjust its definition so as to recover *Pareto optimality*. Given $x \in \mathbb{R}^n$, let $\tilde{x} \in \mathbb{R}^n$ denote the vector obtained from x by writing its coordinates in increasing order. Given $x, y \in \mathbb{R}^n$, x is **lexicographically greater than** y if $\tilde{x}_1 > \tilde{y}_1$ or $[\tilde{x}_1 = \tilde{y}_1$ and $\tilde{x}_2 = \tilde{y}_2]$ or, more generally, for some $k \in \{1, 2, \ldots, n-1\}$, $[\tilde{x}_1 = \tilde{y}_1, \ldots, \tilde{x}_k = \tilde{y}_k$ and $\tilde{x}_{k+1} = \tilde{y}_{k+1}]$. Now, given $S \in \Sigma_0^n$, its **lexicographic egalitarian solution** outcome of S, $E^L(S)$, is the point of S that is lexicographically maximal. (See Figure 10).

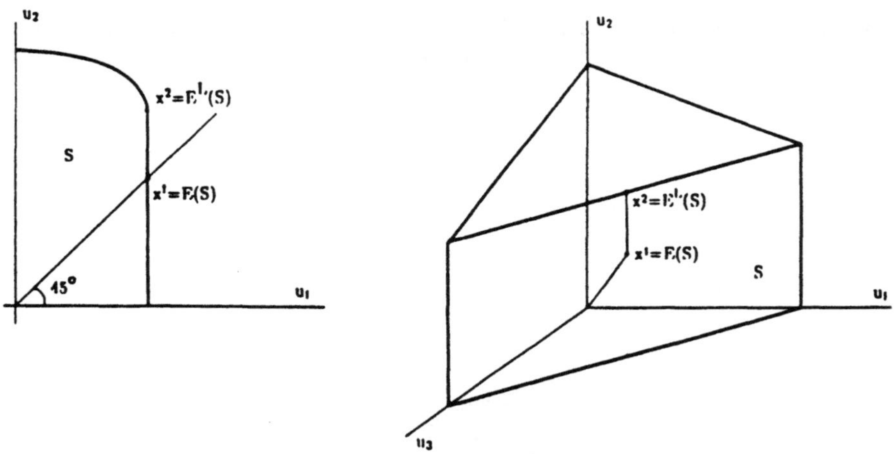

(a) (b)

Figure 10: The lexicographic egalitarian solution (a) a two agent case (b) a three agent case

Note that the lexicographic egalitarian solution is *Pareto optimal* even if the domain is not comprehensive.

The next property requires a solution to be immune to positive affine transformations of the utility functions. (Recall that von Neumann-Morgenstern utilities are unique up to positive affine transformations).

Let $\Lambda_0^n : \mathbb{R}^n \to \mathbb{R}^n$ be the class of independent person by person, positive linear transformations: $\lambda \in \Lambda_0^n$ if there is $k \in \mathbb{R}_{++}^n$ such that for all $x \in \mathbb{R}^n$, $\lambda(x) = (k_1 x_1, \ldots, k_n x_n)$. Given $\lambda \in \Lambda_0^n$ and $S \subset \mathbb{R}^n$, $\lambda(S) = \{x' \in \mathbb{R}^n \mid \exists x \in S \text{ with } x' = \lambda(x)\}$.

Scale invariance: $\lambda(F(S)) = F(\lambda(S))$.

Both the Kalai-Smorodinsky solution and the Nash solution are *scale invariant*. However the egalitarian solution is not.

One of the problems with *strong monotonicity* is that it requires everybody to gain even if opportunities expand in a way that really "favors" one of the agents. This is the motivation for the following property. It requires that if opportunities expand in a direction favorable to an agent, in the sense that the maximal utilities of other agents remain the same, he (weakly) gains.

Individual monotonicity: For all $S, S' \in \Sigma_0^2$, if $S' \supseteq S$ and $a_j(S') = a_j(S)$ for all $j \neq i$ then $F_i(S') \geq F_i(S)$. (See Figure 11).

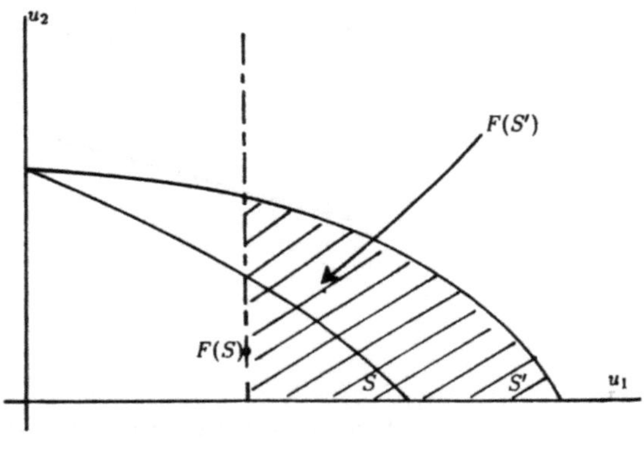

Figure 11: *Individual monotonicity*: if the opportunities expand in a direction favorable to agent 1 then he (weakly) gains

The Kalai-Smorodinsky solution is *individually monotonic*. (See Figure 12a). The advantage of weakening *strong monotonicity* is we recover *scale invariance*. However we recover it essentially in a unique way as revealed by the following theorem.

Theorem 2 (Kalai and Smorodinsky, 1975): The Kalai-Smorodinsky solution is the only solution satisfying *weak Pareto optimality, symmetry, scale invariance*, and *individual monotonicity*.

Proof: It is easy to show that the Kalai-Smorodinsky solution satisfies *weak Pareto optimality, symmetry, scale invariance*, and *individual monotonicity*.

Conversely let F satisfy the four axioms. Let $S \in \Sigma_0^2$ be such that $K(S)$ is on the 45° line. This is without loss of generality as otherwise we can transform S by a positive affine transformation and use *scale invariance*. We need to show that $F(S) = K(S)$. Let $S' = cch\{(a_1(S), 0), K(S), (0, a_2(S)\}$. Note that $a_1(S) = a_2(S)$ as $K(S)$ is on the 45° line. Hence S' is a symmetric problem. Therefore $F(S') = K(S)$ by *weak Pareto optimality* and *symmetry*. But $S \supseteq S', a_1(S) = a_1(S')$, and $a_2(S) = a_2(S')$. Therefore using *individual monotonicity* twice we have $F_1(S) \geq F_1(S') = K_1(S)$ and $F_2(S) \geq F_2(S') = K_2(S)$. But $K(S)$ is the only feasible point to satisfy these two inequalities and hence $F(S) = K(S)$. (See Figure 12b). Q.E.D.

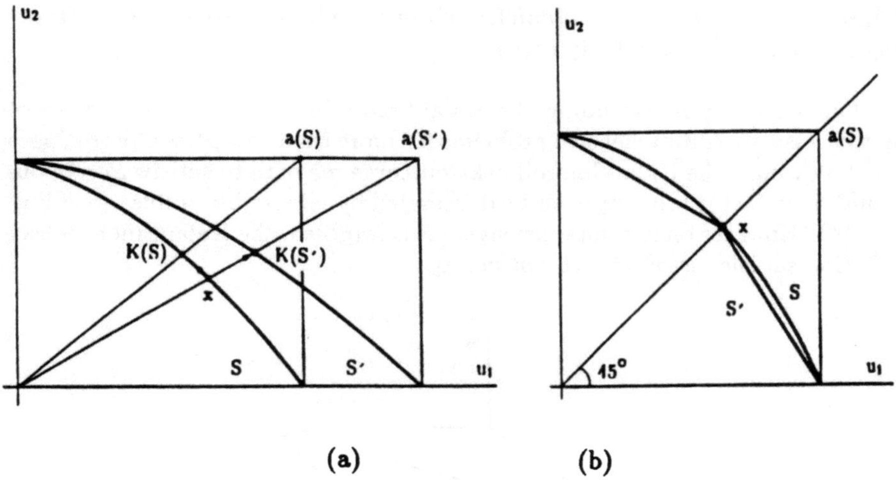

Figure 12: (a) The Kalai-Smorodinsky solution is *individually monotonic* (b) a characterization of the Kalai-Smorodinsky solution on the basis of *individual monotonicity*

Note that the Kalai-Smorodinsky solution also satisfies *Pareto optimality*.

An alternative monotonicity condition, *restricted monotonicity*, says that if opportunities expand but the maximal utilities remain the same, then all agents should (weakly) gain.

Restricted monotonicity: For all $S, S' \in \Sigma_0^2$ if $S' \supseteq S$ and $a(S') = a(S)$ then $F(S') \geq F(S)$. (See Figure 13).

We may replace *individual monotonicity* with *restricted monotonicity* in Theorem 2 and obtain an alternative characterization of the Kalai-Smorodinsky solution.

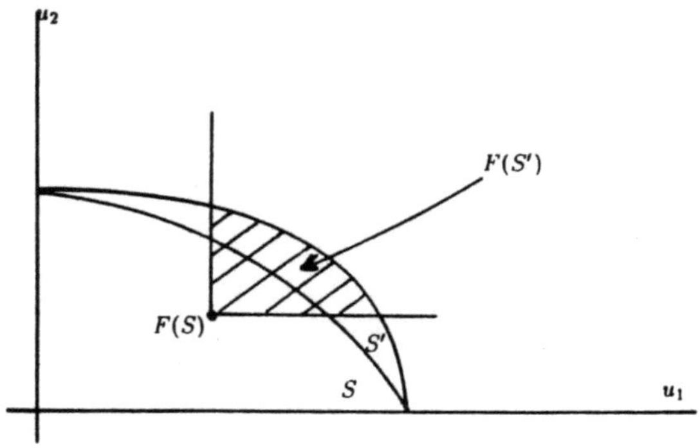

Figure 13: *Restricted monotonicity:* if opportunities expand without favoring anybody then everybody gains

One difficulty in extending the Kalai-Smorodinsky solution to classes of not necessarily comprehensive problems for more than two person problems is the following: The Kalai-Smorodinsky outcome may fail to satisfy *Pareto optimality*; in fact it too may even be dominated by all feasible points. (See Figure 14). However once comprehensiveness is imposed, the Kalai-Smorodinsky solution satisfies *weak Pareto optimality.*

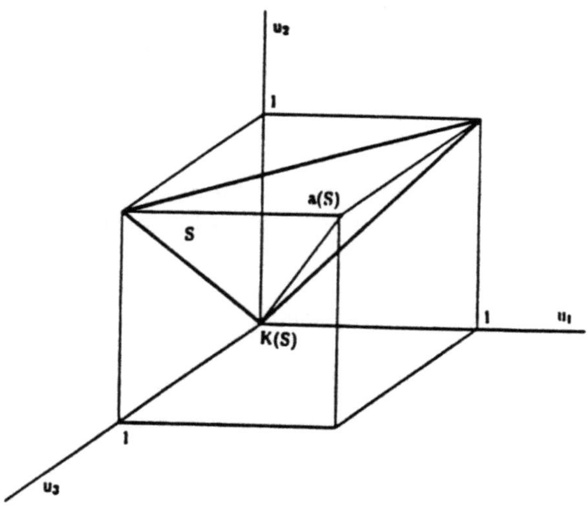

Figure 14: A difficulty in extending the Kalai-Smorodinsky solution for not necessarily comprehensive and more than two person problems: the Kalai-Smorodinsky outcome is dominated by all other points

Next consider the following independence property. It is in the same spirit as Arrow's independence of irrelevant alternatives condition.

Contraction Independence: For all $S, S' \in \Sigma_0^2$ if $S' \subseteq S$ and $F(S) \in S'$ then $F(S') = F(S)$. (See Figure 15).

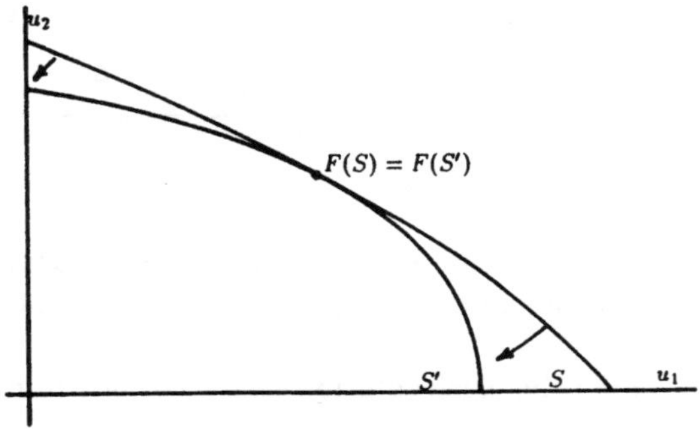

Figure 15: *Contraction independence*: if opportunities shrink leaving the solution outcome feasible, it should still be selected

The Nash solution satisfies *contraction independence.* (See Figure 16a). It is also *Pareto optimal* (even in the case of non-comprehensive problems), *symmetric*, and *scale invariant.* In fact, we have:

Theorem 3 (Nash, 1950): The Nash solution is the only solution satisfying *Pareto optimality, symmetry, scale invariance*, and *contraction independence.*

Proof: It is easy to show that the Nash solution satisfies *Pareto optimality, symmetry, scale invariance*, and *contraction independence.*

Conversely let F satisfy the four axioms. Let $S \in \Sigma_0^2$ be such that $N(S)$ is on the 45° line. This is without loss of generality as otherwise we can transform S by a positive affine transformation and use *scale invariance.* We need to show that $F(S) = N(S)$. Let $T = \{y \in \mathbb{R}_+^2 \mid \sum y_i \leq \sum N_i(S)\}$. The problem T is symmetric and $N(S) \in PO(T)$. Thus $F(T) = N(S)$. But $T \supseteq S$ and $F(T) = N(S) \in S$, therefore $F(S) = F(T) = N(S)$ by *contraction independence.* (See Figure 16b). Q.E.D.

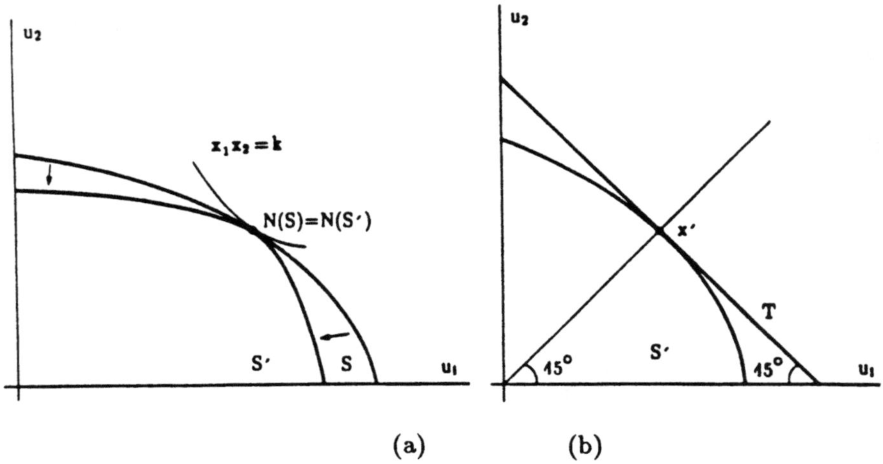

Figure 16: (a) The Nash solution satisfies *contraction independence* (b) a characterization of the Nash solution based on *contraction independence*

The characterization of the Nash solution easily extends to more than two person problems.

References

[1] Nash, J.F. (1950), "The Bargaining Problem," *Econometrica* 28, 155-162.

[2] Kalai, E. (1977), "Proportional Solutions to Bargaining Solutions: Interpersonal Utility Comparisons," *Econometrica* 45, 1623-1630.

[3] Kalai, E. and M. Smorodinsky (1975), "Other Solutions to Nash's Bargaining Problem," *Econometrica* 43, 513-518.

Cooperative Theory of Bargaining II: Modern Development

William Thomson

Department of Economics, University of Rochester
Rochester, NY 14627, USA

1 Variable Number of Agents

The model discussed in Part I is written under the assumption of a fixed number of agents. Here, the number of agents is allowed to vary. How solutions should respond to such changes are formulated as axioms, and additional characterizations of the main solutions are developed. A detailed account of these recent developments can be found in Thomson and Lensberg (1989).

The model is expanded to allow problems involving any finite number of agents. Agents are indexed by the integers. Let $\mathcal{P}$ be the set of all finite subsets of the integers. Given $P \in \mathcal{P}$, the set of problems that the group P could face is denoted by Σ_0^P. The set of all problems is denoted by $\Sigma_0 = \cup_{P\in\mathcal{P}}\Sigma_0^P$. A **solution** is a function $F : \Sigma_0 \rightarrow \cup_{P\in\mathcal{P}} \mathbb{R}_+^P$ such that for all $P \in \mathcal{P}$ and all $S \in \Sigma_0^P$, $F(S) \in S$. Figure 1 represents a 2-agent problem $S \in \Sigma_0^{\{2,3\}}$ involving agent 2 and agent 3.

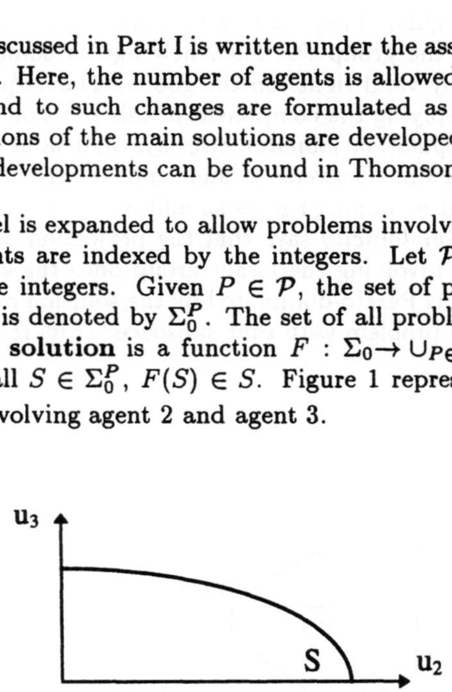

Figure 1: A 2-agent problem involving agent 2 and agent 3

In this more general model it is possible to define solutions by "combining" different solutions across cardinalities: say, the egalitarian solution could be used for 2-agent bargaining problems; the Kalai-Smorodinsky solution for 3-agent problems; the egalitarian solution again for 4-agent problems; and .... One may find it more reasonable using the same solution for all sizes of groups. But even if we decide to do that, how do we compare the solution which uses the Kalai-Smorodinsky solution for all cardinalities and the solution which uses the Egalitarian solution for all cardinalities. We need to develop a theory to help us link the components of solutions across cardinalities. In what follows we propose two principles that can be used for that purpose.

2 Population Monotonicity and the Egalitarian Solution

Consider a group of agents $P \in \mathcal{P}$ that have to divide a bundle of goods on which they have equal rights. They do this by applying the component of a solution pertaining to the group P. Then new agents come in, who are recognized to have equally valid rights on the goods. This requires that the goods be redivided by applying the component of this solution pertaining to the remaining group. *Population monotonicity* says that none of the agents initially present should gain. Conversely, and equivalently, if some agents relinquish their rights, all remaining agents should be better-off. In bargaining theory *population monotonicity* says that the projection of the solution outcome of the problem involving the larger group onto the subspace relative to the smaller group is Pareto-dominated by the solution outcome of the intersection of the larger problem with that subspace. (Figure 2).

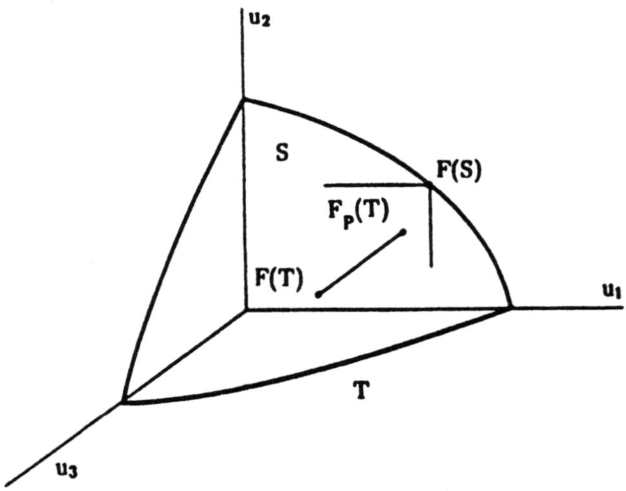

Figure 2: *Population monotonicity* in bargaining theory

Population monotonicity: For all $P, Q \in \mathcal{P}$ with $P \subseteq Q$, if $S \in \Sigma_0^P$ and $T \in \Sigma_0^Q$ are such that $S = T_P$, then $F(S) \geq F_P(T)$.[1]

The egalitarian solution (Figure 3) and the Kalai-Smorodinsky solution satisfy this requirement; but the Nash solution does not. Characterizations of the egalitarian solution and the Kalai-Smorodinsky solution can be obtained with the help of *population monotonicity*.

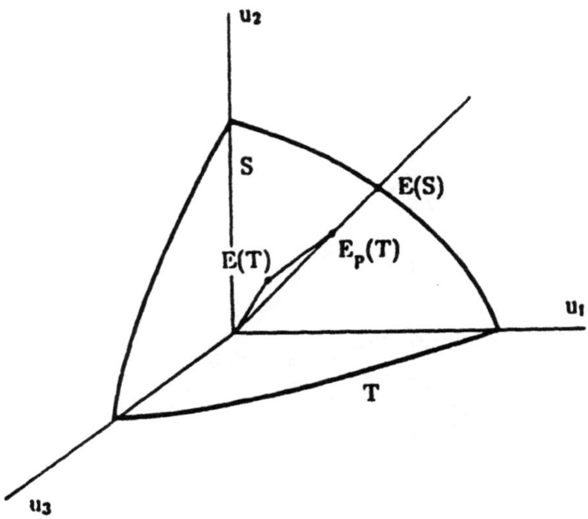

Figure 3: The egalitarian solution is *population monotonic*: the projection of the egalitarian outcome of the problem T onto the coordinate subspace pertaining to agents 1 and 2 is dominated by the egalitarian outcome of the problem S which is the intersection of T with the coordinate subspace pertaining to agents 1 and 2

Theorem 1 (Thomson 1983a): The egalitarian solution is the only solution on Σ_0 satisfying *weak Pareto optimality, symmetry, contraction independence, continuity,* and *population monotonicity.*

Proof (Sketch): It is easy to verify that E satisfies the five axioms (see Figure 3 for *population monotonicity*). Conversely, let F be a solution on Σ_0 satisfying the five axioms. We only show that F coincides with E on Σ_0^2 (Figure 4).

[1]Note that T_P is the intersection of the problem T with the subspace $\mathbf{R}^P$, and $F_P(T)$ is the projection of the outcome $F(T)$ onto the subspace $\mathbf{R}^P$.

For simplicity, let $S \in \Sigma_0^{\{1,2\}}$ be given. Suppose that $E(S) = (\alpha, \alpha)$. Let $T \in \Sigma_0^{\{1,2,3\}}$ be such that $T = \{x \in \mathbb{R}_+^{\{1,2,3\}} | x_1 + x_2 + x_3 \leq 3\alpha\}$. By *weak Pareto optimality* and *symmetry,* $F(T) = (\alpha, \alpha, \alpha)$. Now, let $T' \in \Sigma_0^{\{1,2,3\}}$ be such that $T' = cch\{S, F(T)\}$. Since $T' \subset T$ and $F(T) \in T'$, it follows from *contraction independence* that $F(T') = F(T)$. Suppose $S \subseteq T_{\{1,2\}}$. Then $T'_{\{1,2\}} = S$, so that by *population monotonicity,* $F(S) \geq E(S)$. If $E(S) \in PO(S)$,then we are done. Otherwise we conclude by *continuity.* Next suppose $S \not\subseteq T_{\{1,2\}}$. In this case the previous arguments can be modified by adding more than one agent and obtaining the inclusion. Q.E.D.

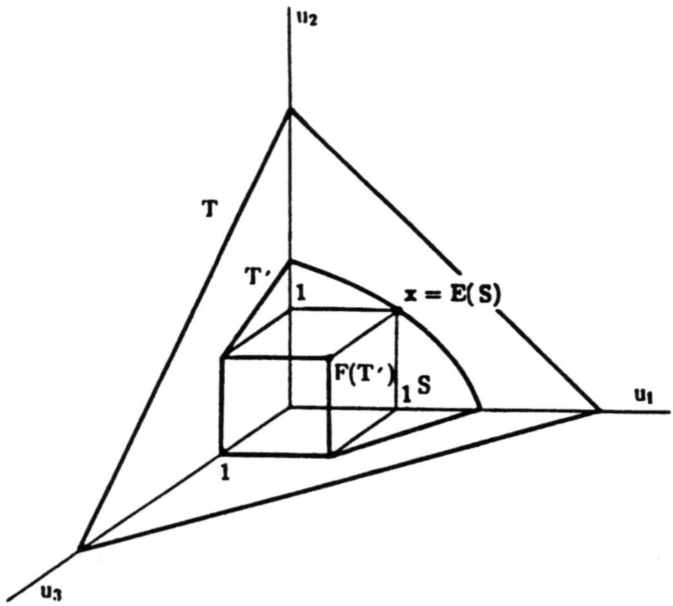

Figure 4: Characterization of the egalitarian solution on the basis of *population monotonicity*

3 Population Monotonicity and the Kalai-Smorodinsky Solution

All of the axioms used in the next theorem except *anonymity* have already been discussed. This property is a strengthening of *symmetry* and it requires the solution to be invariant under exchanges of the names of the agents in each group, as well as the replacement of some of its members by other agents. (See Figure 5).

Anonymity: Given $P, P' \in \mathcal{P}$ with $|P| = |P'|$, $S \in \Sigma_0^P$ and $S' \in \Sigma_0^{P'}$, if there exists a bijection $\gamma : P \to P'$ such that $S' = \{x' \in \mathbb{R}^{P'} \mid \exists x \in S \text{ with } x'_i = x_{\gamma(i)} \text{ for all } i \in P\}$, then $F_{\gamma(i)}(S') = F_i(S)$.

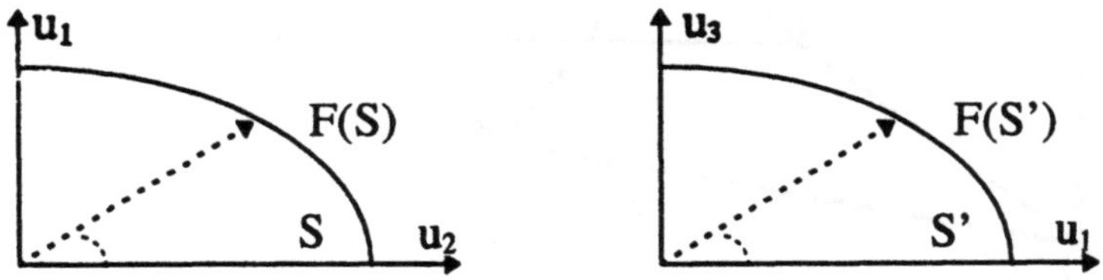

Figure 5: *Anonymity* with a variable number of agents

The following theorem differs from the previous one only in that *scale invariance* is used instead of *contraction independence*.

Theorem 2 (Thomson 1983b): The Kalai-Smorodinsky solution is the only solution on Σ_0 satisfying *weak Pareto optimality, anonymity, scale invariance, continuity,* and *population monotonicity.*

Proof (Sketch): It is straightforward to see that K satisfies the five axioms. Conversely, let F be a solution on Σ_0 satisfying the five axioms. We only show that F coincides with K on Σ_0^2 (Figure 6). Let $S \in \Sigma_0^{\{1,2\}}$ be given. By *scale invariance,* we can assume that S is normalized so that $a(S)$ has equal coordinates. Suppose then that $K(S) = (\alpha, \alpha)$ and let $x = (\alpha, \alpha, \alpha)$. We construct $T \in \Sigma_0^{\{1,2,3\}}$ by replicating S in the coordinate subspaces $\mathbb{R}^{\{2,3\}}$ and $\mathbb{R}^{\{3,1\}}$, and taking the comprehensive hull of S, its two replicas, and of the point x. Since all agents enter symmetrically in the definition of T and $x \in PO(T)$, it follows from *anonymity* and *weak Pareto optimality* that $F(T) = x$. Now, note that $T_{\{1,2\}} = S$, so that by *population monotonicity,*

$F(S) = K(S)$. To prove that F and K coincide for problems of cardinality greater than 2, one has to introduce more agents and *continuity* becomes necessary. Q.E.D.

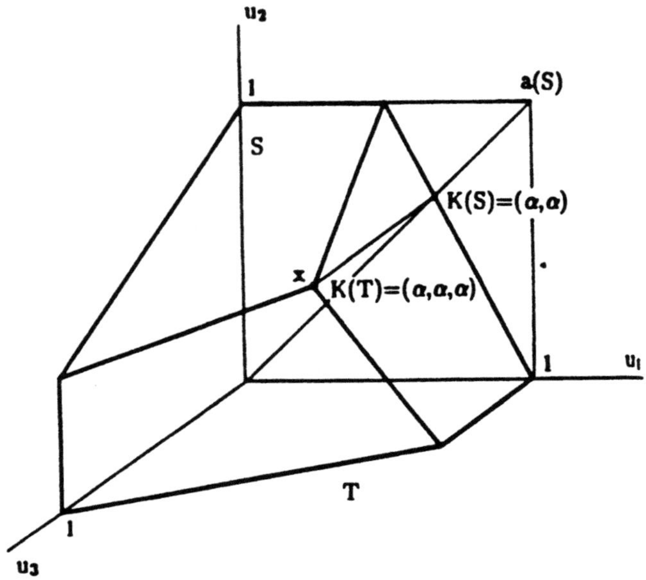

Figure 6: Characterization of the Kalai-Smorodinsky solution on the basis of *population monotonicity*

4 Consistency and the Nash Solution

Instead of allowing the agents to depart empty-handed, we will now imagine them to leave with their payoffs. A solution is *consistent* if the remaining agents receive the same payoffs when some of the agents depart with their payoffs. To be precise, let $Q \in \mathcal{P}$ and $T \in \Sigma_0^Q$, and consider some point $x \in T$ as the candidade compromise for T. Assume that it has been accepted by the subgroup P', and let us imagine its members leaving the scene with the understanding that they will indeed receive their payoffs $x_{P'}$. Now, let us reevaluate the situation from the viewpoint of the group $P = Q \setminus P'$ of remaining agents. It is natural to think as the set $\{y \in \mathbb{R}^P \mid (y, x_{P'}) \in T\}$ obtained from points of T by giving the agents in P' the payoffs $x_{P'}$, as the feasible set for P. Let us denote it by $r_P^x(T)$. Geometrically, $r_P^x(T)$ is the "slice" of T through x by a plane parallel to the coordinate subspace relative to group P. If this set is a well-defined member of Σ_0^P, and if the solution recommends the utilities x_p, then it is *consistent.* (Figure 7).

Consistency: Given $P, Q \in \mathcal{P}$ with $P \subseteq Q$, if $S \in \Sigma_0^P$ and $T \in \Sigma_0^Q$ are such that $x = F(T)$ and $r_P^x(T) = S$, then $F(S) = x_P$.

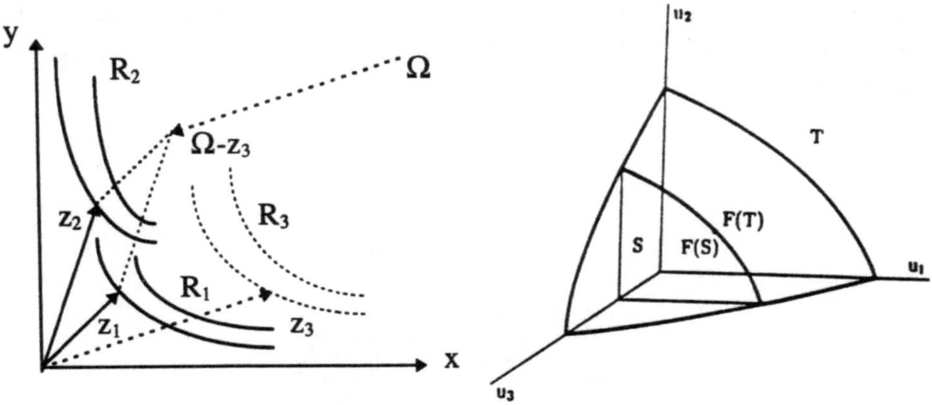

(a) (b)

Figure 7: (a) *Consistency* in exchange economy: if agent 3 leaves with his consumption z_3 the remaining agents should be unaffected. (b) consistency in bargaining theory: if agent 3 leaves with his payoff $F_3(T)$, the remaining agents should be unaffected

Consistency is satisfied by the Nash solution (Harsanyi, 1959) but not by the Kalai-Smorodinsky solution, nor by the egalitarian solution. Violations are usual for the Kalai-Smorodinsky solution but rare for the egalitarian solution; indeed on the class of strictly comprehensive problems, the egalitarian solution does satisfy the condition, and if this restriction is not imposed, it still satisfies the slightly weaker condition obtained by requiring $F(S) \geq x_P$ instead of $F(S) = x_P$. Let us call this weaker condition *weak consistency.* The lexicographic egalitarian solution satisfies *consistency.*

The Nash solution can be characterized on the basis of *consistency:*

Theorem 3 (Lensberg, 1988): The Nash solution is the only solution on Σ_0 satisfying *Pareto optimality, anonymity, scale invariance, continuity,* and *consistency.*

Proof (Sketch): It is straightforward to see that N satisfies the five axioms. Conversely, let F be a solution on Σ_0 satisfying the five axioms. We only show that F coincides with N on Σ_0^2. Let $S \in \Sigma_0^{\{1,2\}}$ be given. By *scale invariance,* we can assume that S is normalized so that $N(S) = (1,1)$. For simplicity, we assume that $PO(S) \supseteq [(1.5,.5),(.5,1.5)]$. (In Figure 8, $S = cch\{(2,0),(.5,1.5)\}$.) Now, we translate S by the third unit vector, we replicate the resulting set twice by having agents 2, 3 and 1, and then agents 3, 1 and 2 play the roles of agents 1, 2 and 3 respectively; finally, we define $T \in \Sigma_0^{\{1,2,3\}}$ to be the convex and comprehensive hull of the three sets so obtained. Since the problem $T = cch\{(1,2,0),(0,1,2),(2,0,1)\}$ is invariant under rotations of the agents, by *anonymity,* $F(T)$ has equal coordinates, and by *Pareto optimality,* $F(T) = (1,1,1)$. But, since $r_P^{(1,1,1)}(T) = S$, *consistency* gives $F(S) = (1,1) = N(S)$.

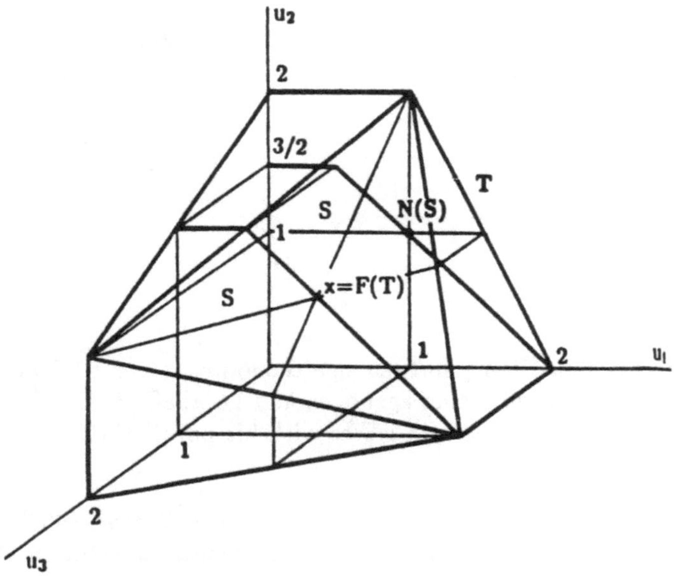

Figure 8: Characterization of the Nash solution on the basis of *consistency*

For the case that $PO(S) \not\supseteq [(1.5,.5),(.5,1.5)]$ and $N(S)$ is contained in the interior of $PO(S)$, the argument can be extended in the same way by introducing more agents. For the remaining case, the argument is completed by using *continuity.* Q.E.D.

5 Conclusion

In the previous chapter we presented the classical results in bargaining theory and in this chapter we presented recent developments concerning variable population problems. These chapters are in no means comprehensive. There has been considerable expansion in bargaining theory in recent years. Some of these recent developments include studies concerning changes in the disagreement point, uncertainty in the feasible set or the disagreement point (or both), bargaining problems with claims, non-convex problems, etc. We refer the reader to Thomson (1994) for an analysis of recent developments.

References

[1] Harsanyi, J.C. (1959). "A Bargaining Model for the Cooperative n-Person Games," in *Contributions to the Theory of Games IV, Annals of Mathematical Studies 40* (A.W. Tucker and R.D. Luce, eds) Princeton University Press, Princeton, NJ, 325-355.

[2] Lensberg, T. (1988). "Stability and the Nash Solution," *Journal of Economic Theory* 45, 330-341.

[3] Thomson, W. (1983a). "Problems of Fair Division and the Egalitarian Solution," *Journal of Economic Theory* 31, 211-226.

[4] Thomson, W. (1983b). "The Fair Division of a Fixed Supply Among a Growing Population," *Mathematics of Operations Research* 8, 319-326.

[5] Thomson, W. (1994). *Bargaining Theory: the Axiomatic Approach*, Academic Press, forthcoming.

[6] Thomson, W. and T. Lensberg (1989). *The Theory of Bargaining with a Variable Number of Agents*, Cambridge University Press.

Classical Cooperative Theory I: Core-Like Concepts

Sergiu Hart [1]

The Hebrew University of Jerusalem

1. Introduction

Pure bargaining games discussed in the previous two lectures are a special case of n-person cooperative games. In the general setup coalitions other than the grand coalition matter as well. The primitive is the *coalitional form* (or, "coalitional function", or "characteristic form"). The primitive can represent many different things, e.g., a simple voting game where we associate to a winning coalition the worth 1 and to a losing coalition the worth 0, or an economic market that generates a cooperative game. Von Neumann and Morgenstern (1944) suggested that one should look at what a coalition can guarantee (a kind of a constant-sum game between a coalition and its complement); however, that might not always be appropriate. Shapley and Shubik introduced the notion of a *C-game* (see Shubik (1982)): it is a game where there is no doubt on how to define the worth of a coalition. This happens, for example, in exchange economies where a coalition can reallocate its own resources, *independent* of what the complement does.

We assume we are given a coalitional function. Let N denote the set of players; a subset $S \subset N$ is called a *coalition*; $V(S)$ is the set of feasible outcomes for S.

How is an outcome defined? Assuming that some underlying utility functions for the players are specified, one can represent outcomes by the players' utilities. We thus use a payoff vector $a=(a^i)_{i \in S}$ in $\Re^S$ to represent an outcome, where a^i is player i'th utility of the outcome. So $V(S) \subset \Re^S$. Usually there are some assumptions made on the set $V(S)$; e.g., comprehensive, closed, convex, etc.

There are two special classes of games:

1) *Pure bargaining* games (PB): In these games only the grand coalition matters. Here $V(S)=\{x \in \Re^S$ such that $x^i \leq 0$ for all $i \in S\}$ for all[2] $S \neq N$.

---

[1] Lecture notes written by Yossi Feinberg.

[2] Sometimes this is relaxed to: $(0,\ldots,0) \in \mathrm{bd}V(S)$ for all $S \neq N$, where "bd" stands for boundary.

2) *Transferable utility* games (TU): (used to be called "games with side payments"). Here one number represents what a coalition can get, and the members of the coalition can arbitrarily divide this amount among themselves. So a TU game is of the form $V(S)=\{x \in \Re^S$ such that $\sum_{i \in S} x^i \leq v(S)\}$ where $v(S)$ is a number for all $S \subset N$. Geometrically these sets are half-spaces with normal vector $(1,...,1)$.

The general games are usually referred as games with *non-transferable utility,* or *NTU-games* for short. The following diagram shows the relationship between the different classes of games.

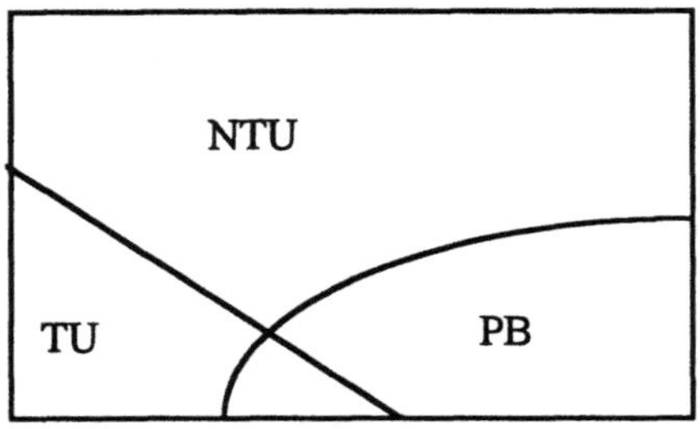

2. Solution Concepts

We will distinguish between two approaches to solution concepts (though the distinction is not always clear cut):

D - definition, description, discussion.

A - axiomatization.

Obviously there are other approaches, e.g., noncooperative, evolutionary, etc.

The D-approach stands for various formal or informal arguments, on how the solution has to look like. In the A-approach, one puts down a set of axioms and gets as a result that these axioms uniquely characterize the solution concept. Most solution concepts started out with the D-approach and only later where axiomatized; the Shapley Value started out with the A-approach.

A p.v. (*payoff vector*) is a vector $x \in \Re^N$. It is *feasible* if $x \in V(N)$, *efficient* (or Pareto optimal) if $x \in \mathrm{bd}V(N)$, and *individually rational* if $x^i \notin \mathrm{int}V(\{i\})$. The set $X := \{x |\ x$ is an efficient and individually rational p.v.$\}$ is called the set of *imputations*. For simplicity we assume that the set of imputations is always non-empty. Thus in the TU case we consider only games which satisfy $v(N) \geq \sum_{i \in N} v(i)$. A solution concept associates payoff vectors (outcomes) with each game.

3. The D-Approach

3.1 Core

The idea of the Core is to look at those payoff vectors which no coalition can improve upon. Let x be a feasible p.v.; a coalition $\emptyset \neq S \subset N$ *can improve upon* x if there is a feasible outcome y for S ($y \in V(S)$) which is better for S, that is $y^i > x^i$ for all i in S (everyone in S must agree that y is better than x); we write $y \succ_S x$. We say that y *dominates* x if there exists a coalition S such that $y \succ_S x$. The *Core* is thus defined as the set of imputations that are not dominated by any p.v..

In the NTU case a feasible p.v. x satisfies: $x \in \text{Core} \Leftrightarrow x^S \notin \text{int } V(S)$ for all[3] S.

In the TU case a feasible p.v. x satisfies: $x \in \text{Core} \Leftrightarrow x(S) \geq v(S)$ for all[4] S.

A first question one poses is the non-emptiness of the Core. This is connected with superadditivity which states that joining two coalitions may only increase their possibilities. The non-emptiness of the core is implied by *balancedness*, which is a generalization of superadditivity. Certain classes of games such as market games turn out to have a non-empty core (under general conditions). In the case of market games, the competitive (Walrasian) equilibrium is always in the Core. On the other hand, unless there is a veto player, voting games have an empty Core.

3.2 von Neumann and Morgenstern Stable Set

Recall that the Core is the set of all feasible p.v. that are not dominated by any p.v..

Consider now the following definition of a solution:

The "Shmore"[5] is the set of all feasible p.v. that are not dominated by any p.v. in the Shmore.

It turns out this concept is indeed well defined. The idea behind it is that the set of "good" p.v. is to be compared against "good" p.v.. The definition of the Shmore can be rewritten by $x \in \text{Shmore} \Leftrightarrow y \not\succ x$ for all $y \in \text{Shmore}$. This can be further restated as follows: Let K=Shmore, then

1) $x, y \in K \Rightarrow x \not\succ y$;

2) $y \notin K \Rightarrow$ there exists $x \in K$ such that $x \succ y$.

---

[3] We write x^S for the projection of x on $\Re^S$, i.e., $x^S = (x^i)_{i \in S}$.

[4] Here we define $x(S) = \sum_{i \in S} x^i$.

[5] This (temporary) name is due to R. J. Aumann.

A set K of efficient p.v. that satisfies these two conditions is called a *von Neumann and Morgenstern Stable Set*. Note that the basic concept here is a set concept, unlike the Core where the payoff vector's properties alone determine if it is in the solution. The Stable Set becomes a "standard of behavior" in the sense that if everyone believes that the solution is in K it will indeed be in K. Note that there may be more than one stable set in a game. Trivially, all von Neumann and Morgenstern Stable Sets contain the core.

There are games for which there are no Stable Sets. The first example was found by Lucas (1968) and was a 13 player game. Unfortunately even in simple cases it is difficult to calculate all (or even one of) the Stable Sets of a game. But finding this solution is very rewarding, it gives a lot of insight. For example, in voting games, where minimal winning coalitions seem important, it turns out that Stable Sets predict actually the formation of minimal blocking coalitions.

3.3 Bargaining Sets

The previous solution concepts are based on the idea that, given a p.v. , some coalitions or players may object to it (using other feasible p.v.). Along this line one can define a *counter objection* by objecting to the p.v. used for the original objection. Then follows the notion of *justified objections*, defined as objections that have no counter objection. Using these definitions one can define the Bargaining Set as the set of efficient p.v. for which there is no justified objection. This solution has many variants and was first conceived by Aumann and Maschler (1964); see Davis and Maschler (1963,1967), and also Mas-Colell (1989) for a new approach. The work on Bargaining Sets led to the following solution concept.

3.4 Nucleolus

This is a one point solution defined for the TU case. There are various suggestions for the generalization of the Nucleolus to the NTU case, but this is not yet settled.

Behind the notion of the Nucleolus is the following interpretation. Given a p.v. x each coalition S looks at $v(S)$-$x(S)$; this number represents the "complaint" of the coalition (it could be positive or negative). The higher the complaint the more loudly the coalition protests against x. Thus we want to minimize complaints under the "budget constraint" (the feasibility of x). We do so starting with the maximal complaint, i.e., we look at $Min_{p.v.\,x}\{Max_{S\subset N}(v(S)-x(S))\}$. Then we minimize the next highest complaint when considering only p.v. which minimized the highest complaint, and so on. What we get is the lexicographic minimum of all complaints. It turns out that we are left with a unique p.v. which is the Nucleolus. This solution concept is due to Schmeidler

(1969). The Nucleolus was applied to various problems, such as an airport landing fees problem in which airlines needed to share the cost of using runways (each coalition of airlines needing its distinct minimal runway length). We remark that when the Core is non-empty there is a feasible p.v. for which all complaints are non positive. Thus, in this case the complaints for the Nucleolus are non positive as well and we have that the Nucleolus is in the Core (it is moreover a special point in the Core, a kind of a symmetry center).

We have presented three kinds of solution concepts: one is a one-point solution (the Nucleolus); the second is a set of points (the Core), and the third is several sets of points (the Stable Sets).

4. The A-Approach

We move now to the second point of view on solution concepts, i.e., the axiomatic approach. One can always use the definition of a concept as its axiomatization, but obviously we would like to have more basic axioms with an intuitive meaning that characterize our concept. It should be noted that these characterizations are comparatively new.

All these axiomatizations have in common the *Consistency axiom* (also called the *Reduced Game Property*). Consistency is based on the following idea: Assume we have a game and its solution. Suppose that a certain set of players agree to the solution. The *reduced game* is the game played by the remaining players, on the remaining payoffs. Consistency requires that the solution of the reduced game be identical to the solution of the original game.

Formally, let (N,V) be an NTU game. Let x be a p.v. and $T \subset N$ be a coalition. We define the game (T,V^*) (where V^* depends on both x and T) by[6] $V^*(T)=\{y^T|(y^T,x^{T^c})\in V(N)\}$; i.e., we give x^{T^c} to the players in T^c and consider what we can give to the players in T so that the whole vector is feasible in the original game. For strict sub-coalitions $S \subset T$ $(S \neq T)$ we define $V^*(S)=\cup_{Q\subset T^c}\{y^S|(y^S,x^Q)\in V(S\cup Q)\}$; that is, we consider all sub-coalitions Q of T^c as those which can complement the members of S and create a feasible p.v. for S through a feasible p.v. for $S \cup Q$ in the original game.

The consistency or reduced game property states:

CONS: If x is a solution of (N,V) then x^T is a solution of (T,V^*) for all T.

It turns out that this definition of consistency yields many results.

6 $T^c=N\backslash T$ is the complement of T.

4.1 The TU Case

A solution associates a set of feasible p.v. for each game, i.e., it is a mapping $(N,v) \rightarrow \sigma(N,v)$ where $\sigma(N,v)$ is a set of feasible p.v. . We shall consider the set of games with non-empty Core. The axioms are:

NE (non-emptiness): $\sigma(N,v) \neq \varnothing$.

IR (individual rationality): In any p.v. in the solution every player gets at least what the game guarantees him, i.e., $x \in \sigma(N,v)$ implies $x^i \geq v(i)$ for all i.

CONS : (as above).

SUPA (superadditivity): $\sigma(N,v)+\sigma(N,w) \subset \sigma(N,v+w)$ where the summation here is a set summation. Note that the set of players is always the same.

SIVA (single valuedness): $|\sigma(N,v)|=1$.

AN (anonymity): If the games (N,v) and (N',v') are isomorphic, i.e., there exists a one-to-one mapping $\Pi: N \rightarrow N'$ such that $v(S)=v'(\Pi S)$ for all S, then $\Pi\sigma(N,v)=\sigma(N',v')$ (Π is a "relabeling of the players").

INV (TU invariance): For all $a>0$ and $b \in \Re^N$, if $w(S)=av(S)+\sum_{i \in S} b^i$ for all S then $\sigma(N,w)=a\sigma(N,v)+b$.

Theorem (Peleg (1986)): The Core is the unique solution concept satisfying NE,IR,CONS,SUPA.

(Note that this result requires a world with 3 players at least.)

Theorem (Sobolev (1975)): The PreNucleolus (defined in the same way as the Nucleolus, with respect to all efficient but not necessarily individually rational p.v. x) is the unique solution concept satisfying SIVA,AN,INV,CONS.

(This result requires a world with an infinite number of players.)

The axiomatization of the Stable Sets is an open problem.

4.2 The NTU Case

Theorem (Peleg (1985)): The Core is the unique solution concept satisfying NE,IR,CONS (under some regularity conditions on the games considered).
(Note that this result requires a world with an infinite number of players.)

References

Aumann, R.J., and M. Maschler 1964. 'The bargaining set for cooperative games', in: M. Dresher, L.S. Shapley and A.W. Tucker, eds., *Advances in games theory*. Princeton, NJ: Princeton University Press. pp. 443-476.

Lucas, W.F., 1968. 'A game with no solution', *Bulletin of the American Mathematical Society*, 74: pp. 237-239.

Lucas, W.F., 1992. 'Von Neumann-Morgenstern stable sets', in: R.J. Aumann and S. Hart, eds., *Handbook of Game Theory with Economic Applications*. Vol. I, ch. 17. Amsterdam, The Netherlands: Elsevier Science Publishers.

Maschler, M., 1992. 'The bargaining set, kernel, and nucleolus', in: R.J. Aumann and S. Hart, eds., *Handbook of Game Theory with Economic Applications*. Vol I, ch. 18. Amsterdam, The Netherlands: Elsevier Science Publishers.

Mas-Colell, A., 1989. 'An equivalence theorem for a bargaining set', *Journal of Mathematical Economics*, 18:pp. 129-139.

Peleg, B., 1985. 'An axiomatization of the core of cooperative games without side payments', *Journal of Mathematical Economics*, 14: pp. 203-214.

Peleg, B., 1988. 'On the reduced game property and its converse', *International Journal of Game Theory*, 15: pp. 187-200.

Peleg, B., 1992. 'Axiomatizations of the core', in: R.J. Aumann and S. Hart, eds., *Handbook of Game Theory with Economic Applications*. Vol I, ch. 13. Amsterdam, The Netherlands: Elsevier Science Publishers.

Schmeidler, D., 1969. 'The nucleolus of a characteristic function game', *SIAM Journal of Applied Mathematics*, 17: pp. 1163-1170.

Shubik, M., 1982. *Game Theory in the Social Sciences: Concepts and Solutions*. Cambridge, MA: The MIT Press (2nd ed., 1983). pp. 130-131.

Sobolev, A.I., 1975. 'The characterization of optimality principles in cooperative games by functional equations', in: N.N. Vorobjev, ed., *Matematischeskie metody v socialnix naukakh*, Proceedings of the Seminar, Issue 6, Vilnius: Institute of Physics and Mathematics, Academy of Sciences of the Lthuanian SSR. pp. 94-151. [in Russian, English summary].

von Neumann, J., and O. Morgenstern, 1944. *Theory of Games and Economic Behavior*. Princeton, NJ: Princeton University Press (2nd ed., 1947).

Classical Cooperative Theory II: Value-Like Concepts

Sergiu Hart [1]

The Hebrew University of Jerusalem

1. Introduction

The *Value* is a solution concept originally due to Shapley (1953). The idea behind the concept is to evaluate how much will a player be willing to pay to participate in a given game. It seeks to represent what the game is worth for a player. In some sense the value captures the expected outcome of the game. We will start with the TU (transferable utility) framework and the axiomatic approach and then consider various extensions to the NTU (non transferable utility) case.

2. TU Games

A TU game (N,v) is defined by associating a real number $v(S)$ to every coalition $S \subset N$ (put $v(\emptyset)=0$). $v(S)$ is referred to as the worth of the coalition S, i.e., what the members of the coalition S can divide between them. The value will associate one p.v. (payoff vector) with each game; i.e., the value of a game will be denoted $\varphi(N,v) \in \Re^N$ where $\varphi^i(N,v)$ is the value of player i in the game (N,v).

Shapley (1953) started out with the following set of axioms.

EFF (Efficiency) (also called PO-Pareto optimality):
$\varphi(N,v)$ satisfies $\quad \Sigma_{i \in N} \; \varphi^i(N,v)=v(N)$.

ET (Equal Treatment) (usually called symmetry):
Two players $i,j \in N$ are called substitutes in (N,v) if $v(S \cup \{i\})=v(S \cup \{j\})$ for all coalitions S such that $i,j \notin S$. The axiom states that if i,j are substitutes in the game (N,v) then $\varphi^i(N,v)=\varphi^j(N,v)$.

---

[1] Lecture notes written by Yossi Feinberg.

NP (Null Player) (or Dummy):
A player i is a null player in (N,v) if $v(S\cup\{i\})=v(S)$ for all S. The axiom states that if i is a null player in (N,v) then[2] $\varphi^i(N,v)=0$.

ADD (Additivity):
For all players i in N, $\varphi^i(N,v)+\varphi^i(N,w)=\varphi^i(N,v+w)$, where the game $(N,v+w)$ is defined by $(v+w)(S)=v(S)+w(S)$ for all coalitions[3] S.

Theorem (Shapley (1953)):
a) There exists a unique φ satisfying EFF, ET, NP, ADD on the class of all TU games.
b) φ is given by $\varphi^i(N,v)=E_{\{S|i\notin S\}}[v(S\cup\{i\})-v(S)]$.

The formula of φ given in b) can be explained as follows. We look at $v(S\cup\{i\})-v(S)$ which is the marginal contribution of player i to the coalition S, and average the marginal contributions according to a distribution over S. This distribution can be defined by looking at the players in a random order, where the players preceding player i in a given order form the coalition S. One may think of the players as entering a room one by one in a random order, and averaging the contribution of player i, when he enters the room, to the players already in the room.

There is some distant indication of the idea of marginality in the NP (Null Player) axiom. There, a player which always contributes marginally zero gets zero value, but this is still very far from the marginal contributions that appear in the formula. It is easy to see that the four axioms are satisfied by $\varphi^i(N,v)=E_{\{S|i\notin S\}}[v(S\cup\{i\})-v(S)]$ (to see that EFF is satisfied notice that if we take the marginal contributions of all players for a given order the terms in the summation cancel each other out and we are left with $v(N)$ which is averaged over all orders). We now sketch the proof in the other direction, i.e., that the axioms imply the Shapley value. Consider a unanimity game u_T for a given set of players T, i.e., u_T is defined so that all coalitions S that contain T have $u_T(S)=1$ and all the other coalitions get 0. By NP, players outside T must get 0, and by EFF and ET all players in T must get $1/\#T$. These unanimity games form a basis of the linear space of games. Thus by[4] ADD we get that the solution has to be the Shapley value since both are linear and agree on the basis of unanimity games.

As was mentioned above, the underlying notion here is "marginality", i.e., the value of a player is only a function of his marginal contributions. Thus we introduce a new axiom.

---

[2] In the original paper by Shapley this axiom was combined with the EFF axiom.

[3] Sometimes a different version of this axiom is used. It corresponds to playing the average game, and it requires that $1/2\varphi^i(N,v)+1/2\varphi^i(N,w)=\varphi^i(N,(v+w)/2)$.

[4] Note that the value of cu_T is c times the value u_T for all constants c, again by NP, EFF and ET.

MARG (Marginality):
Let (N,v) and (N,w) be two games (with the same set of players N). Let $i \in N$. If $v(S \cup \{i\})-v(S)=w(S \cup \{i\})-w(S)$ for all S such that $i \notin S$, then $\varphi^i(N,v)=\varphi^i(N,w)$.

It turns out that ADD which is considered as a strong axiom and NP can be replaced by MARG.

Theorem (Young (1985)):
The Shapley value is the unique solution concept satisfying EFF, ET, MARG.

One can obtain the Shapley value through yet another approach. Consider what player j contributes to the *value* of player i, i.e., by how much the value of i changes when j "drops out": $\varphi^i(N,v)-\varphi^i(N\backslash\{j\},v)$, (in the second term we consider the subgame of v with player set $N\backslash\{j\}$). It turns out that the Shapley value satisfies that j'th contribution to i'th value always equals i'th contribution to j'th value, i.e., $\varphi^i(N,v)-\varphi^i(N\backslash\{j\},v)=\varphi^j(N,v)-\varphi^j(N\backslash\{i\},v)$. Moreover these equations for all i,j together with EFF characterize uniquely the Shapley value (Myerson (1980) and Hart and Mas-Collel (1989)). The marginal contributions above have a structural resemblance to derivatives and the requirement of equal contributions reminds us of the mixed derivatives condition of Frobenius. Taking this line of thought even further implies the existence of a "potential function" whose "gradient vector" is the value p.v. . One defines a real valued function P on games as a *potential* if it satisfies $\sum_{i \in N} [P(N,v)-P(N\backslash\{i\},v)]=v(N)$ for all games (N,v).

Theorem (Hart and Mas-Collel (1989))
There exists a unique function P satisfying $\sum_{i \in N} [P(N,v)-P(N\backslash\{i\},v)]=v(N)$; moreover, its "derivative" is the Shapley value, i.e., $\varphi^i(N,v)=P(N,v)-P(N\backslash\{i\},v)$.

In the first part of this lecture where we discussed core-like concepts the consistency axiom was shown to play a major role in axiomatization. Using a related definition of consistency yields another characterization of the Shapley value.

Theorem (Hart and Mas-Colell (1989))
EFF, ET, INV for[5] 2 players games and CONS characterize the Shapley value[6].

Here CONS is defined as follows: let σ be a one point solution concept. Let (N,v) be a game and T a subcoalition of N, we define the game (T,v^*) by

---

[5] INV is the axiom of covariance with respect to linear transformation (see part 1 of the lecture).

[6] Notice the similarity between this characterization and Sobolev's characterization of the Nucleolus; the difference hinges on the definition of CONS.

$v^*(S)=v(S\cup T^c)-\sigma(S\cup T^c,v)(T^c)$. Here we assume that all the players in T^c agree that σ is a good solution concept. We look at the subgame $(S\cup T^c,v)$ and subtract what the solution gives to T^c from what the grand coalition in this game gets. The axiom states that the solution of the reduced game (T,v^*) is the restriction to T of the solution to the game (N,v).

Unlike the previous definition of CONS, we use here the solution concept itself - for subgames - to define the reduced game. It turns out that in cost allocation problems this form of reduced games has a more natural appeal and indeed the Shapley value is used in such problems. The notion of value has been also extensively applied to voting games (weighted majority games). In such games the core is often empty and there may be many stable sets (some are hard to find). Clearly, the total number of votes a coalition has is usually different from its value. In these games the Shapley value (known as the Shapley-Shubik index) is best viewed as the probability that a player is pivotal. For example, if we have one big party and many small parties, the value of the large party is higher than its share of votes. But with two large parties and many small ones the power of the large parties is greatly diminished and is lower than their actual share of the votes. These phenomena implied by the value are frequently observed in practice. The value is easy to apply and is very tractable, thus it becomes a most applied solution concept. Note that the value gives us a kind of an expected outcome. Furthermore the value is linear unlike the core and the nucleolus which are only piecewise linear.

3. NTU Games

An NTU game (N,V) is defined by associating a set $V(S)\in\Re^S$ for every coalition S. Consider the following diagram of classes of games (see previous chapter).

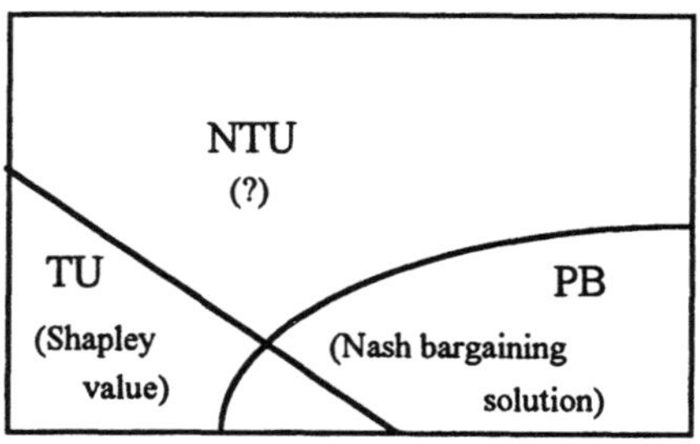

If we take the classical solutions: the Shapley value in the TU-case, and the Nash bargaining solution in the PB-case, one would like to extend these solutions to the whole space of NTU games.

The Nash solution to 2-person PB games is characterized by the "equal angles" property, i.e., the marginal rate of substitution between the players' utilities at the solution point is equal to the ratio of the payoffs. This is shown in the following diagram.

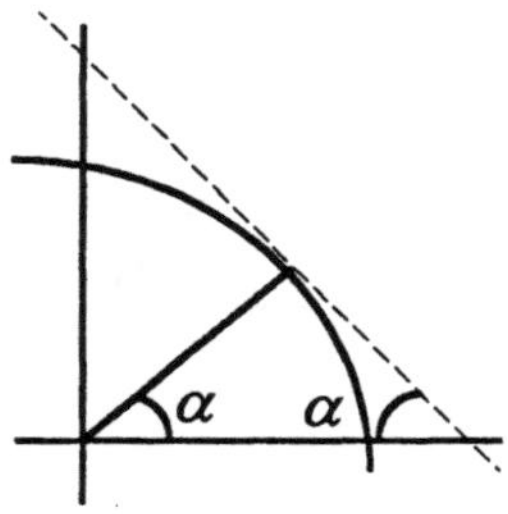

There are two underlying criteria that appear here. Let c be a positive number. The first is the Utilitarian "local efficiency" criterion which means that we are on the Pareto frontier and the marginal rates of substitution there are precisely c The second is the Egalitarian criterion which requires the payoffs to be distributed according to the ratio given by c. These two criteria are jointly satisfied at the Nash solution: i.e., they are satisfied for the same c. We would like to extend these criteria to the general NTU case. To do so, the first step is to define an egalitarian solution relative to a vector λ of utility comparison weights (where λ is a strictly positive vector in $\Re^N$). This solution is called the *λ-egalitarian* solution. It tries to capture the idea that the gains from cooperation are split equally among the players (hence comparison weights are needed). The second step consists in endogenizing the determination of the comparison weights λ. This is done by demanding that λ be such that the λ-egalitarian solution be also λ-utilitarian, i.e., that it maximizes the sum of λ-rescaled payoffs. A fixed point theorem asserts that there are weights which will yield such a value. This approach is due to Harsanyi (1963).

A different approach was given by Shapley (1969). He looked at the induced TU game (N, v_λ), given a vector λ of utility comparison weights: $v_\lambda = Max\,\{\sum_{i \in S} \lambda^i x^i \,|\, x \in V(S)\}$.

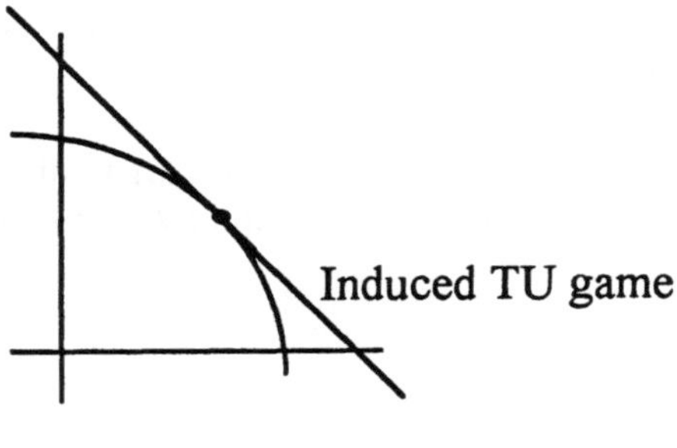

The Shapley value of the TU-induced game will be somewhere on the hyper-plane and may well be non feasible in the NTU game as is shown in the diagram below.

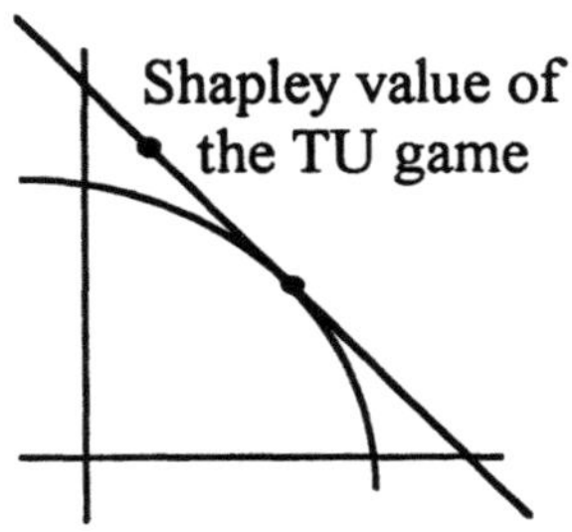

If the value of the TU game is feasible (thus the two points in the diagram above coincide) we get the Shapley NTU value. Again a fixed point argument ensures that the NTU value exists. Here we start with the marginal rate of utility substitution and check if it corresponds to the induced TU Shapley value.

There are other NTU values and some applications of values (e.g., in market games), there are also axiomatizations of the Shapley NTU value (Aumann(1985)) and of the Harsanyi NTU value (Hart(1985)).

When comparing the two approaches, i.e., Shapley's vs. Harsanyi's, it seems that Shapley's approach considers more the effect of the grand coalition on the expanse of smaller subcoalitions, whereas the Harsanyi's approach does the opposite. This can be clearly seen in their axiomatizations.

One should also mention a third NTU value. This is the consistent Maschler and Owen (1989,1992) NTU value. The natural extension of CONS to NTU games is self contradicting (no solution satisfies it), so they defined an "average" reduced game. Thus they have an NTU value with a notion close in spirit to the CONS.

This entire lecture has considered the traditional approach alone. It will be seen in a later lecture how these solutions may emerge from the non cooperative approach.

References

Aumann, R.J., 1985. 'An axiomatization of the non-transferable utility value', *Econometrica* 53: pp.599-612..

Harsanyi, J.C., 1963. 'A simplified bargainning model for the n-person cooperative game', *International Economic Review* 4, pp. 194-220.

Hart, S., 1985. 'An axiomatization of Harsanyi's non-transferable utility solution', *Econometrica* 53, pp. 1295-1313.

Hart, S., 1990. 'Advances in value theory', in: T. Ichiishi, A. Neyman, and Y. Tauman, eds., *Game Theory and Applications*. pp. 166-175. San Diego, CA. Academic Press, Inc.

Hart, S., and A. Mas-Colell, 1989. 'Potential, value, and consistency', *Econometrica* 57, pp. 589-614.

Maschler, M., and G. Owen, 1989. ' The consistent Shapley value for hyperplane games', *International Journal of Game Theory* 18, pp. 389-407.

Maschler, M., and G. Owen, 1992. ' The consistent Shapley value for games without side payments', in: *Rational Interaction*, R. Selten ed., Springer-Verlag, pp. 5-12.

Myerson, R., 1980. 'Conference structure and fair allocation rules', *International Journal of Game Theory* 9, pp. 169-182.

Shapley, L.S., 1953. 'A value for n-person games', in: *Contributions to the Theory of Games II (Annals of Mathematics Studies* 28), H. Kuhn and A. Tucker, eds., Princeton, NJ: Princeton University Press. pp. 307-317.

Shapley, L.S., 1969. 'Utility comparison and the theory of games', in: *La Decision*. Paris: Editions du CNRS, pp. 251-263.

Young, H.P., 1985. 'Monotonic solutions of cooperative games', *International Journal of Game Theory* 14: pp. 65-72.

Cooperative Theory with Incomplete Information

Beth Allen

Department of Economics, University of Minnesota, Minneapolis, MN 55455 USA and Federal Reserve Bank of Minneapolis, 250 Marquette Avenue, Minneapolis, MN 55480 USA

Abstract. This paper surveys cooperative game theory when players have incomplete or asymmetric information, especially when the *TU* and *NTU* games are derived from economic models. First some results relating balanced games and markets are summarized, including theorems guaranteeing that the core is nonempty. Then the basic pure exchange economy is extended to include asymmetric information. The possibilities for such models to generate cooperative games are examined. Here the core is emphasized as a solution, and criteria are given for its nonemptiness. Finally, an alternative approach is explored based on Harsanyi's formulation of games with incomplete information.

Keywords. Asymmetric information, incomplete information, core, market games, *NTU* games, *TU* games

1 Introduction

This paper considers the incorporation of uncertainty and information into cooperative game theory. In particular, we examine the cooperative games that arise from economies with asymmetric information. To simplify, we focus on the case of general equilibrium models of perfectly competitive pure exchange economies.

One frequently encounters the opinion that cooperative game theory cannot easily be adapted to include informational considerations. In fact, economists' interest in asymmetric information is sometimes cited as an important reason for the recent emphasis on noncooperative models of strategic behavior, especially in fields such as industrial organization and corporate finance. I disagree with this viewpoint—one *can* put asymmetric information into cooperative games, albeit at the expense of certain complications which may lead to somewhat surprising results.

Since we stress cooperative games that are derived from economies with asymmetric information, we first digress to present a more general, brief survey of the relationships between cooperative games and perfectly competitive exchange economies. After summarizing these results on market games in the next section, we proceed to introduce information in the following section. Section 4 is devoted to Wilson's article on the core with asymmetric information. The derivation of cooperative games from economies with asymmetric information is examined in Section 5, as preparation for analysis of the core and the value in the following two sections. Section 8 concludes by presenting an alternative approach based on Harsanyi's formulation of noncooperative games with incomplete information.

2 Market Games

To fix notation, let N be the (finite) set of traders in the economy (or players in the game), and denote a typical agent by $i \in N = \{1, \ldots, n\}$ (where n is the cardinality of the set N). Suppose that the number of commodities present in the economy is the finite positive integer ℓ, and take $\mathbb{R}^{\ell}_{+}$ to be the consumption set of each trader $i \in N$. Traders are specified by initial endowment vectors and utility functions where, for each $i \in N$, $e_i \in \mathbb{R}^{\ell}_{+}$ and $u_i : \mathbb{R}^{\ell}_{+} \to \mathbb{R}$ is a continuous (or, more generally, upper semicontinuous) and concave function representing the preferences of trader $i \in N$. By definition, a *coalition* is a nonempty subset of agents; the *grand coalition* N is a coalition as is any nontrivial collection of agents. Each coalition induces a smaller economy containing only those traders who belong to the coalition; such subeconomies are called *submarkets*, and they induce *subgames*.

An n-player *cooperative game with transferable utility* (or *TU* game) is a function $v : 2^N \to \mathbb{R}$ with $v(\emptyset) = 0$, where 2^N denotes the set of all subsets of $N = \{1, \ldots, n\}$. The *TU* cooperative game induced from the pure exchange economy as above, in which each trader $i \in N$ has consumption set $\mathbb{R}^{\ell}_{+}$, initial endowment $e_i \in \mathbb{R}^{\ell}_{+}$, and utility function $u_i : \mathbb{R}^{\ell}_{+} \to \mathbb{R}$ which is assumed to be upper semicontinuous (so that maxima in the definition below are well defined), is given by $v : 2^N \to \mathbb{R}$ with $v(\emptyset) = 0$ and, for all $S \subseteq N$ with $S \neq \emptyset$, $v(S) = \max\{\sum_{i \in S} u_i(x_i) | x_i \in \mathbb{R}^{\ell}_{+}$ for all $i \in S$ and $\sum_{i \in S} x_i \leq \sum_{i \in S} e_i\}$. [With sufficient monotonicity, the inequality sign can be replaced by an equality.] In words, $v(S)$ is the maximum total utility that the players in S can achieve by redistributing their own resources; if members of coalition S were to pool all of their initial endowments and redistribute these goods, so as to maximize the total utility of the entire coalition S, the resulting sum would equal $v(S)$.

The *TU core* of the n-person *TU* game v is defined to be the set of all payoff vectors $w = (w_1, \ldots, w_n) \in \mathbb{R}^n$ such that (1) $(w_1, \ldots, w_n)$ is feasible (for N): $\sum_{i \in N} w_i \leq v(N)$, and (2) $(w_1, \ldots, w_n)$ is not *blocked* by any coalition: $\sum_{i \in S} w_i \geq v(S)$ for all $S \subseteq N$. Feasibility of $w \in \mathbb{R}^n$ says that $(w_1, \ldots, w_n)$

is an *imputation* for v. The second property is sometimes described by the statement that no coalition can improve upon w.

Transferable utility games with nonempty cores [note that the core always exists] can be characterized by a balancedness condition. If N is any finite set, a balanced family $\mathcal{B}$ of subsets of N is $\mathcal{B} \subseteq 2^N$ for which there exist balancing weights $\{\gamma_S\}_{S\in\mathcal{B}}$ with $\gamma_S \geq 0$ ($\gamma_S \in \mathbb{R}$) such that for each $i \in N$, $\sum_{S\ni i}\gamma_S = 1$. Obvious examples of balanced families include N itself (with weight $\gamma_N = 1$) and $\mathcal{B} = \{\{1\},\ldots,\{n\}\}$ (again with each balancing weight equal to one). For a nontrivial example, consider the two-player coalitions of $N = \{1,2,3\}$ and set $\gamma_S = 1/2$ for $S = \{1,2\}$, $S = \{1,3\}$, and $S = \{2,3\}$. A *TU* game on N is said to be ***balanced*** if, for all balanced collections $\mathcal{B}$ of subsets of N and all collections of associated balancing weights $\{\gamma_S\}_{S\in\mathcal{B}}$, we have $\sum_{S\in\mathcal{B}}\gamma_S v(S) \leq v(N)$. [Note that the left-hand side of this inequality differs from the operation of taking convex combinations in that we could have $\sum_{S\in\mathcal{B}}\gamma_S > 1$.]

Theorem 1 *A finite TU game has a nonempty core if and only if the game is balanced.*

This result was discovered independently by Bondareva (1962) and Shapley (1967). Its proof involves demonstrating that a certain system of linear inequalities has a solution precisely when the constraints defining balancedness are satisfied.

For an example of a game that fails to be balanced, again let $N = \{1,2,3\}$, and define (with an obvious abuse of notation) $v(123) = v(12) = v(13) = v(23) = 1$ and $v(S) = 0$ otherwise. Then v is not balanced, since examination of the balanced family of two-player coalitions would require $v(12)/2 + v(13)/2 + v(23)/2 \leq v(123) = 1$, an obvious contradiction. Intuitively, we know that the *TU* core of this game is empty, because any two-player coalition that does not include the best treated player(s) can block any (feasible) imputation. The game describes a situation ("three men and a trunk") in which three people discover buried treasure which can be removed from the jungle only if at least two individuals carry it.

The Bondareva-Shapley Theorem is of particular interest because it applies to all market games as described above. Moreover, there is an equivalence between games satisfying a stronger balancedness property and those games that can be derived from pure exchange economies satisfying the conditions stated above. A ***totally balanced*** game is one for which every subgame is balanced.

Theorem 2 *Every market game derived from a finite pure exchange economy in which each trader i has consumption set $\mathbb{R}^\ell_+$, initial endowment $e_i \in \mathbb{R}^\ell_+$, and utility function $u_i : \mathbb{R}^\ell_+ \to \mathbb{R}$, which is upper semicontinuous and concave, is totally balanced. Conversely, every n-player totally balanced TU cooperative game can be generated by a pure exchange economy as above with $\ell = n$.*

This result was discovered by Shapley and Shubik (1969) in their classic study of market games. The proof in one direction uses the balancing weights to define feasible allocations that are convex combinations of allocations available to smaller coalitions. Concavity of utilities then implies, by Jensen's inequality, that total utility cannot be forced to decrease in the larger coalition. For the converse, Shapley and Shubik (1969) construct very special economies in which each player's payoff essentially depends only on the player's allocation of one commodity. [Note that the relations between exchange economies and totally balanced games cannot be described by a one-to-one correspondence because the space of n-player games can be identified with a Euclidean space of dimension $2^n - 1$, whereas the space of n-agent exchange economies parameterized by endowments and utilities must be infinite-dimensional. More specifically, changing a trader's utility function off of the compact set of feasible allocations cannot alter the TU game generated by the economy.]

Cooperative games with nontransferable utility (or NTU games) can similarly be derived from economies. Of course, NTU games are preferable in general for economics, as they do not require one to impose the assumption that each agent's preferences are representable by a quasilinear utility function in order to justify the addition of payoffs of different agents. [A quasilinear utility is a function of the form $u(x) + m$, where x can be a vector of goods and m denotes the quantity of a commodity—such as money—in which side payments are made.]

Recall that an NTU cooperative game with player set $N = \{1, \ldots, n\}$ is a correspondence $V : 2^N \to \mathbb{R}^n$ such that, for all $S \subseteq N$, the sets $V(S)$ are nonempty, closed, and comprehensive [i.e., $V(S) \supseteq V(S) - \mathbb{R}^n_+$], and, moreover, the $V(S)$ sets are ***cylinder sets*** in that if $u = (u_1, \ldots, u_n) \in V(S)$ and if $u' = (u'_1, \ldots, u'_n)$ is such that $u_i = u'_i$ for all $i \in S$, then $u' \in V(S)$. I follow the convention that $V(\emptyset) = \mathbb{R}^n$. Define the ***projections*** of the $V(S)$ sets into the subspace of payoffs for players in S by $V(S)_S = \{u \in V(S) | u_j = 0$ if $j \notin S\}$. Note that $V(N)_N = V(N)$ and $V(\emptyset)_\emptyset = \{0\}$. In addition, for each $S \subseteq N$, $V(S)_S$ generates the cylinder set $V(S)$.

A cooperative game $V : 2^N \to \mathbb{R}^n$ with nontransferable utility is *balanced* if, for all balanced collections $\mathcal{B}$ on N with associated weights γ_T for $T \in \mathcal{B}$, $V(N) \supseteq \sum_{T \in \mathcal{B}} \gamma_T V(T)_T$. [Note that since $\mathcal{B} = \{N\}$ with $\gamma_N = 1$ is a balanced collection, taking the union over all balanced collections on the right-hand side gives a subset of $\mathbb{R}^n$ which precisely equals $V(N)$.] This definition of balancedness is well suited for economies with concave utilities. An alternative definition, which Billera (1974) terms "quasibalancedness," is weaker. Say that an NTU game $V : 2^N \to \mathbb{R}^n$ is *quasibalanced* if, for all balanced collections $\mathcal{B}$ on N, $\bigcap_{T \in \mathcal{B}} V(T) \subseteq V(N)$. Every balanced game is quasibalanced. As in the case of transferable utility, games with nontransferable utility are said to be *totally (quasi)balanced* if all of their subgames are (quasi)balanced.

The core of an *NTU* game is defined to be the set of feasible imputations that cannot be blocked—or improved upon—by any coalition. Formally, $u = (u_1, \ldots, u_n) \in \mathbb{R}^n$ belongs to the *core* of $V : 2^N \to \mathbb{R}^n$ if and only if $u \in V(N)$ and there does not exist a coalition $S \subseteq N$ (with $S \neq \emptyset$) and a payoff vector $u' = (u'_1, \ldots, u'_n) \in V(S)$ such that $u'_i > u_i$ for all $i \in S$. Note that, by definition, the core always exists—every game has a core, although it may be empty. Of course, we are interested in games with nonempty cores. One rationale for the core as a solution concept is the observation that, although not all points in the core may be attractive solutions, whenever the core is nonempty we may be justified in eliminating noncore outcomes from further consideration.

Theorem 3 *Every quasibalanced NTU game has a nonempty core.*

This result was proved by Scarf (1967). The implication holds in one direction only; in contrast to the *TU* case, one does not have equivalence. Moreover, because every balanced game is quasibalanced, Scarf's Theorem implies that every balanced game (as defined above) has a nonempty core.

Now let us return to our model of an exchange economy and show how it generates a well-behaved game with nontransferable utility. As before, we permit each coalition to redistribute its own resources provided that every coalition member receives an allocation belonging to the consumption set $\mathbb{R}^\ell_+$. Accordingly, define $V : 2^N \to \mathbb{R}^n$ by $V(\emptyset) = \mathbb{R}^n$ and for each nonempty $S \subseteq N$, $V(S) = \{(w_1, \ldots, w_n) \in \mathbb{R}^n |$ there exists $(x_1, \ldots, x_n) \in \mathbb{R}^{\ell n}$ with $\sum_{i \in S} x_i \leq \sum_{i \in S} e_i$, $x_i \in \mathbb{R}^\ell_+$ for each $i \in S$, and $w_i \leq u_i(x_i)$ for all $i \in S\}$.

By definition, the $V(S)$ are comprehensive cylinder sets. They're compactly generated (and, hence, closed as the sum of a closed set and a compact set) by the upper semicontinuity of utility functions. This implies that each $V(S)$ set is bounded above in all of the coordinates corresponding to players in S or, equivalently, that the $V(S)_S$ sets are bounded above. Moreover, concavity of utility functions implies that each $V(S)$ or $V(S)_S$ set is convex.

Finally, the economic model specified above gives rise to an *NTU* game which is totally balanced. The proof uses convex combinations of feasible allocations and concavity of utilities. This implies the following desirable property.

Theorem 4 *A finite pure exchange economy, having n agents $i = 1, \ldots, n$ with consumption sets $\mathbb{R}^\ell_+$, initial endowments $e_i \in \mathbb{R}^\ell_+$, and utilities $u_i : \mathbb{R}^\ell_+ \to \mathbb{R}$ which are assumed to be concave and upper semicontinuous, generates a totally balanced NTU game so that the game and all of its subgames have nonempty cores.*

Billera (1974), Billera and Bixby (1974), and Mas-Colell (1975) examine whether totally balanced NTU games satisfying the properties mentioned above can be generated by economies. The results are less sharp than those for the TU case and require technical restrictions which are not discussed here.

An extremely useful reference for much of this material is the book by Hildenbrand and Kirman (1976). The Shapley and Shubik (1969) article is also accessible. Needless to say, all students interested in cooperative game theory should read the following papers relating balancedness to the property of having a nonempty core: Bondareva (1962), Shapley (1967), and Scarf (1967).

3 Economies with Asymmetric Information

This section explains how one can add information to the basic model of a pure exchange economy. We are interested in situations in which different agents may initially possess different information. Moreover, the information must matter to traders.

To model these phenomena, we begin with an arbitrarily given abstract set Ω of states of the world. Elements ω of the set Ω are assumed to completely describe the relevant uncertainty in the universe. A σ-field $\mathcal{F}$ of measurable subsets of Ω is also given. Subsets of Ω that belong to $\mathcal{F}$ are also termed *events*. Technically, $(\Omega, \mathcal{F})$ is a measurable space. Finally, $(\Omega, \mathcal{F})$ is endowed with a (σ-additive) probability measure μ. [This could be generalized to permit agents to have different subjective probabilities regarding the *ex ante* likelihood of various events in Ω, provided that all agree about the null events—those which occur with probability zero.]

The information of trader $i \in N$ is given by a sub-σ-field $\mathcal{G}_i$ of $\mathcal{F}$. Notice that information becomes an *ex ante* concept, in that it means the capacity to condition one's actions on a particular sub-σ-field, where the agent knows which sub-σ-field can be used. Thus, information is like an entire random variable (or measurable function from Ω to $\mathbb{R}$), rather than a single observation of the random variable (or a real number which equals the function evaluated at a specific $\bar{\omega} \in \Omega$). Another analogy is that one should think of information as access to an instrument or measuring device, not as a measurement which is the output of the instrument. In particular, information is not equivalent to the fact that a certain state $\bar{\omega}$ has actually occurred. Note that asymmetric information is sometimes called *differential information,* while *incomplete information* properly refers to situations in which $\mathcal{G}_i$ is smaller than $\mathcal{F}$, regardless of whether the $\mathcal{G}_i$ may be different for different agents. Symmetric information is a special case of asymmetric information, and complete information is a special case of incomplete information.

A simpler model which captures most of the main ideas starts from a finite set Ω of states of the world, where each state occurs with strictly positive probability. Agents' information is specified by partitions of Ω. When state ω occurs, the agent learns the (unique) element of the partition containing ω.

States of the world can also be interpreted as signals (about some underlying fundamental states of the world). However, rather than using dual terminology to include this case, I prefer to think of a state of the world as an n-tuple of the signals that have been received by each agent.

The state of the world can affect traders' endowments and utilities. We formalize this by two measurable functions defined on Ω. For each $i \in N$, trader i's initial endowments are given by $e_i : \Omega \to \mathbb{R}_+^\ell$ which is $\mathcal{G}_i$-measurable. The restriction to $\mathcal{G}_i$-measurability (rather than $\mathcal{F}$-measurability) means that trader i must know his or her own initial endowment; the endowment vector can depend only on the trader's own information. If Ω is infinite, we assume further that each e_i is uniformly bounded almost surely in order to avoid technicalities; this condition is automatically satisfied for finite Ω. State-dependent utilities are frequently written as functions $u_i : \mathbb{R}_+^\ell \times \Omega \to \mathbb{R}$ which are continuous on $\mathbb{R}_+^\ell$ and $\mathcal{F}$-measurable on Ω (so that they're jointly measurable). We use $\mathcal{F}$-measurability instead of $\mathcal{G}_i$-measurability, because we envision that traders eventually learn their true utilities upon consumption of their allocations, but they may not fully know their state-dependent utilities when they make trades or choose strategies. The essential uncertainty here pertains to one's own preferences. We assume (again, to avoid potential difficulties of a technical nature) that for almost all $\omega \in \Omega$ and all $i \in N$, the utility functions $u_i(\cdot;\omega) : \mathbb{R}_+^\ell \to \mathbb{R}$ are not only continuous, but also strictly concave and strictly monotone. [For technical reasons (based on the fact that proper regular conditional probability distributions are defined only up to null sets) state-dependent utilities should be specified by $\mathcal{F}$-measurable functions $U_i : \Omega \to C(\mathbb{R}_+^\ell, \mathbb{R})$, where the space $C(\mathbb{R}_+^\ell, \mathbb{R})$ of continuous functions from $\mathbb{R}_+^\ell$ to $\mathbb{R}$ is endowed with the Borel σ-field corresponding to the topology of uniform convergence on compact subsets, which makes $C(\mathbb{R}_+^\ell, \mathbb{R})$ into a Frechet space. All conditional expectations are taken with respect to the induced image measure on $C(\mathbb{R}_+^\ell, \mathbb{R})$, not on the abstract probability space $(\Omega, \mathcal{F}, \mu)$. Here one assumes that for all $i \in N$ and for almost all $\omega \in \Omega$, $U_i(\omega)$ is strictly monotone and strictly concave; it may also be convenient or necessary to assume that the (unconditional) distribution on $C(\mathbb{R}_+^\ell, \mathbb{R})$ has compact support for all $i \in N$.]

An important conceptual problem with asymmetric information models is that one must carefully delineate those actions (i.e., trades or strategies) among which agents may choose. Radner (1968) considers the question of what people can do in a market when they have asymmetric information. He proposes that one should be able to verify one's own (net) trades. For example, you will never pay a strictly positive amount to sign a contract with me stating that I will give you $100 if I do not have a headache tomorrow morning. If you do, I can always tell you that I have a headache, and you can never know that I'm lying. [You also can't prove to a third party that I'm lying, which exemplifies the issue of verification rather than asymmetric information.] In competitive equilibrium models, agents trade impersonally with the market, which means that one's own net trade should depend only on information available to the agent at the time the market meets. Hence, the individual excess demand of agent i should be $\mathcal{G}_i$-measurable, which implies (because e_i is $\mathcal{G}_i$-measurable) that i's allocation is also $\mathcal{G}_i$-measurable. Radner (1968) demonstrates that, in such models in which consumers have different consumption sets because of the restrictions to subspaces of $\mathcal{G}_i$-measurable functions from Ω to $\mathbb{R}^\ell_+$, competitive equilibria exist, provided that Ω or all of the $\mathcal{G}_i$ are finite.

However, the appropriate informational restrictions in cooperative games are less clear-cut. Do agents share their information freely within a coalition, or can coalitions only make binding agreements based on information which is common to all members? The same problem arises when one attempts to define Pareto optimality in asymmetric information models. [A different approach involving interim efficiency is explored by Holmstrom and Myerson (1983) and more recently by Forges (1990, 1991).]

4 Wilson's Article

In a seminal article, Wilson (1978) examines the core of an economy with asymmetric information. He focuses on the need to define the information of players in a coalition when they (initially) have access to different information.

The analysis is performed in a pure exchange environment with finitely many states. Initial endowments are assumed to be always measurable for every agent in every coalition. Wilson (1978) first defines the abstract concept of *communication structures* and then focuses on two special extreme cases: the *coarse core*, defined by the condition that the information for all players in coalition S is precisely the sub-σ-field $\bigwedge_{i\in S} \mathcal{G}_i$ of information that they have in common, and the *fine core*, defined by giving every member of coalition S the sub-σ-field $\bigvee_{i\in S} \mathcal{G}_i$ of pooled information, so that the coalition can use any information that was initially available to any of its members.

Wilson (1978) then examines whether these two cores are nonempty. The intuition is as follows: The use of only common information renders blocking difficult, so that the coarse core is expected to be nonempty. However, blocking is easy with pooled information, so that the fine core may be empty. To prove nonemptiness of the coarse core, Wilson (1978) argues that a game he defines, in which each "player" consists of a state-player pair, is balanced. For the fine core, he provides a counterexample with three states and three players in which any feasible, efficient allocation for the grand coalition can be blocked in some state by some coalition.

However, when one contemplates these results in light of the market games literature and the relationships between balanced games and those with nonempty cores, the intuition about easy versus difficult blocking seems problematic. Potential blocking allocations should be compared to the set of allocations available to the grand coalition. As in the earlier market games literature, one examines convex combinations (with balancing weights) of feasible allocations. Concavity guarantees that the utilities of convex combinations dominate the average utilities of original allocations.

For the coarse core, this argument seems to fail. Convex combinations of $\bigwedge_{i\in S}\mathcal{G}_i$-measurable and $\bigwedge_{i\in T}\mathcal{G}_i$-measurable functions need not be $\bigwedge_{i\in S\cup T}\mathcal{G}_i$-measurable. However, the paradox is solved when one notices that Wilson's (1978) model uses $\bigwedge_{i\in S}\mathcal{G}_i$ as the information for subcoalitions $S \subset N$ with $S \neq N$ and reverses this logic to take $\bigvee_{i\in N}\mathcal{G}_i$ as the information for the grand coalition N. This explains the apparent inconsistency of the two strategies for proving that the core is nonempty.

In contrast, the argument that market games are balanced seems to apply to the fine core, as all convex combinations are measurable with respect to the information $\bigvee_{i\in N}\mathcal{G}_i$ of the grand coalition. Yet, detailed examination of Wilson's (1978) counterexample indicates that the blocking he employs to show that the core is empty must occur *ex post*. In Wilson's (1978) argument, some coalition blocks a given feasible allocation by dominating it in some particular state of the world. This would be consistent with a parallel state-by-state definition of the feasible and efficient allocations, so that in this case the economy with asymmetric information essentially reduces to three distinct economies in which all agreements and all trades take place after agents learn their information about the particular state of the world that has occurred.

Kobayashi (1980) obtains some results extending Wilson's (1978) coarse core using the concept of common knowledge. He also permits the set Ω of states of the world to be infinite.

5 Market Games with Asymmetric Information

In order to study cooperative solution concepts for economies with asymmetric information more systematically, one must derive the TU or NTU games that are generated by such economies. Standard results from game theory then apply, provided that the induced games are well defined and satisfy the necessary assumptions. Moreover, failures of certain solution concepts—such as the potential emptiness of the core—can be understood in terms of the game theoretic hypotheses that are violated as a consequence of asymmetric information. My formulation of market games with asymmetric information is based on *ex ante* agreements within coalitions. In particular, blocking can occur only before agents learn about the state of the world that has occurred, so that all payoffs in the resulting games consist of (unconditional) expected utilities (of information-conditional allocation functions). An advantage of this approach is that it enables agents to engage in risk-sharing trades and to write contracts that Pareto dominate those available with *ex post* agreements and *ex post* blocking.

Before the market games can be defined, one must specify the information available to every agent in every coalition. Let $\mathcal{H}_i^S$ denote the information that agent $i \in S$ can use as a member of coalition S. Assume that $\mathcal{H}_i^S$ is a sub-σ-field of $\mathcal{F}$ and that $e_i : \mathbb{R}_+^\ell \to \mathbb{R}$ is measurable with respect to $\mathcal{H}_i^S$ for all $S \ni i$ and all $i \in N$. Note that all members of a coalition need not be restricted to the same information; $\mathcal{H}_i^S \neq \mathcal{H}_j^S$ is permitted. A natural assumption is that $\mathcal{H}_i^S = \mathcal{G}_i$ for all $S = \{i\}$ and all $i \in N$.

Given an economy with asymmetric information as modeled in Section 3 and given the sub-σ-fields $\mathcal{H}_i^S$ for all $S \subseteq N$ ($S \neq \emptyset$) and all $i \in S$, define the induced cooperative game $v : 2^N \to \mathbb{R}$ with transferable utility by $v(\emptyset) = 0$ and $v(S) = \max\{\sum_{i\in S} \int_\Omega u_i(x_i(\omega);\omega)\, d\mu(\omega) |$ for all $i \in S$, $x_i : \Omega \to \mathbb{R}_+^\ell$ is $\mathcal{H}_i^S$-measurable and $\sum_{i\in S} x_i(\omega) = \sum_{i\in S} e_i(\omega)$ *a.s.*$\}$ for all $S \subseteq N$ with $S \neq \emptyset$.

Theorem 5 *The induced TU game v defined above is well defined.*

To show that the game is well defined requires proving that the maximum exists. Doing so gives rise to technical difficulties (to be discussed briefly below) whenever Ω is finite.

Similarly, one can define the derived *NTU* games. Let $V : 2^N \to \mathbb{R}^n$ be defined by $V(\emptyset) = \mathbb{R}^n$ and, for $S \subseteq N$ with $S \neq \emptyset$, $V(S) = \{(w_1, \ldots, w_n) \in \mathbb{R}^n |$ there exist $\mathcal{H}_i^S$-measurable functions $x_i : \Omega \to \mathbb{R}_+^\ell$ with $\sum_{i\in S} x_i(\omega) = \sum_{i\in S} e_i(\omega)$ *a.s.* such that $w_i \leq \int_\Omega u_i(x_i(\omega);\omega)\, d\mu(\omega)$ for all $i \in S\}$.

Theorem 6 *The induced NTU game is well defined. Moreover, for all $S \subseteq N$, the $V(S)$ sets are convex, and they are compactly generated whenever $S \neq \emptyset$.*

Closedness of the $V(S)$ sets is roughly equivalent to existence of the maxima for *TU* games; this is difficult when Ω fails to be finite. Convexity follows from concavity of utilities, while the property of being compactly generated comes from the uniform boundedness of initial endowments and upper semicontinuity of utilities.

If Ω is infinite, the argument exploits the characterization of weakly and strongly compact convex subsets in $\mathcal{L}^1$ spaces, especially the theorem of Dunford and Pettis (1940). [See Dunford and Schwartz (1958) or Rudin (1973) for technical background material.] Details appear in Allen (1991a, 1991b, 1991c). Page (1993) extends these theorems to allow the underlying commodity space $\mathbb{R}^\ell$ to be replaced by an infinite-dimensional space.

6 Cores with Asymmetric Information

In this section, we utilize balancedness conditions on the derived *TU* and *NTU* games to obtain nonempty cores with asymmetric information. A summary of the finite state case appears in Allen (1994), while the basic general references are Allen (1991b, 1991c).

Theorem 7 *A sufficient condition for balancedness of the derived TU or NTU game is that for all coalitions $S \subseteq N$ and all agents $i \in S$, $\mathcal{H}_i^S \subseteq \mathcal{H}_i^N$. Total balancedness holds if $\mathcal{H}_i^S \subseteq \mathcal{H}_i^T$ whenever $i \in S \subseteq T \subseteq N$. In particular, if $\mathcal{H}_i^S \subseteq \mathcal{H}_i^N$ whenever $i \in S$, then the core is nonempty, while $\mathcal{H}_i^S \subseteq \mathcal{H}_i^T$ whenever $i \in S \subseteq T \subseteq N$ implies that the cores of all submarkets are nonempty.*

The proof is based on the observation that sums of functions measurable with respect to different sub-σ-fields are measurable with respect to the smallest σ-field generated by all of the sub-σ-fields. The second statement follows from Bondareva (1962) and Shapley (1967) or Scarf (1967).

Theorem 7 implies that the fine information [i.e., $\mathcal{H}_i^S = \sigma\left(\bigcup_{j\in S} \mathcal{G}_j\right)$ whenever $i \in S$] core in the sense defined here with *ex ante* blocking is nonempty. It also implies that the private information [i.e., $\mathcal{H}_i^S = \mathcal{G}_i$ for all $i \in S$ and all coalitions $S \subseteq N$] core is nonempty. It does not apply to the coarse information [$\mathcal{H}_i^S = \bigcap_{j\in S} \mathcal{G}_j$ for $i \in S \subseteq N$] core and, in fact, counterexamples are not too difficult to find. However, a consequence of the theorem is that Wilson's coarse core [$\mathcal{H}_i^S = \bigcap_{j\in S} \mathcal{G}_j$ for $i \in S$ if $S \neq N$ and $\mathcal{H}_i^N = \sigma(\bigcup_{j\in N} \mathcal{G}_j)$] is necessarily nonempty. Of course, many other specifications for information sharing within coalitions are possible, and the theorem provides sufficient (but not necessary) conditions for such models to yield nonempty cores.

Yannelis (1991) shows that exchange economies have private information core allocations. Allen (1992) provides a different proof, under somewhat different assumptions, which follows directly from the market games approach. In Allen (1993a), private information sharing is related to a condition, termed *publicly predictable information*, stating that any single agent's information can always be deduced from the pooled information of all other coalition members.

7 Values with Asymmetric Information

Having derived cooperative games from economies with asymmetric information, one can apply any of the myriad of alternative (*TU* or *NTU*) solution concepts, provided that the requisite hypotheses are satisfied by the derived game. The value is only one of many solution concepts, albeit it is one that has nice properties and has proved to be extremely useful for many problems in economics. Thus, it is discussed here for illustrative purposes.

Theorem 8 *The TU Shapley value of the derived game exists and is unique. The NTU value exists if* $\mathcal{H}_i^S \subseteq \mathcal{H}_i^N$ *whenever* $i \in S$.

The qualification for *NTU* games is needed for monotonicity of the λ-transfer games. See Allen (1991a) for details, which use results from Aumann and Shapley (1974, Appendix A) and Shapley (1969).

Krasa and Yannelis (1994) show via a direct argument that economies with asymmetric information have value allocations. They do not examine all of the information sharing possibilities that are covered by the above argument.

8 The Harsanyi Approach

A different framework for the analysis of information in cooperative game theory is based on Harsanyi's (1967-68) formalization of noncooperative games with incomplete information. Recall that a noncooperative game is specified by a player set $N = \{1, \ldots, n\}$, strategy sets $\mathcal{S}_i$ for each $i \in N$, and payoff functions $\mathcal{U}_i : \Pi_{i \in N} \mathcal{S}_i \to I\!R$ for each $i \in N$. To capture the notion of incomplete information, Harsanyi (1967-68) replaces the single payoff function for each player by payoff functions that are parameterized by a type space. The basic idea is that a ***type*** for player i is taken to consist of the player's own payoff function and his or her beliefs about the payoff functions of other players, which are distributions (possibly depending on the player's own type) over the product of other players' type spaces.

If one contemplates this approach in the context of cooperative theory, problems arise even with transferable utility. For instance, when i learns his or her own type with certainty, does player i then know the entire game $v : 2^N \to I\!R$ in characteristic function form—in which case, either beliefs are inconsistent or there is no asymmetric information—or does player i only have some belief about the correct distribution over possible characteristic functions? How do players make enforceable agreements within coalitions when they believe they're playing different games—i.e., when their beliefs over $v(S)$ are different? Even if coalition members agree about v, they may disagree about how they can actually achieve the maximal worth of their coalition. For example, you and your spouse could both believe that you can double your household wealth, but you may disagree over whether this can be done by buying orange juice futures or by shorting Singapore stocks. What, then, is the worth of such a coalition?

These problems can be interpreted as suggesting the need to include functions from strategy sets or actions to payoffs as part of the primitive description of a cooperative game with incomplete information. Under complete or symmetric information, cooperative game theory usually supresses any explicit notion of strategies or actions, although implicitly, when we say that $v(S)$ is the worth of coalition S, we really mean that the players of coalition S together have some (feasible) joint strategy that enables them to earn $v(S)$. Under incomplete information, the strategies should be explicitly included in our cooperative games. Observe that this issue did not arise for market games with asymmetric information because we did write the strategies in the definition of $v(S)$ and $V(S)$; the strategies were state-dependent net trades (or state-dependent allocations) for each player in the coalition.

A pathbreaking article by Myerson (1984) explores such a formulation of cooperative games with incomplete information. For each coalition S, let $\mathcal{D}_S$ denote the set of actions available to S, and assume that $\mathcal{D}_S \times \mathcal{D}_T \subseteq \mathcal{D}_{S \cup T}$ when $S \cap T = \emptyset$. [This is a superadditivity condition.] Take the set $\mathcal{T}_i$ of types for player i to be a finite set for all $i \in N$; assume that all combinations of types [i.e., all n-tuples $(t_1, \ldots, t_n) \in \Pi_{i \in N} \mathcal{T}_i$] occur with strictly positive probability. Write $\mathcal{T}_S = \Pi_{j \in S} \mathcal{T}_j$ for the set of profiles of types in the coalition S. Let $\mathbf{u}_i(d, t_1, \ldots, t_n)$ be the payoff to player $i \in N$ if the grand coalition N chooses strategy $d \in \mathcal{D}_N$ when players' types are $(t_1, \ldots, t_n) \in \mathcal{T}$. This model permits externalities, although there is no obvious way to define subgames except by having the subgames depend on some given type realization and action of the complementary coalition. [This situation is worse than in cooperative games with complete information in that, while a coalition can perhaps observe the action of its complement, the coalition may have no way to ascertain the type drawings of players who do not belong to the coalition.] Myerson (1984) further assumes the consistency condition of Harsanyi (1967-68) that there exists a probability p on $\mathcal{T}$ such that its conditional distributions satisfy $p_i(t_{)i(} | t_i) = p(t) / \sum p(t_i, s_{)i(})$, where the summation is taken over $s_{)i(} \in \mathcal{T}_{)i(}$. Then a *cooperative game with incomplete information* and player set N is defined by $\left((\mathcal{D}_S)_{S \subseteq N, S \neq \emptyset}, (\mathcal{T}_i, \mathbf{u}_i)_{i \in N}, p\right)$ satisfying the above assumptions. Myerson (1984) studies bargaining solutions in such games.

This model forms the basis for recent research by Allen (1993b), Ichiishi and Idzik (1992), and Rosenmüller (1990), among others. As this work focuses on issues of incentive compatibility and, hence, relates to implementation, I do not discuss it further here.

Acknowledgements

This work was supported by the National Science Foundation (through grant SBR93-09854), the Federal Reserve Bank of Minneapolis, and the Curtis Carlson Chair in Economics at the University of Minnesota. The views expressed herein are those of the author and not necessarily those of the Federal Reserve Bank of Minneapolis or the Federal Reserve System. Mr. Xiaodong Huang of the University of Minnesota was the notetaker for the lecture on which this paper is based.

References

Allen, Beth (1991a) Market games with asymmetric information: The value. Center for Analytic Research in Economics and the Social Sciences Working Paper #91-08, Department of Economics, University of Pennsylvania

Allen, Beth (1991b) Market games with asymmetric information and nontransferable utility: Representation results and the core. Center for Analytic Research in Economics and the Social Sciences Working Paper #91-09, Department of Economics, University of Pennsylvania

Allen, Beth (1991c) Transferable utility market games with asymmetric information: Representation results and the core. Center for Analytic Research in Economics and the Social Sciences Working Paper #91-16, Department of Economics, University of Pennsylvania

Allen, Beth (1992) Market games with asymmetric information: The private information core. Center for Analytic Research in Economics and the Social Sciences Working Paper #92-04, Department of Economics, University of Pennsylvania

Allen, Beth (1993a) Market games with asymmetric information: Verification and the publicly predictable information core. Hitotsubashi Journal of Economics **34** (Special Issue in Honor of Lionel McKenzie), 101-122

Allen, Beth (1993b) Supermechanisms. Mimeo, Department of Economics, University of Minnesota

Allen, Beth (1994) Market games with asymmetric information: The core with finitely many states of the world. In: Munier, B., Machina, M.J. (eds.) Models and experiments in risk and rationality. Proceedings of the 6th Foundations and Applications of Utility, Risk and Decision Theory Conference, Cachan (Paris), 1993. Theory and Decision Library, Kluwer Academic Publishers, Dordrecht, The Netherlands, 377-411

Aumann, R.J. and Shapley, L.S. (1974) Values of Nonatomic Games. Princeton University Press

Billera, Louis J. (1974) On games without side payments arising from a general class of markets. Journal of Mathematical Economics **1**, 129-139

Billera, Louis J. and Bixby, Robert E. (1974) Market representations of n-person games. Bulletin of the American Mathematical Society **80**, 522-526

Bondareva, Olga N. (1962) Theory of core in an n-person game. Vestnik Leningradskogo Universiteta, Seriia Matematika, Mekhaniki i Astronomii, **13**, 141-142

Dunford, N. and Pettis, B.J. (1940) Linear operators on summable functions. Transactions of the American Mathematical Society **47**, 323-392

Dunford, Nelson and Schwartz, Jacob T. (1958) Linear Operators, Part I: General Theory. Interscience, New York

Forges, Françoise (1990) Some thoughts on efficiency and information. CORE Discussion Paper #9071, Center for Operations Research and Econometrics, Université Catholique de Louvain, Louvain-la-Neuve, Belgium

Forges, Françoise (1991) Posterior efficiency. CORE Discussion Paper #9145, Center for Operations Research and Econometrics, Université Catholique de Louvain, Louvain-la-Neuve, Belgium

Harsanyi, John D. (1967-68) Games with incomplete information played by "Bayesian" players. Management Science **14**, 159-182, 320-334, 486-502

Hildenbrand, W. and Kirman, A.P. (1976) Introduction to Equilibrium Analysis. North-Holland, Amsterdam
Holmstrom, Bengt and Myerson, Roger B. (1983) Efficient and durable decision rules with incomplete information. Econometrica **51**, 1799-1819
Ichiishi, Tatsuro and Idzik, Adam (1992) Bayesian cooperative choice of strategies. Mimeo, Department of Economics, Ohio State University
Kobayashi, Takao (1980) Equilibrium contracts for syndicates with differential information. Econometrica **48**, 1635-1665
Krasa, Stefan and Yannelis, Nicholas C. (1994) The value allocation of an economy with asymmetric information. Mimeo, Department of Economics, University of Illinois at Champaign-Urbana
Mas-Colell, Andreu (1975) A further result on the representation of games by markets. Journal of Economic Theory **10**, 117-122
Myerson, Roger B. (1984) Cooperative games with imcomplete [*sic*] information. International Journal of Game Theory **13**, 69-96
Page, Frank H. Jr. (1993) On Allen's approach to economies with differential information. Mimeo, Department of Finance, University of Alabama at Tuscaloosa
Radner, Roy (1968) Competitive equilibrium under uncertainty. Econometrica **36**, 31-58
Rosenmüller, Joachim (1990) Fee games: $(N)TU$-games with incomplete information. Institut für Mathematische Wirtschaftsforschung Working Paper Nr. 190, Universität Bielefeld
Rudin, Walter (1973) Functional Analysis. McGraw-Hill, New York
Scarf, Herbert E. (1967) The core of an n person game. Econometrica **35**, 50-69
Shapley, Lloyd S. (1967) On balanced sets and cores. Naval Research Logistics Quarterly **14**, 453-460
Shapley, Lloyd S. (1969) Utility comparisons and the theory of games. In: La Décision. CNRS, Paris, 251-263
Shapley, Lloyd S. and Shubik, Martin (1969) On market games. Journal of Economic Theory **1**, 9-25
Wilson, Robert B. (1978) Information, efficiency and the core of an economy. Econometrica **46**, 807-816
Yannelis, Nicholas C. (1991) The core of an economy with differential information. Economic Theory **1**, 183-198

PART B

NON-COOPERATIVE APPROACHES

Bargaining Games

Andreu Mas Colell

Department of Economics, Harvard University, Cambridge, MA 02138, USA

This is a close rendition of the delivered lecture. The manuscript has been written and prepared by Shira Lewin, currently a doctoral candidate at Harvard University. The lecturer wishes to acknowledge and thank her for a very well done job.

1 Introduction

The topic to be reviewed in this lecture is included in what Bob Aumann described in his lecture as the bridges between cooperative and noncooperative theory. If I had all the time in the world, I would begin by presenting the basics of noncooperative game theory, but I cannot possibly do this. I will therefore remain very elementary, and I will be somewhat loose about the noncooperative concepts.[1] The flavor of what I will be doing today consists in writing down or describing game procedures, understood as noncooperative mechanisms for interaction, discussion, and the formulation of agreements about how to split things. These bargaining procedures will be set in a context which will stay very close to the frameworks presented by earlier lecturers. We will then see how the noncooperative solutions of the bargaining procedures relate to the axiomatic procedures presented earlier by others.

In interpreting these procedures, there are two positions that we can take — not quite two points of view, but two sources of light with which we can look at this sort of theory. The first is the descriptive source of light, and the second is the prescriptive.

The descriptive view In the descriptive view, the noncooperative procedure comes first, not only logically but also conceptually and theoretically. We are discussing bargaining procedures, and when we analyze these procedures, we may discover that the equilibria exhibit some relationship with an axiomatically based solution. Then, if we wish, we may call the bargaining procedure under discussion the noncooperative foundation of the axiomatic

[1] For general references on game theory, see Myerson (1991) or Osborne & Rubinstein (1994).

solution. But we certainly view the noncooperative approach as the conceptual starting-point.

The prescriptive approach The prescriptive approach relates more to implementation theory. (The elements of this theory will be presented in a forthcoming lecture.) Here, the point of view is to think of the axiomatic solutions as well founded on the axiomatic grounds on which they are presented. However, one recognizes that to reach them, it may be necessary to design devices — call them bargaining procedures — that will yield the axiomatic solutions as noncooperative equilibria. So, logically, the cooperative part comes first, and we really think of the noncooperative part of the theory as an instrument with which to obtain the cooperative result.

As I said, the distinction is very much there, but I would not want to trace a rigid boundary for the purposes of this lecture.

In this first part of the lecture, I will discuss two player games, and in the second part, I will say something about N-player games.

2 Two-player games

In this section, we have two players, those in $N = \{1,2\}$. I will start with transferable utility (TU) games, but I will move to the non-transferable utility (NTU) case very soon.

2.1 TU case

2.1.1 Cooperative approach

Think of two players that have to split a pie. If they cooperate, then they will get a total amount of utils, or dollars or whatever, equal to $v(N)$. If they do not cooperate, then they will get a certain point $(c_1, c_2) = c \in \mathbf{R}$.

It would be tempting to adhere to the exact framework of a characteristic function, and write, say, $c_1 = v(1)$ and $c_2 = v(2)$. We could do this, but we will not. It is not clear that we should really think of c_1 and c_2 as if they were exactly what $v(1)$ and $v(2)$ would be in a cooperative framework. There is no need to regard c_1 as what 1 could get by himself, and similarly for c_2. We only require that the combination $c = (c_1, c_2)$ is what would happen if there were no cooperation.

We do assume that

$$c_1 + c_2 < v(N)$$

so that there is some reason to cooperate. Graphically, we present this in figure 1. Here, the segment AB is the utility possibility frontier $\{u_1, u_2 \,|u_1 + u_2 = v(N)\}$. Now, let us make a slight conceptual jump, and let us associate the vector c with the threat point of a bargaining problem (like those presented in W. Thomson's lecture). Then we see that each of

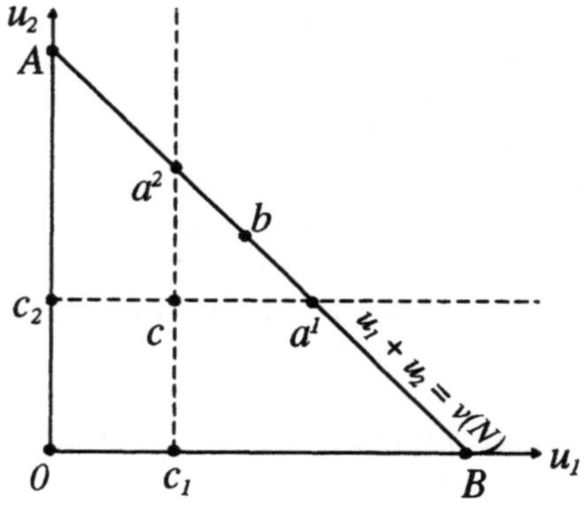

Figure 1: Standard solution in the TU case

the solutions that we have discussed from an axiomatic perspective has the property that it splits the surplus above c, rather than, say, giving each player $\frac{1}{2}v(N)$. Thus, we end up at the midpoint between a^1 and a^2 shown above in figure 1. We will call this split the **standard solution**:

$$b = \left(c_1 + \frac{v(N) - c_1 - c_2}{2}, c_2 + \frac{v(N) - c_1 - c_2}{2} \right) = \frac{1}{2} \left(a^1 + a^2 \right)$$

where $v(N) - c_1 - c_2$ is the surplus that is split.

2.1.2 Noncooperative approach

There is a very simple way to obtain the standard solution noncooperatively: the all-or-nothing (or take-it-or-leave-it) mechanism. Consider the following bargaining procedure:

- Choose one player by tossing a coin. Call this player the proposer.
- The proposer proposes a split of $v(N) : (u_1, u_2)$.
- The respondent accepts or rejects.
  - acceptance $\Rightarrow (u_1, u_2)$.
  - rejection $\Rightarrow (c_1, c_2)$.

To solve this (extensive form) game, the natural noncooperative solution concept is backward induction (or, if you prefer, you can say that I am choosing the perfect equilibrium of this game). If 1 is the proposer, then how will 1 reason? He will say, "If I don't offer 2 at least c_2, then 2 will reject, because by rejecting she can get c_2." Therefore, 1 will propose the split that gives him the maximum amount compatible with 2 getting c_2, and this is the point a^1 in figure 1. (We always assume that player 2 is cooperative enough with player 1 that she breaks ties in his favor, so I won't have to worry about little ϵ's.) Thus, 1 will propose a^1, and similarly, when 2 is the proposer, she will propose a^2 and get all the surplus herself. If we accept the von Neumann-Morgenstern expected utility theory, then, in expectation, the outcome is $\frac{1}{2}\left(a^1 + a^2\right)$, which is exactly the standard solution defined above.

So, are we done? Have we implemented the standard solution noncooperatively? The answer is, not quite. Why not? Well, the procedure just described has some drawbacks. One is very apparent, and the other will become clear momentarily when we move to the NTU case. The apparent drawback is that we get the correct outcome only in expectation. It is true that, ex ante, each player i gets b_i, but the proposals that actually take place in the game are not the standard solution. They are either a^1 or a^2. (Actually, in the current TU case, there is an easy fix for this problem — just perform the procedure twice instead of once — but as we will now see, the matter is not always so simple.)

2.2 NTU case

2.2.1 Failure of the take-it-or-leave-it procedure

We draw the problem graphically as before (figure 2). Note that, with a^1

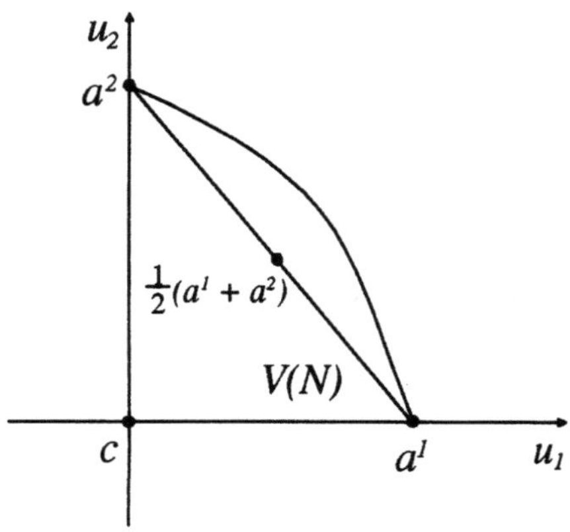

Figure 2: Failure of the one-stage mechanism in the NTU case

and a^2 defined as before, the 50/50 expectation of the two outcomes is not even efficient. Of course, we could get efficiency by not randomizing, and by simply imposing that 1 begin, or 2 begin, but then we would not implement the standard solution. While this is a very simple example, it illustrates an issue that tends to be a general problem. If you are in the TU case, then you can implement (in expectation) by relatively short bargaining procedures; but these procedures are not likely to be efficient if you have an NTU problem.

2.2.2 A multi-stage procedure

At this juncture, it seems logical to argue that if we want to get better outcomes, perhaps we should work with more elaborate bargaining procedures — in particular, bargaining procedures that keep repeating themselves, so that if somebody rejects, then this is not the end of the world; another round of negotiation may yet take place. I will now present a particular, but typical, multi-stage procedure. It goes as follows:

- As before, a player is chosen by tossing a coin, and she makes a (feasible) proposal $(u_1, u_2) \in V(N)$.
- The other player can accept or reject.
  - Acceptance $\Rightarrow (u_1, u_2)$.
  - Rejection $\Rightarrow$
    * With probability $\rho < 1$, the game repeats.
    * With probability $1 - \rho$, the players get c.

Note that in the case of rejection, the probability of breakdown is not 1, but only $1 - \rho < 1$. This number (which is a parameter of the problem) could be large or small, but I want you to think of it as small, so that if there is persistent rejection, then with high probability, the procedure will not terminate with breakdown immediately, but will do so only quite far in the future. A typical interpretation (but not the one that I want to emphasize here), is to think of $1-\rho$ as a rate of time-discounting. (In this case, we should interpret c as the utility that will be obtained if there never is agreement.) The point is that there is a cost of delaying agreement one round. This cost may be that of time passing, or it may be something else. For example, in the case of implementation, the designer can set a device which incorporates the possibility of breakdown.

The procedure just described is not the only possible one. Note, in particular, that it is time-stationary. We could, for example, also have a non-time-stationary rule. Fix a horizon $T < \infty$, such if there is no agreement by time T, everything stops and players get c. However, assume that up to time T there is no cost of delaying agreement. This is somewhat discontinuous (and nonstationary), and it doesn't lend itself to a simple analysis, whereas the device proposed above does.

2.2.3 Stationary Perfect Equilibrium

Which solution concept will we adopt? We have a game that will terminate with probability 1, but which is, in principle, infinite. There is no end of time, and therefore I cannot use backward induction. Instead, I will adopt a very simple solution concept called *stationary perfect equilibrium*. *Perfect* implies that we are still within the framework of backward induction. *Stationary* means that we are focusing on equilibria where chosen strategies do not depend on history or on calendar time. Proposals will be independent of whatever has happened in the past. Similarly, the responses will depend only on the proposal received, and not on past proposals.

Finally, note that we are still talking about an *equilibrium*. That is, I am referring to an equilibrium, in which the strategies happen to be stationary ones, but it is a true equilibrium. In particular, there is no restriction on the strategy set, and players do contemplate the possibility of every sort of complicated nonstationary deviation. In the universe of perfect equilibria — where every equilibrium is as good as any other — the equilibria that are most descriptively simple are the stationary ones. So just as, when one looks at a dynamical system, one first looks at the rest points, it makes some sense to look at the stationary equilibria first.

2.2.4 Graphical solution by equilibrium equations

I'm going to try to solve the equilibrium problem graphically. The treatment is not meant to be rigorous. Focus on a particular stationary perfect equilibrium. Call $b = (b_1, b_2)$ the expected payoffs at $t = 0$ when this equilibrium is played. Since the utility possibility set is convex, b must be a feasible point, but it does not need to be at the boundary of $V(N)$, and in figure 3, it is shown in the interior. Remember that we do not know a priori that efficiency is guaranteed (and in fact, as we will see, it is not).

Now suppose that player 1 is chosen to be the proposer. What will 1 propose? He will try to evaluate how much it costs 2 to reject. Well, if 2 rejects 1's proposal, then with probability ρ, everything is repeated, and because of stationarity, we come back to b. With probability $1-\rho$, we go to c. So the expected payoff vector is $\rho b+(1-\rho)c$. Hence, by rejecting 1's proposal, 2 can guarantee herself $\rho b_2 + (1-\rho)c_2$. Player 1 will therefore propose the point a^1 (shown in figure 3) that maximizes his own payoff subject to 2 getting at least $\rho b_2 + (1-\rho)c_2$, the minimum payoff that guarantees 2's acceptance. Similarly, 2 will propose the point a^2 that maximizes her payoff, subject to 1 getting at least $\rho b_1 + (1-\rho)c_1$, which guarantees that 1 accepts as well. Note that along the equilibrium path, there will be no rejection.

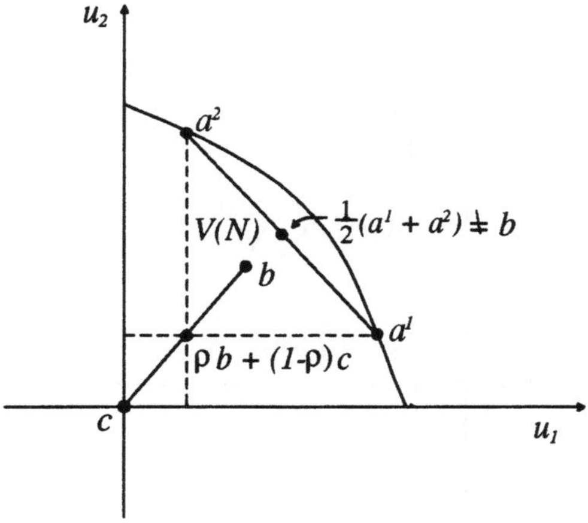

Figure 3: Graphical analysis of the mult-stage procedure

We conclude that, if 1 is the proposer, the outcome is a^1, and if 2 is the proposer, the outcome is a^2. So far, we have not brought any consistency conditions into the analysis. To close the system, we notice that the expected outcome of the equilibrium is b. Therefore, we must have

$$b = \frac{1}{2}\left(a^1 + a^2\right)$$

That is, the vector b must lie at the midpoint of the line segment between a^1 and a^2. This is a real condition. If I start with an arbitrary b, and then I construct a^1 and a^2 (as indicated above) and take their midpoint, I need not come back to b, and in figure 3, I do not. If I do happen to come back to b, then I have found an equilibrium. This is the case in figure 4.

The stationary equilibrium payoff vector b depends on ρ, but it can be verified that, given ρ, b is unique. Note that it is not efficient.

At this point, we can observe something very interesting. Normalize c to $(0,0)$ (this is just for convenience). Take the straight line through a^1 and a^2, and extend it until it hits the axes (figure 5). The two triangles BOA and a^1Da^2 are similar, and the line Ob splits them in half. Now, since b is the midpoint of the hypotenuse of a^1Da^2, it follows that b is also the midpoint of the hypotenuse of BOA. Now imagine that $1-\rho$ is very small, so that the triangle a^1Da^2 is also very small. Then the slope of AB is almost equal to the slope of the boundary of $V(N)$ near a^1 and a^2 (assuming that $V(N)$ has a smooth boundary).

So, for $1-\rho$ very small, we have, almost, the following property: The equilibrium payoffs b are efficient and are such that when we take the tangent

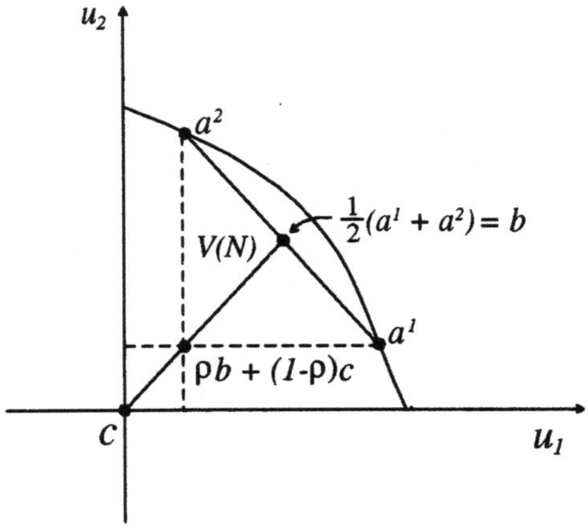

Figure 4: Equilibrium condition for the multi-stage procedure

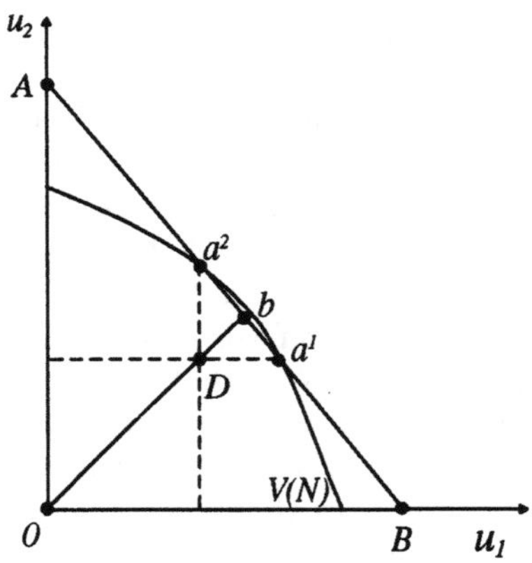

Figure 5: Approximate efficiency for large ρ

to the boundary of $V(N)$ at the equilibrium payoffs, b falls at the midpoint of this tangent (more precisely, the midpoint of the segment of this tangent that lies between its intersections with the axes). We should recognize this as the defining property of the Nash bargaining solution. Recall that for a utility possibility set as illustrated below (figure 6), if the vector b has the property that the segments Ab and bB have the same length, then it follows that b is the Nash solution. This comes right out of the axioms of the Nash

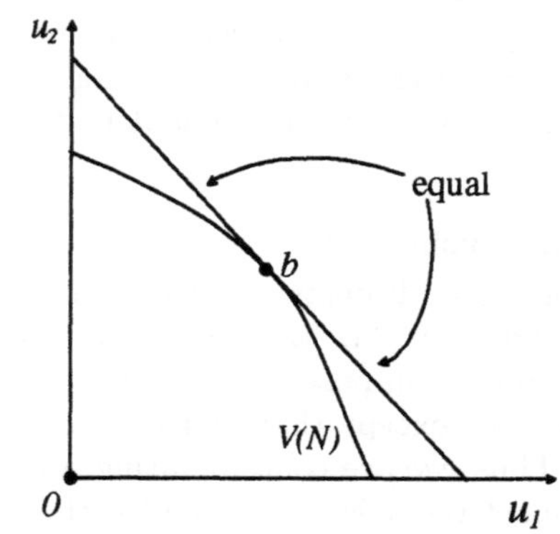

Figure 6: The Nash solution

solution. Consider first an economic budget set — a straight line (See figure 6). Then if we maximize the product of the coordinates on this line, we come to the midpoint. So b is the Nash solution for the budget set. But then the contraction independence axiom implies that, since when we go from the economic budget set to $V(N)$, we only make the utility possibility set smaller while b remains feasible, the solution remains b after the change.

We conclude that if the cost of renegotiation is very low, then at the first stage of the stationary perfect equilibrium of the bargaining procedure, the proposer will propose a payoff which is close to the Nash bargaining solution, and the respondent will accept. We emphasize that:

1. We obtain this result because, in principle, negotiation can go on for a very long time. But in fact, it will not go on for long. It will end in the first round.

2. The proposer does not really matter. Both agents will propose almost the same outcome.

2.3 Some Remarks

2.3.1 On the stationarity restriction

I have been talking about stationary perfect equilibria. In fact, there is a notable result due to Rubinstein (1982), which asserts that, for this model, to get the equilibrium payoffs we derived above, the stationarity restriction is actually not required. Every perfect equilibrium of this procedure has exactly the characteristic that we just described: On the equilibrium path, one player (whoever is the proposer) makes some proposal (the same for all equilibria), and this proposal is accepted. This is a very singular result, but I will not emphasize it. It is quite remarkable, but something of a jewel — admirable and beautiful, but hard to replicate. In particular, it does not generalize to more than two players.

2.3.2 A dynamic analysis

Associated with the previous discussion there is a nice dynamic analysis. I can not be precise here, but M. Maschler and collaborators have done much research on this topic. Suppose that you take exactly the bargaining procedure I have presented, except that you truncate it at period $T \gg 0$, when the world ends. Thus, we are contemplating a nonstationary problem such that the disagreement point is, say, 0, and such that at $t = 0$, we have a utility possibility set $V_0 = V(N)$. Then, there is a contracting sequence of utility possibility sets $V_1, V_2, \ldots V_T$ (figure 7) such that each $V_t = \rho^t V(N)$ is the set of feasible expected payoffs if there is no agreement before time t. Assume also that T is so large that ρ^T is nearly equal to 0.

This problem can be solved by backward induction. You just think about what would happen at the end of the world, and then given that, you look at stage $T-1$, etc. Figure 7 illustrates the construction of the equilibrium expected payoffs $b(T-1)$ at $t = T-1$ from the equilibrium expected payoffs $b(T) = \frac{1}{2}\left[a^1(T) + a^2(T)\right]$ in the last period. You can proceed in this manner until, finally, you derive $b(0)$. The process will begin to look like a differential equation. We can then make the jump to real differential equations, so that the backward induction yields a system of differential equations which, as it turns out, converges to the Nash solution.

2.3.3 Variation in the breakdown point

I have assumed that the breakdown point c is given independently of the history that leads to breakdown. I could consider (why not?) a more complicated model in which the breakdown point depends, for example, on who has been responsible for the breakdown, perhaps the refuser or perhaps the last proposer. One can think of many variations. But let me focus on one.

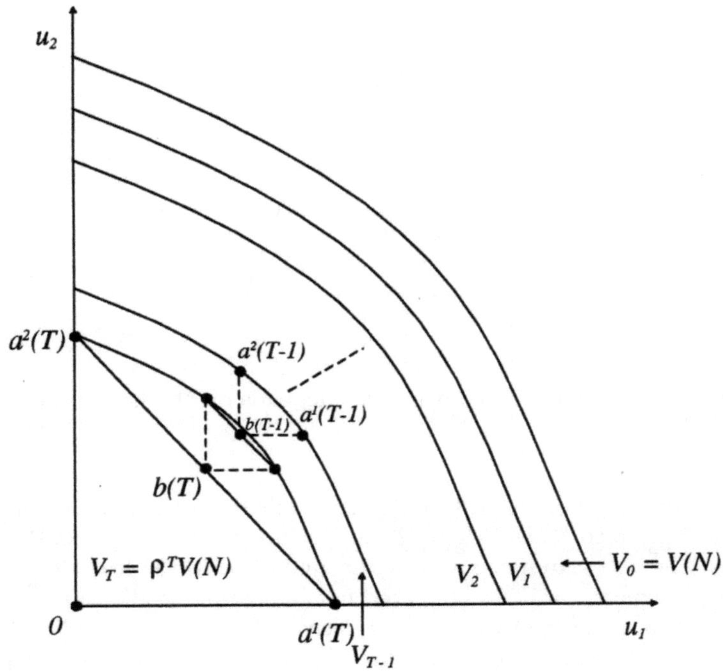

Figure 7: Truncated multi-stage problem

Suppose that $V(N)$ is as before, but that the breakdown point depends on who the last proposer was before breakdown. Otherwise, the bargaining procedure is as before. Again we have a ρ. The only change is that as we go through time, if player 1 makes a proposal, 2 rejects it, and there is breakdown, then we go to some point c^1. If the breakdown happens after 2 proposes and 1 rejects, then we go to c^2 instead (figure 8). Note that if you think of ρ in terms of time discounting then this doesn't make sense, but for other interpretations, it does make sense. When 1 evaluates the utility 2 gets from rejecting, he should consider c^1. If 2 rejects, then c^1 occurs with probability $1-\rho$, and play continues with probability ρ. It turns out (this is very easy to check) that to solve this model, we can proceed by constructing a kind of fake disagreement point (shown in figure 8): $c = \left(c_1^2, c_2^1\right)$. Then we can continue exactly as we did before. Taking c as the disagreement point, if ρ is large, the outcome will be nearly the Nash solution calculated from this fake disagreement point. Note that this disagreement point has no reality. Outcome c will never occur. What can occur is c^1 or c^2. But the theory can still make use of the fake disagreement point.

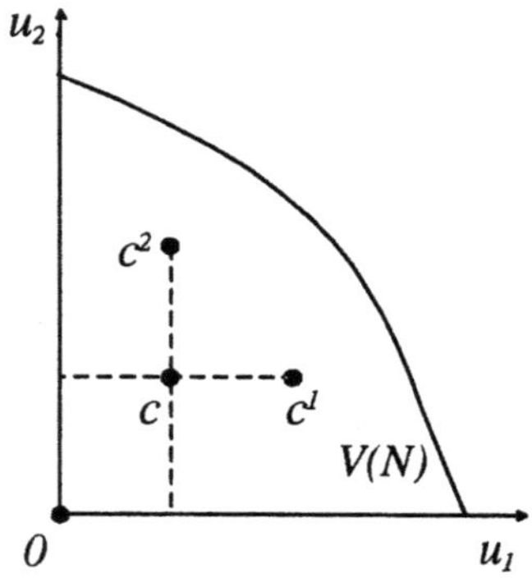

Figure 8: The breakdown point depends on the last proposer

One interesting point that I would like to mention is that there is no reason why things have to be as I drew them in figure 8. In fact, it could be as in figure 9. The only requirement is that c^1 and c^2 be in the feasible set $V(N)$. But, as constructed in figure 9, c need not be feasible itself. But we can still look at the bargaining procedure and derive its stationary equilibria. What will we get? The equilibrium will have to satisfy exactly the same equations as before. Given an equilibrium payoff vector b, we construct a^1 and a^2 as we did earlier, and it must be the case that b is at their midpoint. (See figure 9.) If $1-\rho$ is small, then to find something that is almost the solution, you look at c as a disagreement point, and you look at the point b on the boundary of $V(N)$ such that when you take the tangent at b, b is at the midpoint of the line segment AB shown in figure 10. Two things are worth noting about this construction. First, it amounts to guaranteeing the first order conditions (but not the second!) of the Nash product "maximization problem." Second, the solution need not now be unique.

2.3.4 What about Kalai-Smorodinsky?

My entire discussion has led us to the Nash bargaining solution. You heard in W. Thomson's lecture that the Kalai-Smorodinsky solution is as important as Nash's, so you may ask whether I can get Kalai-Smorodinsky's solution by a bargaining procedure similar to the one I have described above. I cannot give you an affirmative answer to this question. I could offer you some bargaining procedures, but these would be of a very different character. However, I can offer you some insight by obtaining a solution which is in the spirit of Kalai-Smorodinsky.

We do this as follows: Put $\rho = 1$, so that there is no cost of delay. To avoid being degenerate, also assume a fixed time horizon $T < \infty$; i.e. we repeat only T times. We now apply backward induction. Consider the last period. If 1 is the proposer, then 1 will offer $a^1(T)$; similarly, 2 will offer $a^2(T)$, as shown

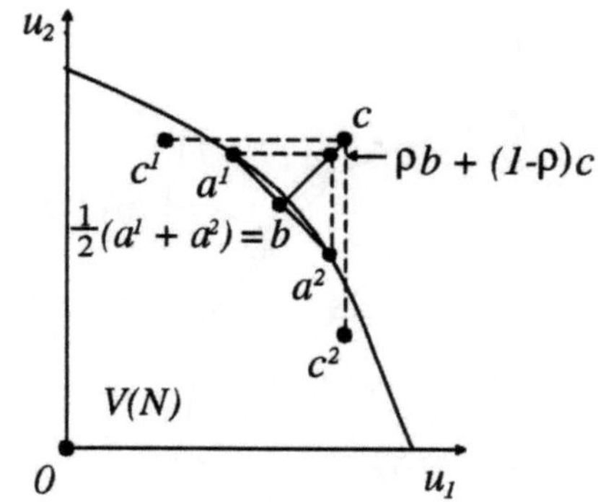

Figure 9: Disagreement point outside feasible set

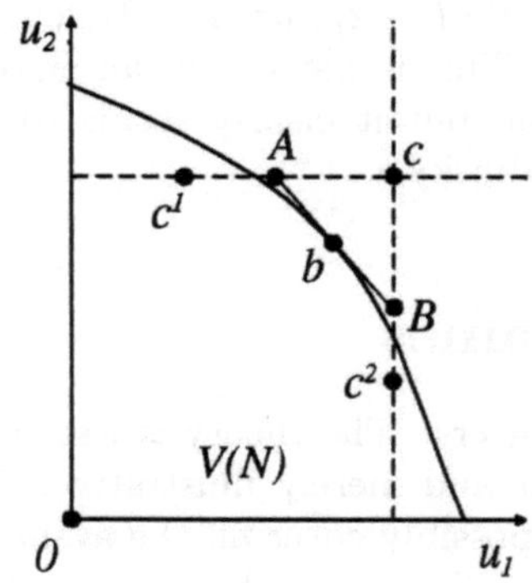

Figure 10: Near efficiency with a fake disagreement point

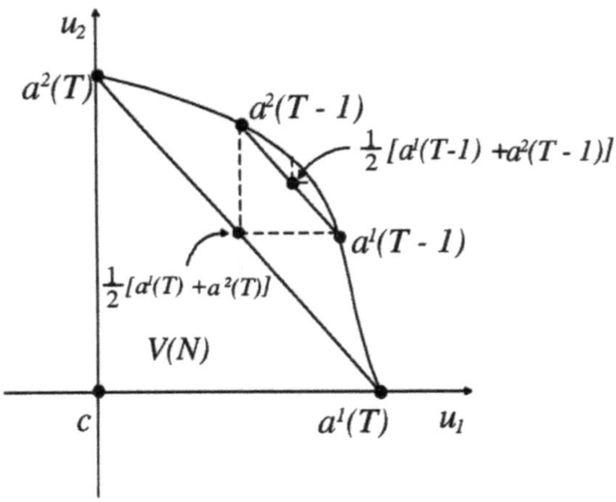

Figure 11: The Raiffa solution

in figure 11. Therefore, if there is rejection at $T-1$, the expected payoff is the midpoint $b(T) = \frac{1}{2}\left[a^1(T) + a^2(T)\right]$. What will happen in period $T-1$? Both players know that if the other rejects, then they get $b(T)$, so player 1 proposes $a^1(T-1)$ shown above, and 2 proposes $a^2(T-2)$. We continue this construction to get $a^1(T-2)$, $a^2(T-2)$, etc. As T grows large, we approach the boundary. This is not the Kalai-Smorodinsky solution. It is called the Raiffa solution, but it clearly seems to be in the same general category as Kalai-Smorodinsky's.

3 N-Player Games

I will now move to N players. The theory is less settled here, and so I will be much more particular and merely illustrative. There is much work on this topic, and I cannot possibly cover all the available results. Fortunately, the lecture by P. Reny will also touch on this general area, and he will complement our discussion quite well. The presentation from now on is in the spirit of implementation theory. Thus, I will just present instances of how, under certain restrictions, this or that solution concept can be supported by a noncooperative procedure. But I will not discuss whether such a procedure is sufficiently descriptive of "real bargaining."

3.1 Background

Since I want to relate my discussion to cooperative game theory, let me begin by describing the physical situation that is being contemplated in terms of the formalism of the characteristic function. I assume that there are N players and that for all $S \subset N$, $V(S)$ is the attainable set for S. I want to be a bit more precise here. Thus, in the spirit of an economic approach, let us think of this V as describing an economic situation in which people are endowed with resources and where there are no externalities of any kind, so that $V(S)$ is the set of utility possibilities that the group S can reach by using the resources of its members (this set is independent of whatever the players not in S eventually get).

I am focusing on this particular scenario because it is a very simple one in which there is no complication whatsoever in interpreting what the characteristic function is. In general I think that for the type of discussion that we are carrying out, it is indispensable to have an interpretation, a particular story about this characteristic function. Otherwise it is very hard to know what one is talking about. For example, in a model with externalities and other interactions, the characteristic function may have been constructed taking into account a certain number of strategic considerations that concern how the members of a coalition and its complement will behave. In this case, it would be artificial to analyze the problem using other strategic considerations. If we are focusing on the core, then the kind of strategic considerations that go into its definition may also be strategic considerations that lead us to a certain kind of characteristic function; while for the Shapley Value, say, we may be led to another characteristic function. I think, therefore, that you cannot separate the construction of the characteristic function from the solution concept that you are going to use, except in completely natural cases like pure resource problems. Since matters are already complicated enough here, we will stick to this case.

3.2 Two approaches to the N-player case

In the spirit of what I have done so far, I want to generalize the bargaining procedures presented earlier, and in a manner that fits in with the lessons of cooperative game theory. This already points one in certain directions. For example, in the $N = 2$ case, I focused on the Nash solution. That means I have already focused on a single-valued solution. It would not be natural now to move in the direction of the core, which is multi-valued even in the case of two players, and which certainly does not equal the Nash solution. In fact, the approach I am taking is directing me towards value-type solutions, and I will not resist this direction.

Let me remark that, in my view, it is not yet well understood what distinguishes the types of bargaining solutions that take us to the core from those that take us to value. Very vaguely, my impression is that the distinction has something to do with the meeting technology of the players — that is, with how different players meet. If the choice of players that meet is very strategic — namely, a player chooses and looks around for partners — then I think that we are pushed towards core-like notions; while if the meeting technology is that people meet at random somehow, so that people have very little choice about how they find their companions, then I think we are pushed more towards value solutions.

From now on, we have N players, and they need to meet to bargain over how to split some sort of pie. To analyze this situation further, it is critical to establish how players meet. In the value-oriented literature, we find two varieties of what we could call "meeting technologies." Because I am more familiar with one of them I will follow that one, but I must mention both. They both generalize the formulation we adopted in the two-player case.

The first is the technology of pair-wise meetings. The meeting of pairs (the "buyer" and the "seller") is pervasive throughout economics (cf. Gale 1986, Rubinstein & Wolinsky 1985). For the current problem, the technology of pair-wise meetings has been used by Gul (1989). In his important paper, there is a collection of people who have resources and who meet at random in pairs. When they meet, one proposer is chosen at random. The proposer makes a proposal to buy the resources of the respondent. The respondent may accept, in which case he disappears from the game with the payment, or he may not accept, in which case both members of the pair go back to the pool of players and negotiation continues in this manner.

The second technology, which is the one I will adopt, is that of multilateral meetings. More precisely, I mean by this that at any point in time, there is an assembly of all the bargainers, and the proposer (chosen in some manner) addresses the entire assembly. I will follow this approach, but I also recommend that you look at the Gul paper.

3.3 An illustrative example

3.3.1 Setup

I am going to present, as an example, a bargaining procedure which is a generalization of the previous (2-player) bargaining procedure. It is taken from Hart & Mas-Colell (1992). The key feature of this procedure is that players may drop out throughout the negotiation process.

The procedure is as follows: Assume that $S \subset N$ is the set of players still involved in negotiation. Initially, we will just have $S = N$.

- Choose a player i at random from S using a uniform distribution.

- Player i proposes a payoff vector $u \in V(S)$.

- Other players are asked (sequentially) if they agree or dissent.
  - All agree $\Rightarrow u$ is implemented.
  - Any player dissents $\Rightarrow$
    * With probability ρ, the game repeats.
    * With probability $1-\rho$, breakdown occurs.

What does breakdown mean? I don't want it to mean, as before, that everything is ended and that we go to some disagreement point $c \in V(S)$. I avoid this meaning because I want subcoalitions to matter. There are a multitude of other meanings of "breakdown" that one could consider. I encourage you to consider some. For example, "breakdown" could mean that some player is at risk of disappearing. It could be that there is some $\delta < 1$ such that each player disappears with his own resources with probability δ. If δ is small, then the probability of two players disappearing simultaneously is negligible, and so breakdown means that one of the players, chosen at random, will disappear, and we will go on with a smaller game.

To be specific, I will focus on another particular meaning of breakdown. I choose this particular example purely because it fits with the analysis of the Nash solution I presented before, and because I want to tie this analysis to the Shapley Value. Thus, I will present a breakdown technology which has the feature that if I look at the equilibria, then in the pure bargaining case I get the Nash solution, and in the case of transferable utility — another leading case for analysis — I get the Shapley Value.

The breakdown technology which I will use, and which, I could argue, is the only technology which works for this purpose, is the following: With probability $1-\rho$, the proposer disappears (taking with him and consuming his own resources). The game then repeats with only the players in $S\setminus\{i\}$. So proposers that are frivolous enough to invite rejection run the risk of being out of the game. At the same time, of course, they are not always thrown out of the game because they have resources that the other players value.

As before, I will look at the stationary perfect equilibrium. If I could do with perfect equilibrium, I would be happier, but unfortunately, in the games that we are analyzing, the set of perfect equilibria is large (if $N > 2$).

3.3.2 The equilibrium conditions

How do we analyze problems like this? We already know how to determine the stationary perfect equilibrium equations. For the case $N = 2$, I drew a picture (figure 4). Now, since there are many more than two equations, I cannot draw a picture, but I can still write down the equilibrium equations without any difficulty. The logic is the same.

The equilibrium objects are the proposals. For every foreseeable coalition S of players still in the game and for every $i, j \in S$, let $a_j^{S,i}$ represent the proposal that i makes to j if i is the proposer. Let's see now what sort of consistency conditions these numbers need to satisfy. Define the expected payoff of the players in S to be

$$a^S = \frac{1}{|S|} \sum_{i \in S} a^{S,i} \in V(S)$$

This is the average of all the payoffs that result if all players' offers are accepted. There will be two equilibrium conditions:

1. For all S and $i \in S$, $a^{S,i} \in \partial V(S)$; i.e. $a^{S,i}$ is efficient, and is therefore on the boundary of $V(S)$ and not in the interior.

2. If $i \in S$ is the proposer, then he will offer j the minimum possible payoff, which is what j would get if she rejected. He must make sure that j will not reject. If j rejects, then with probability ρ, everything will be repeated, and she will get a_j^S, and with probability $(1-\rho)$. i will be thrown out of the game, and j will expect to get $a_j^{S \setminus \{i\}}$ instead. Thus,
$$a_j^{S,i} = \rho a_j^S + (1-\rho)\, a_j^{S \setminus \{i\}} \tag{1}$$

We can then solve the system using conditions 1 and 2. Notice that the result is a stationary perfect equilibrium.

3.3.3 A Remark

If ρ is very close to 1, then the term $(1-\rho)\, a_j^{S \setminus \{i\}}$ is very small. So, the proposals to j will depend on who the proposer i is, but in fact, no matter who i is, this proposal will be very close to the average a_j^S, which does not depend on i. Therefore, the proposals of all the players in S will lie close together on the boundary of $V(S)$, so that the average of these proposals, a^S, will be almost efficient, as illustrated in figure 12. If we examine the equilibrium equations, then as long as ρ is close to 1, we see that, first, it won't matter very much who the first proposer is and, second, the average proposal will be approximately efficient.

3.3.4 The TU case

Do we know anything else about the case in which ρ is large? Well, in the pure bargaining case for two players, once we defined the disagreement point, the stationary equilibrium of this model was the Nash solution. The same argument generalizes to N players if the strict subcoalitions cannot generate gains from trade (the pure bargaining case). But we are interested in the general situation in which the worth of a subcoalition matters.

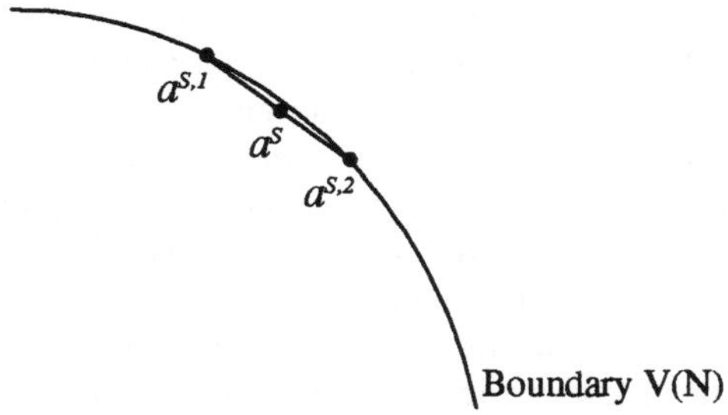

Figure 12: Approximate efficiency for large ρ

Consider the well understood TU case. Here, it turns out that, for any ρ, the stationary equilibrium payoffs are the Shapley values. This is, incidentally, why I chose this particular breakdown technology. While this result holds for all ρ, remember that in order to make sure that the proposals themselves (not just their average) are the Shapley values , we do need ρ to be close to 1. I will try to provide some intuition for this result. If you are familiar enough with the Shapley value, then you look at the system of equations and say, "Of course." But let me argue directly from the axioms. The Shapley value is characterized by four axioms.

1. Efficiency This is guaranteed for equilibrium payoffs because all the equilibrium proposals are on the boundary and the boundary is flat, so that the expectation is also on the boundary.

2. Equal Treatment I have never distinguished any particular player from any other. Thus, clearly, the "stationary perfect equilibrium payoffs" solution must be symmetric.

3. Additivity (Linearity) Write down the system of equations. In the linear (TU) case we simply have, for all S and $i \in S$,

$$\sum_{j \in S} a_j^{S,i} = v(S) \tag{2}$$

and therefore

$$a_i^{S,i} = v(S) - \sum_{\substack{j \in S \\ i \neq j}} a_j^{S,i} \tag{3}$$

A quick examination of (1) and (3) reveals that we can solve these for all the $a_j^{S,i}$'s recursively. We will thus obtain some complicated expression, but this expression will be linear in the $v(S)$'s.

4. Dummy Axiom This is where our particular breakdown technology comes into play. Intuitively, this technology implies that when some $i \in S$

makes a proposal to the players in S, if she has nothing to contribute to the other players, then these players do not pay any cost when they force a delay by rejecting her offer. Either the game is repeated, or i is kicked out, which makes no difference since the pie remains the same. Thus, the procedure gives no power to a dummy player. In a bargaining procedure, a player can have two types of power: one that derives from her resources, and another that derives from her ability to prevent agreement. A dummy has no resources, and this bargaining procedure makes the delays caused by her harmless to the other players. She therefore has no power of either kind.

Here is a formal proof that a dummy player gets nothing in a stationary perfect equilibrium. The proof is by induction. Suppose that the claim holds for all games up to size $N-1$. I will prove that it also holds for games of size N. Suppose that i is a dummy. I need to show, first, that when i proposes, he proposes 0 for himself, and, second, that when another player proposes, the proposal to i is 0.

Suppose that i is the proposer. How much will i propose to the other players? Adding the equilibrium equations (1) above, we get:

$$\sum_{j\neq i} a_j^{N,i} = \sum_{j\neq i} \left[\rho a_j^N + (1-\rho)\, a_j^{N\setminus\{i\}}\right]$$

Then substituting using (3), we get

$$v(N) - a_i^{N,i} = \rho\left[v(N) - a_i^N\right] + (1-\rho)\underbrace{v(N\setminus\{i\})}_{=v(N)} = v(N) - \rho a_i^N$$

since i is a dummy. Therefore, i proposes for himself

$$a_i^{N,i} = \rho a_i^N$$

Now suppose that $j \neq i$ is the proposer. Then, again using (1), we have

$$a_i^{N,j} = \rho a_i^N + (1-\rho)\, a_i^{N\setminus\{j\}}$$

Because i is a dummy, and $N\setminus\{j\}$ has only $N-1$ players, the induction hypothesis tells us that $a_i^{N\setminus\{j\}} = 0$. Therefore, in parallel to what we derived above, we have

$$a_i^{N,j} = \rho a_i^N$$

and this is true for all $j \neq i$.

Hence, whether i is the proposer or not, he gets ρ times his expected value. So by definition,

$$a_i^N = \frac{1}{N}\sum_{j\in N}\underbrace{a_i^{N,j}}_{=\rho a_i^N} = \rho a_i^N$$

which implies that

$$a_i^N = 0 \text{ because } \rho < 1$$

Therefore $a_i^{N,j} = \rho \cdot 0 = 0$ for all $j \in N$. ■

We have now shown that all the axioms of the Shapley value are satisfied. Therefore, the outcome must be the Shapley value.

3.3.5 Closing Remarks: The NTU Case

I repeat that what I have shown you her is only an example. However, it prompts a question: We have considered a bargaining procedure which in familiar cases gives very familiar solutions. Can we now do some conceptual boot-strapping, so that, after using cooperative theory to motivate a non-cooperative procedure, we can then go back to cooperative theory and discover what the particular non-cooperative procedure yields in the general NTU case? Interestingly enough, for $\rho \approx 1$, this procedure yields the "consistent solution" introduced by Maschler & Owen (1992). It does not yield either of the two more familiar solutions: the Shapley NTU value solution or the Harsanyi solution. Note that our particular bargaining procedure was not designed to yield the consistent solution, and that Maschler and Owen derived it from very different consistency-like requirements. Now you can ask, "What is the consistent solution?". I could spell it out for you, and I could also add:

Theorem: *When ρ is close to one, the stationary perfect solution of the bargaining procedure is close to the consistent solution.*

But since I have only –2 minutes left, I am going to simplify matters by transforming a theorem into a definition, and I will answer your question by saying that the consistent solution is the limit as $\rho \to 1$ of the stationary perfect equilibrium of the particular bargaining procedure that I have described!

References

Gale, D. (1986) Bargaining and computation, Part I: Characterization. *Econometrica* 54, 785-806.

Gul, F. (1989) Bargaining foundations of Shapley value. *Econometrica* 57, 81-95.

Hart, S., Mas Colell, A. (1992) Bargaining and value. *Econometrica,* forthcoming.

Maschler, M., Owen, G. (1992) The consistent Shapley value for games without side payments. In: R. Selten (ed.) *Rational Interaction.* Berlin: Springer-Verlag.

Myerson, R. B. (1991) *Game Theory.* Cambridge: Harvard University Press.

Osborne, M., Rubinstein, A. (1994) *A Course in Game Theory.* Cambridge: MIT Press.

Rubinstein, A. (1982) Perfect equilibrium in a bargaining model. *Econometrica* 50, 97-109.

Rubinstein, A., Wolinsky, A. (1985) Equilibrium in a market with sequential bargaining. *Econometrica* 53, 1133-1150.

Two Lectures on Implementation Under Complete Information: General Results and the Core

Philip J. Reny

Department of Economics, University of Pittsburgh, Pittsburgh PA 15260

1. Some General Results

1.1 Introduction

What is implementation theory all about? To answer this question we shall follow Moore (1991) by describing a classic problem known as "King Solomon's Dilemma." As it happened, two women approached Solomon with a newborn child. Each claimed to be the child's mother. It was up to Solomon to decide which one was telling the truth. In his wisdom, Solomon had the women lay the child before him. He drew his sword and announced that he would settle the dispute by cutting the child in half. However, just before he brought down his sword the true mother begged that he spare the child's life and give it to the impostor. Knowing that only the real mother would be willing to give up the child rather than allow it to die, Solomon gave it to her. Such is the nature of an implementation problem which we now describe in rather general terms.

An *implementation problem* consists of the following:

1. A finite set of agents $N = \{1, 2, ..., n\}$.
2. A finite set of states of the world, S.

A state may include whatever is relevant for the problem at hand. This may include only the preferences of the n agents while it might also include their endowments etc.

3. A finite set of social choices, C. (For King Solomon, a choice specifies which woman gets the baby.)

4. A social choice correspondence (SCC) $f : S \to 2^C$. (For King Solomon, it is single-valued; i.e. a function. An economic example is the Pareto correspondence.)

This framework covers problems involving the provision of public goods, optimal taxation, auction design, monopoly pricing, voting theory, bargaining, contract theory, agency theory, etc.

In general, a "planner" wishes to induce, for each state of the world (hereafter simply a state), s, a social choice (hereafter simply a choice) in $f(s) \subseteq C$. The difficulty is that the planner does not know the true state s. In our complete information setting we assume that the agents $1, 2, ..., n$ know the true state. When S includes all agents' preferences over C this is a very strong assumption.

In some circumstances this assumption is justified (the two mothers in King Solomon's dilemma, for example). When it is not, the theory of implementation under incomplete information is the relevant framework within which to analyze the problem. We shall stick with the complete information case however.

What does it mean for a planner to "induce a choice in $f(s)$?" How can he do this? One way is to have the agents (who know s) participate in a game (designed by the planner) in which they are driven to reveal their private information through the course of playing the game "rationally." Since the planner must design the game in ignorance of s, *the rules of the game cannot depend on the true state in any way.*

1.2 Nash Implementation

A (simultaneous) game form (rules of a game), denoted G, endows each agent i with a set of messages, M^i, and specifies for each n-tuple of messages chosen by the agents, a choice in C. The latter is represented by a function $g : M^1 \times \ldots \times M^n \to C$.

The issue then is this: Given a SCC, f, does there exist a game form, G, such that

(*) for all $s \in S$ the set of "equilibrium" outcomes of the game (G, s) is precisely $f(s)$. (Note that the set of equilibrium outcomes of the game may change as s changes since s determines the agents' preferences.)

Up to now I have avoided specifying an equilibrium concept. We'll say that a SCC is *Nash implementable* if there is a game form satisfying (*) where equilibrium there is Nash equilibrium. We similarly define *subgame perfectly implementable* SCC's, *dominant strategy implementable* SCC's, etc.

Let's now return to King Solomon. Does Solomon's scheme of threatening to cut the child in half meet the demands we have set forth above? Specifically, has Solomon succeeded in Nash implementing the desired outcome? No. By keeping quiet, the impostor failed to get the baby. Clearly the impostor can do no worse by mimicking the true mother. What would Solomon have done if both women had begged him to give the child to the other? Unfortunately, the biblical account is silent on this issue.

Can we fare any better than Solomon at solving his dilemma? Is King Solomon's SCC Nash implementable? The following analysis is taken from Moore (1991). The elements of the implementation problem are: $N = \{\text{Ann}, \text{Bess}\}$, $C = \{a, b, c, d\}$, $S = \{\alpha, \beta\}$, $f(\alpha) = a$, $f(\beta) = b$, where the choices a, b, c, d are respectively, Ann gets the baby, Bess gets the baby, the baby is cut in half, death to all; and the state α (β) denotes that Ann (Bess) is the true mother. In each state we assume that getting the baby is the best choice and death to all is the worst choice for both Ann and Bess. In state α (when Ann is the mother) we assume (as did Solomon) that Ann prefers b to c, and that Bess prefers c to a. Their preferences are reversed in state β.

We'll now argue that the SCC f above is not Nash implementable. If it were, then some game form would do the job. Let the game form be represented by a matrix whose entries are the outcomes and where Ann chooses the row and

Bess chooses the column. For simplicity, consider the 3x3 matrix below, where most entries are intentionally left unspecified.

	Bess		
	a	x	?
Ann	?	?	?
	?	?	?

Since the matrix game implements f, it must yield precisely choice a as a Nash equilibrium in state α. Thus choice a must occur as one of the entries, which it does. But if this corresponds to a Nash equilibrium, then the entry labelled x (and indeed every other entry in the first row) must either be a or d. Otherwise Bess would deviate in state α. But this implies that a remains an equilibrium outcome in state β, even though $f(\beta) = b$. Consequently, this game does not Nash implement f. This argument is completely general and therefore establishes that Solomon's SCC is not Nash implementable.

What is it about Solomon's problem that renders it nonimplementable? The key observation is this. When the state switches from α to β, choice a moves up (weakly) in both agents' preferences. Hence regardless of the game form, if a is a Nash equilibrium in state α, then it remains an equilibrium in state β. However, $a \notin f(\beta)$ and so f cannot be Nash implemented. This important insight was first recognized by Maskin (1977) where a distinction is drawn between SCC's that are *monotonic* and those that are not. Formally, a SCC $f : S \to C$ is *monotonic* if whenever $a \in f(s)$ and $a \notin f(t)$, there is an agent i and a choice b such that i weakly prefers a to b in state s, but strictly prefers b to a in state t. Clearly Solomon's SCC is not monotonic and this accounts for our inability to Nash implement it. Thus our argument above in the case of Solomon's problem has established the following general result.

Lemma 1.2.1 (Maskin (1977)): *If a SCC is Nash implementable, then it is monotonic.*

For the moment, we'll leave King Solomon and very briefly consider another SCC, namely the Pareto correspondence. Consider an exchange economy with a fixed set of agents and endowments. Let $f(s)$ denote the set of Pareto efficient allocations in state s, where the agents' preferences are determined by the state s. Is f monotonic? Yes. Consider $a \in f(s)$ such that $a \notin f(t)$. Then the latter relation implies that there is a choice b such that every agent strictly prefers b to a in state t. However, the former implies that there is at least one agent i who weakly prefers a to b in state s. Thus the Pareto correspondence satisfies this necessary condition for Nash implementability. But is the Pareto correspondence Nash implementable? If monotonicity were a sufficient condition for Nash implementation then the answer would certainly be, yes. However, monotonicity is not sufficient for Nash implementation (see Maskin (1985) for an example), but it is almost enough. Maskin (1977) proves the following remarkable result.

Theorem 1.2.2 (Maskin (1977)): *When there are three or more agents, any monotonic SCC satisfying no veto power is Nash implementable.*

A SCC f satisfies *no veto power* if whenever choice a in state s is top ranked for all but perhaps one agent, then $a \in f(s)$. Together with Lemma 1.2.1, this provides nearly a full characterization.[1] Since the Pareto correspondence satisfies no veto power and monotonicity, Theorem 1.2.2 establishes that it is Nash implementable when there are at least three agents.

Proof. The proof is constructive. Let $R_i(s)$ and $P_i(s)$ denote agent i's preference relation and strict preference relation, respectively, over C in state s. The following game form implementing f is slightly modified from that of Moore (1991) whose presentation is based on Repullo (1987).

Each agent announces a state in S, a choice in C, and an integer.

1. If all agents announce the same state s and the same choice $a \in f(s)$, then the outcome is a.
2. If all agents but one agree on s and $a \in f(s)$, then a is still the outcome of the game unless the remaining agent i announces a choice b such that $aR_i(s)b$, in which case the outcome is b.
3. Otherwise the outcome is the choice announced by the agent who announced the highest integer. (Ties are broken in any previously specified manner.)

We first show for every a and s with $a \in f(s)$, that a is an equilibrium outcome when the state is s. Indeed consider the common announcement $(s, a, 1)$. If all agents make this announcement in state s, then any individual agent i can change the outcome from a to b only if $aR_i(s)b$. Consequently, no agent can profitably deviate. Thus, it is a Nash equilibrium to make the common announcement above in state s. This then yields the outcome a in state s.

To complete the proof we must argue that if a is an equilibrium outcome in state s, then $a \in f(s)$. So, let a denote the equilibrium outcome in state s, and assume by way of contradiction that $a \notin f(s)$. Then, by no veto power, a is not top ranked in state s by at least two agents. Consequently, all agents must be announcing the choice a and the same state, t say, where $a \in f(t)$. Otherwise, by part 3 of the game form one of the two agents could deviate (by announcing the highest integer) and obtain his top ranked choice. Now, since $a \in f(t)$ and $a \notin f(s)$, monotonicity implies that there is an agent i and a choice b such that $aR_i(t)b$ and $bP_i(s)a$. But this means, by part 2 of the game form, that agent i can deviate by announcing $(s, b, 1)$ and render the outcome of the game b which he strictly prefers in state s to a. But this contradicts our assumption that a is an equilibrium outcome in state s. ■

Although ingenious, the game form used to implement the SCC f in the proof above has a drawback, namely the integer game that comes into play in

[1] For a complete characterization see Moore and Repullo (1990).

part 3 of the description of the game form. The integer game is included in order to rule out unwanted equilibria. However, in practice when confronted with the opportunity to play an integer game, the fact that a Nash equilibrium fails to exist might not be reason enough not to choose to play it. For instance consider a situation in which two players can choose to play an integer game at a cost of \$1 each. The status quo prevails if one of them decides not to play. If both decide to play, then the one who names the highest integer receives \$1000.00. The only Nash equilibrium of this game is for both players to decide not to play. But is it irrational to decide to play? How confident would you be that in practice intelligent players would always decide not to play? The integer game in the proof of Maskin's theorem rules out unwanted equilibria in precisely this fashion. Thus although technically the game form provided above does Nash implement f, one might well be suspicious of the practical success such a game form would enjoy. For an excellent critique of integer games and the like see Jackson (1992).

1.3 Subgame Perfect Implementation

Up to this point we have only considered Nash implementation. There are a variety of reasons for considering other solution concepts, in particular subgame perfection. Indeed, some SCC's that are not Nash implementable are subgame perfectly implementable. In addition, the use of subgame perfection rather than Nash equilibrium sometimes permits the use of simpler game forms and more compelling solutions. We now illustrate these possibilities.

In this example we provide a nonmonotonic (and therefore non Nash implementable; by Theorem (1.2.2) SCC that is subgame perfectly implementable. There are two agents 1 and 2, and two states s and t. There are three social choices, a, b, and c. Agent 1 ranks the choices in both states in the following strictly decreasing order: b, a, c. Agent 2's ranking (again in strictly decreasing order) is a, c, b in state s, and a, b, c in state t. The SCC is single valued and given by $f(s) = a$, and $f(t) = b$. It is straightforward to verify that f is not monotonic. However the following extensive game form subgame perfectly implements f. Agent 1 can immediately choose a (in which case the outcome of the game is a), or he can decline and give the move to agent 2. In the latter case 2 is informed of this and can choose the outcome of the game to be either b or c. It is easy to see that in state s the unique subgame perfect equilibrium outcome is a, while in state t it is b. Consequently, this extensive game form implements f as claimed. Note that it fails to Nash implement f (as it must) since in state t, a remains a Nash equilibrium outcome (agent 1 chooses a, and agent 2 if called upon to play chooses c).

Finally we demonstrate that subgame perfection sometimes implements the desired SCC in a more compelling fashion than does the Nash equilibrium concept. For this we return to King Solomon's dilemma but now we allow Solomon to exact fines. For concreteness, suppose that the true mother values the child at 20, the impostor values the child at 10 (the reader can choose the units as he/she deems appropriate), and that Solomon can either fine both women 15 or fine neither. Letting A denote the woman whose name is Ann and B Bess, the social choices are $(A,0), (B,0), (A,15), (B,15)$, where the first component of each ordered pair denotes who gets the baby, and the second denotes the amount both women are fined. As before there are two states α and β, where Ann is the true mother in the first and Bess the true mother in the second. Solomon's SCC, f, is given by $f(\alpha) = (A,0)$ and $f(\beta) = (B,0)$. The preferences of Ann and Bess are summarized in the following table. Social choices in higher rows are strictly preferred to those in lower rows.

State α:	Ann	Bess	State β:	Ann	Bess
	$(A,0)$	$(B,0)$		$(A,0)$	$(B,0)$
	$(A,15)$	$(A,0)$		$(B,0)$	$(B,15)$
	$(B,0)$	$(B,15)$		$(A,15)$	$(A,0)$
	$(B,15)$	$(A,15)$		$(B,15)$	$(A,15)$

Evidently, f is now monotonic. For instance, when the state switches from α to β, Bess's preferences between $(A,0)$ and $(B,15)$ are reversed. Similarly, when the state switches from β to α, Ann's preferences between $(B,0)$ and $(A,15)$ are reversed. Consequently, f in this setting with fines may well be Nash implementable. Note that we cannot appeal to Theorem 1.2.2, since there are only two agents here. Nonetheless, f is Nash implementable by the following matrix game in which Ann chooses the row and Bess the column.

$(A,0)$	$(B,15)$	$(A,15)$
$(B,15)$	$(A,15)$	$(B,0)$

In state α, the only Nash equilibrium is for Ann to choose the top row and Bess the leftmost column. In state β, the only (pure strategy) Nash equilibrium is for Ann to choose the bottom row and Bess the rightmost column. Thus, the matrix game implements f in (pure strategy) Nash equilibrium.

Note that in our whole discussion up to this point, we have said nothing about the agents' preferences over lotteries over social choices. Consequently we have restricted attention up to now to pure strategy equilibria of the game forms introduced. The advantage of this approach is that it requires fewer assumptions about the agents' preferences (in particular the von Neumann-Morgenstern axioms need not hold) and consequently (in our complete information setting) weaker assumptions about what each agent knows about the other agents' preferences.[2] The disadvantage is that the agents may well have von Neumann-Morgenstern preferences over lotteries and then there may be natural mixed strategy equilibria that upset the implementation result[3] The matrix

[2] I am grateful to Motty Perry for pointing this out to me.

[3] We will see that when we restrict preferences to the von Neumann-Morgenstern class, every social choice function is (virtually) implementable. This striking reult is due to Abreu and Matsushima (1992a). They also provide a nice discussion of a number of shortcomings of the previous implementation literature.

game just presented is a good example. Were Ann and Bess endowed with von Neumann-Morgenstern utility functions assigning in each state the number 4 to the top ranked choice, the number 3 to the next, down to 1 to the lowest ranked choice, there would in state β be an additional equilibrium in which Ann mixes between the two rows with equal probability, and Bess mixes between the left and right columns giving right weight 3/4. This equilibrium is not at all pathological and upsets the implementation result. This difficulty can be avoided by considering the remarkably simple extensive game form described below and employing subgame perfection rather than Nash equilibrium.

The game proceeds as follows (see Glazer and Ma (1989)). Ann moves first and can choose either to give the baby to Bess, in which case the outcome is $(B, 0)$, or she can claim that the baby is hers. If she claims that the baby is hers, Bess is informed of this and it is Bess' turn to move. Bess can then choose either to give the baby to Ann, in which case the outcome is $(A, 0)$, or she can claim that the baby is hers in which case the outcome is $(B, 15)$. It is easy to check that in state α the unique subgame perfect equilibrium outcome is $(A, 0)$ and that in state β it is $(B, 0)$. Moreover, this remains the case even if Ann and Bess have von Neumann-Morgenstern preferences and mixed strategies are considered.[4]

For general results on subgame perfect implementation see Moore and Repullo (1988) and Abreu and Sen (1990). We now turn to one of the most important recent developments in the theory of implementation. It is due to Abreu and Matsushima (1992a).

1.4 Virtual Implementation

1.4.1 The Setup

In this section we shall not rule out the use of lotteries as we've done for the most part up to now. Moreover, we shall assume that the agents have von-Neumann-Morgenstern utilities over lotteries over the social choices in each state. Let A denote the set of lotteries over the given finite set of social choices. Thus A might as well be the unit simplex in $\Re^n$, where the ith unit vector represents the lottery assigning probability one to the ith social choice. Let $u_i(a, s_i)$ denote agent $i's$ von Neumann-Morgenstern utility of lottery a in state $s = (s_1, ..., s_n) \in S \subseteq \times_{i=1}^n S_i$. We shall restrict attention to single-valued SCC's. Thus, f is a social choice function (SCF) mapping the set of states, S, into lotteries over social choices, A. Finally, we shall be concerned with implementation in iteratively strictly undominated (IU) strategies. This is a very weak solution concept in that any outcome supported by any refinement of Nash equilibrium survives iterative removal of strictly dominated strategies.

[4]Moore (1991) generalizes this result to the case in which each woman knows the value the other places on the child, but Solomon only knows that the true mother's value is the higher of the two. Perry and Reny (1994b) extend this further to the case in which the women's values are purely private information but that each knows who has the higher value (i.e. each knows who the true mother is). The latter paper implements Solomon's SCC in iteratively (weakly) undominated strategies.

But there is an advantage in this weakness. One need assume much less about the agents in order to conclude that they will play an IU strategy as compared to what must be assumed to conclude that they will play a Nash equilibrium. Thus from a practical point of view there is much to be gained from considering implementation in IU strategies. The real surprise is that IU implementation is not hopeless.

A number of significant ideas will be brought together here to produce an extraordinarily permissive implementation result. The first of these is the idea that one ought to be content if one can implement the desired social choice function up to any prespecified degree of accuracy.

Definition 1.4.1: *The SCF f is virtually implementable in iteratively (strictly) undominated strategies if $\forall \epsilon > 0, \exists$ an SCF g with $\|g(s) - f(s)\| < \epsilon$ $\forall s \in S$ such that g is implementable in iteratively strictly undominated strategies.*[5]

We shall maintain the following assumption throughout this section.

Assumption A: For all agents i, and all pairs of distinct s_i, $t_i \in S_i$, there is a pair of lotteries, $a, b \in A$, such that $u_i(a, s_i) > u_i(b, s_i)$ and $u_i(a, t_i) < u_i(b, t_i)$.[6]

Insight 1: Under Assumption A, every SCF is virtually monotonic in the following sense. For every SCF f, and every $\epsilon > 0$, there is a monotonic SCF g that is ϵ-close to f. This is a consequence of the self-selection result below. Thus monotonicity, the necessary condition for Nash implementation, is automatically (virtually) satisfied when lotteries are available.

Lemma 1.4.1 (Self-Selection): *Under Assumption A, $\forall i$, $\exists d_i : S_i \to A$ such that $u_i(d_i(s_i), s_i) > u_i(d_i(t_i), s_i)$, $\forall$ distinct $s_i, t_i \in S_i$.*

Proof. For each agent i, let A_i denote the union of all lotteries a and b as in Assumption A when all distinct pairs s_i, t_i are considered. Without loss of generality we may assume that $u_i(\cdot, s_i)$ strictly orders A_i for all $s_i \in S_i$. Let $p_1, p_2, ..., p_{\#A_i}$ be a strictly decreasing sequence of positive probabilities whose sum is one. For each s_i, define $d_i(s_i)$ to be the lottery giving the jth ranked member of A_i (according to s_i) probability p_j. ■

To see that self-selection yields *Insight 1* above, consider the social choice function

$$g(s) = \epsilon d_1(s_1) \oplus \epsilon d_2(s_2) \oplus ... \oplus \epsilon d_n(s_n) \oplus (1 - n\epsilon) f(s), \tag{1}$$

---

[5] $\|\cdot\|$ denotes Euclidean distance in $\Re^n$. Consequently, the definition requires that in every state, the probability that g assigns to any particular social choice is within ϵ of that assigned by f . In this case, we'll say that g is ϵ−close to f.

[6] Abreu and Matsushima (1992a) show that Assumption A is a consequence of the following two conditions: (i) no agent is indifferent over all of A in some state, and (ii) different states in S_i for agent i induce different preferences over A for i.

where for lotteries a and b, and $p \in [0,1]$, $pa \oplus (1-p)b$ denotes the compound lottery in which a occurs with probability p and b with probability $1-p$. For ϵ small enough, g is arbitrarily close to f. Moreover, with the $d_i(\cdot)'s$ chosen as in the self-selection Lemma, g is easily seen to be monotonic.

Given the SCF f, our object will be to implement the SCF g above in iteratively undominated strategies for $\epsilon > 0$ arbitrarily small.

Insight 2: If ϵ were large enough in (1), then we could implement g (which would then be very much like a random dictator SCF) in dominant strategies. Thus when the probability that each player is a dictator (namely ϵ) looms large relative to the probability assigned to f, there is no problem. Of course we are concerned with the case in which ϵ is small. An ingenious way of getting the right relative weights is to break the probability that f occurs into many (k say) small (relative to ϵ) pieces. So write g as

$$g(s) = \epsilon d_1(s_1) \oplus \epsilon d_2(s_2) \oplus \ldots \oplus \epsilon d_n(s_n) \oplus \frac{(1-n\epsilon)}{k} f(s) \oplus \ldots \oplus \frac{(1-n\epsilon)}{k} f(s) \quad (2)$$

In order for this to have any real effect, the game form must be designed so that the agents can affect each of the last k outcomes of the lottery in g independently of one another. Once this is done, agent $i's$ incentive to change any *one* of the last k components of the lottery $g(s)$ will be outweighed by his incentive to obtain $d_i(s_i)$ in the event he is chosen as dictator on A_i since the latter event is much more likely to occur than the former. The game form we shall construct below (due to Abreu and Matsushima (1992a)) takes full advantage of this idea.

In order to simplify matters we make the following assumption.[7]

Assumption B: Small personal taxes can be levied. Moreover, agent $i's$ utility of lottery a together with a tax of $\tau > 0$ in state s is $u_i(a, s_i) - \tau$.

1.4.2 The Game form

Each agent i (simultaneously) sends a message consisting of $k+1$ cells. The first cell is a personal state $s_i \in S_i$. The remaining k cells are each members of the state space S. Thus a typical message is of the form $m_i = (m_i^0, m_i^1, \ldots m_i^k)$, where $m_i^0 \in S_i$, and $m_i^j \in S$ for all $j = 1, 2, \ldots k$. The joint message $(m_1, m_2, \ldots m_n) \in \times_{i=1}^{n}[S_i \times S^k]$ determines the components $a_1, \ldots a_n, b_1, \ldots b_k \in A$ of the lottery

$$\epsilon a_1 \oplus \ldots \oplus \epsilon a_n \oplus \frac{1-n\epsilon}{k} b_1 \oplus \ldots \oplus \frac{1-n\epsilon}{k} b_k \quad (3)$$

as follows:

$$a_i = d_i(m_i^0)$$

$$b_j = \begin{cases} f(s), \text{ if } m_l^j = s, \text{ for all but perhaps at most one } l = 1, \ldots n \\ a^*, \text{ otherwise,} \end{cases}$$

[7] Abreu and Matsushima (1992a) do not assume this. Instead they assume that each agent can be punished (if only slightly) independently of other agents.

where a^* is some previously specified lottery. Note that the last k cells of agent $i's$ message independently affect the last k components of the lottery in (3), and the first cell determines the outcome in the event that agent i is given the opportunity to dictate over the elements of A_i. Recall *Insight 2* above for the significance of this. Also note that if all agents report honestly in state s, then the outcome will be $g(s)$.

In addition to determining the outcome of the above lottery, the joint message determines the agents' personal taxes. Before describing these, we first provide an appropriate choice of the parameter k to be employed in (3) as well as two other parameters, τ and δ, where the first is used to compute the personal taxes and the second provides a lower bound on the utility gain available from switching to an honest report from a false one when agent i is chosen to be dictator on A_i. Specifically choose k, τ, and δ all positive such that:

$$u_i(d_i(s_i), s_i) - u_i(d_i(t_i), s_i) \geqslant \delta \text{ for all distinct } s_i \text{ and } t_i \in S_i \tag{4}$$

$$\tau < \epsilon\delta \tag{5}$$

$$k > \frac{\Delta(1 - n\epsilon)}{\tau} + 1 \tag{6}$$

where $\Delta = \max[u_i(\bar{a}, s_i) - u_i(\underline{a}, s_i)]$, and where the maximum is taken over all agents i, personal states s_i, and lotteries $\bar{a}$ and $\underline{a}$. Note that the Self-Selection Lemma ensures that (4) can be satisfied.

There are two kinds of personal taxes that are potentially imposed, an *outlier tax*, and a *false report tax*. These are determined by the joint message $(m_1, ...m_n)$ as follows.

Outlier Tax: Let τ_o denote the outlier tax. Set $\tau_o = \tau/k$. The outlier tax is levied on an agent whenever that agent is the only one to disagree with the reports of the others in a particular cell (beyond the zeroth) of his message. That is, agent i is taxed τ_o for every cell $l = 1, ..., k$ such that $m_1^l = m_2^l = ... \neq m_i^l$.

False Report Tax: Let τ_F denote the false report tax. Set $\tau_F = \tau$. The false report tax is levied at most once per agent when that agent in some cell of his message does not announce the (induced) state announced in the combined zeroth cells of all agents when all previous cells of all agents did so. That is, let j be the first cell such that $m_{i'}^j \neq (m_1^0, ...m_n^0) \in S$ for some agent i'. Then agent i is taxed τ_F if $m_i^j \neq (m_1^0, ..., m_n^0)$.

Each agents' total tax is the sum of his false report tax and all his outlier taxes. This completes the description of the game form.

1.4.3 The Result

Theorem 1.4.2: *The above game form implements g defined by (1) in iteratively strictly undominated strategies.*

Proof. Suppose that the state is $s = (s_1, ...s_n)$. It suffices to show that for each agent i the only iteratively undominated message is $(s_i, s, ..., s)$. To do so we shall show that the first round of elimination of dominated messages implies that the first cell of agent $i's$ message must be s_i, and that the second and subsequent rounds imply that the second and subsequent cells must be s.
First Round: Suppose that the first cell in $i's$ message is not s_i. Then the gain from switching only the first cell of his message to s_i is at least $\epsilon\delta$ (in expected utility) by (4). The loss in making such a switch is at most τ_F since the outlier tax levied on i is unaffected by the switch. Consequently, the net gain of the switch is at least $\epsilon\delta - \tau_F$, which is positive by definition of τ_F and (5). Thus all messages not involving honest reporting in the first cell are removed in the first round of elimination.
Second Round: From the first round all agents' first cells are honest reports of their personal state. Thus, $(m_1^0, ..., m_n^0) = s$. Suppose that some agent $i's$ second cell is not equal to s.
Case A. Consider the case in which some other agent's second cell is also not equal to s. If agent i switches only the second cell of his message to s, then he gains τ_F in utility since he no longer pays the false report tax. This switch may affect the lottery that enters in the place of b_1 in (3) and it may also result in i paying an additional outlier tax (since it may be that all other agents' second cells are equal and distinct from s). Thus $i's$ loss (in expected utility)from making such a switch is at most $\tau_o + \frac{1-n\epsilon}{k}\Delta$. Consequently, $i's$ net gain from this switch is at least $\tau_F - (\tau_o + \frac{1-n\epsilon}{k}\Delta) = \frac{k-1}{k}\tau - \frac{1-n\epsilon}{k}\Delta$, which is positive by (6).
Case B. Consider the alternative case in which all other agents report s in the second cell of their message. The gain to agent i from switching the second cell of his message to s is now $\tau_o + \tau_F$, since both the outlier and false report taxes are avoided for this cell. Since all other agents' second cells matched, $i's$ second cell report has no impact on the lottery b_1 in (3). Hence the only potential loss in making the switch is that i might incur the false report tax for a different cell. Thus $i's$ net gain from the switch is at least τ_o which is positive.

Hence after the first round of elimination, it is strictly dominant for each agent to report s in the second cell of his message. The argument for the second cell can be repeated to conclude that for each agent i the only message surviving iterative elimination of strictly dominated strategies is $(s_i, s, ..., s)$. ■

Two features of the game form are well worth noting: (i) it involves no integer game, and (ii) mixed strategies are not ruled out of the analysis. It is remarkable what has been achieved here. Not only is the implementation result completely permissive (i.e. all SCF's are virtually IU implementable), but the equilibrium concept is extraordinarily weak. Indeed it is enough that expected utility maximizing behavior be common knowledge among the players in order that only iteratively undominated strategies survive. Moreover, two fundamental difficulties inherent in previous implementation results (namely the use of integer games and the ban on consideration of mixed equilibria) have been entirely overcome. It should also be noted that the total taxes levied can be made arbitrarily small both on and off the equilibrium path. Finally, we remark that the full force of the expected utility hypothesis is not needed. In addition to continuity, the only required property is the following: Suppose that the probabilities assigned to two outcomes of a compound lottery are switched. If the newly created compound lottery now gives higher weight to the (strictly) more preferable of the two outcomes, then the new compound lottery is strictly preferred to the old.

In our view, the result due to Abreu and Matsushima (1992a) reviewed here necessitates a fundamental change in the direction of research in implementation theory. Their result has more or less completely settled the question "What SCF's can be implemented?" from a *theoretical* standpoint. In practice, the number of rounds of elimination required in the Abreu-Matsushima game form depend upon how closely one wishes to approximate the original SCF. The number of rounds increases (without bound) as the approximation improves. If one is uneasy about long chains of iterative elimination being made by individuals in practice, then one might well hesitate to actually employ the Abreu-Matsushima game form. See Glazer and Rosenthal (1992) for further comments on this issue as well as the reply in Abreu and Matsushima (1992b). Thus there remains an important avenue for research, namely the search for "simple" and "compelling" game forms that are capable (in practice) of implementing "interesting" and "important" SCF's.

2. The Core

2.1 Introduction

There is little doubt that the core is a central idea both in game theory and economics. In the sequel I shall focus on two questions.

1. Under what conditions can a planner ensure any (and only) core outcomes?
2. Are core outcomes inevitable?

The second question is the familiar one of Edgeworth (1881) and it lies at the heart of the positive (as opposed to the normative) interpretation of the core. Put somewhat differently, "Are outcomes not in the core necessarily unstable?"

We will begin by focussing our attention on the first question. To the extent that the core has normative appeal, this question is clearly of interest. In the language of the previous sections it asks whether or not the core can be implemented. It is straightforward to show that the core correspondence, viewed as a SCC, is monotonic. Moreover on economic domains it satisfies no veto power.[8] Consequently, by Theorem 1.2.2 the core can be Nash implemented.

[8] In an exchange economy setting in which all commodities are desirable, no veto power is vacuously satisfied since the allocation top ranked for a consumer is that which gives him everything and hence it cannot be top ranked for anyone else.

Since our concerns at this point are largely normative, it is rather important to have confidence that the game form employed will actually perform in practice as it does according to the theory. With this in mind, it makes sense to search for game forms having rather simple message spaces. The message space employed in the proof of Maskin's Theorem is potentially complex. Indeed it is the entire state space and more. Thus in particular, each agent is asked to report the preferences of all other agents over the entire set of social choices. In specific cases it is sometimes possible to simplify the message space substantially. Again we emphasize that this has at most practical, not theoretical, consequences. We now present a game form implementing the core in subgame perfect equilibrium with rather simple message spaces. The presentation is based on Serrano and Vohra (1993).

2.2 An Implementation of the Core

Let (v, N) be a TU game. Thus, N is a finite (with cardinality n) set of players, and $v(\cdot)$ is a function mapping nonempty subsets of N (coalitions) into non-negative numbers. For a payoff vector $x \in \Re^n$, $x(S)$ denotes the sum of the components x_i of x for $i \in S$. A *feasible proposal* is a pair (x, S) where $x \in \Re^n$, $S \subseteq N$ and $x(S) \leq v(S)$. A payoff vector x is in the *core* of (v, N) if $x(S) \geq v(S)$ for every $S \subseteq N$, with equality for $S = N$.

Of course, the planner, who wishes to implement the core of (v, N), does not know the characteristic function v. Otherwise he would simply impose an outcome in the core. We shall however make the following assumptions about the planner.

1. The planner can extract fines (measured in payoff units) from each player.
2. The planner can verify the feasibility of any payoff vector.
3. The planner can impose a coalition structure.

The content of (1) beyond the obvious is that the planner is assumed to know a lower bound on the resources of each agent. Without this, an agent might claim to have no resources in order to avoid paying a fine. Assumption (2) says that for each coalition $S \subseteq N$, and any payoff vector x, the planner can verify whether or not $x(S)$ can be attained by the members of S. This does not require the planner to know $v(S)$.[9] The last assumption is self explanatory. We now describe a two stage game implementing the core of (v, N) in subgame perfect equilibrium.

[9] For instance, in an exchange economy setting the planner may know the players' preferences but be unaware of their endowments. The planner can nonetheless verify the feasiblity of any payoff vector by simply requiring the players to display the required amounts of commodities that they claim to have. On the other hand the planner may not know the players' preferences but may know their endowments. Although the utility vector now cannot be verified, the planner can verify the feasibility of any allocation for any coalition. The mechanism below can be easily modified to implement the core in this (perhaps more natural) case as well.

Stage 1: All players i simultaneously announce a feasible proposal of the form (x^i, N),[10] and a natural number m_i.

1. If $x^i \neq x^j$ for some i and j, then player $i' = \sum_{k \in N} m_k \bmod(n)$ is fined. The coalition structure in which all players are alone is then imposed.

2. If $x^i = x$ for all $i \in N$, then all players are informed of all the announcements (x^i, m_i) and the game proceeds to Stage 2.

Stage 2: Player $i' = \sum_{k \in N} m_k \bmod(n)$ from Stage 1 makes a feasible proposal (y, S).[11] All members of S are informed of this proposal and respond to it (by saying "accept" or "reject") sequentially in some predetermined order. If all members of S accept then the game ends with $i \in S$ obtaining y_i, and $i \notin S$ obtaining $v(i)$. If at least one member of S rejects the proposal, then the grand coalition is imposed and each $i \in N$ obtains x_i, the status quo from Stage 1.

Proposition 2.2.1: *The above game form implements the core of (v, N) in pure strategy subgame perfect equilibrium.*

Proof. We leave it to the reader to verify that any core payoff x can be supported as an SPE. The equilibrium path has all players announcing the payoff vector x and some natural number in stage one; the player chosen as proposer in stage two proposing (x, N); and all players accept.

To prove the converse, we suppose that z is a pure strategy SPE outcome.
STEP 1: Each players' first stage payoff vector announcement must be the same. Otherwise player i' could change his announced number and avoid the fine and increase his payoff. Let this common stage 1 payoff announcement be x.
STEP 2: Each player i can, by changing his announced number, become the proposer in stage 2 and there propose (x, N). Subsequently, regardless of how the others respond, the outcome will be x. Since the outcome is in fact z, this implies that $z_i \geq x_i$ for all players i.
STEP 3: Suppose by way of contradiction that z is not in the core of (v, N). Then for some payoff vector y and coalition S with $y(S) \leq v(S)$, we have $y_i > z_i$ for all $i \in S$. Choose $j \in S$ and consider a deviation by j rendering j the proposer in stage 2 (the status quo then remains x) and in which j proposes (y, S). Clearly subgame perfection demands that every $i \in S$ accept this proposal since $y_i > z_i \geq x_i$ for all $i \in S$. Hence j can profitably deviate, a contradiction. ■

Remark 2.2.1: The restriction to pure strategies is essential. The presence of the "modulo" game in stage 1 creates mixed strategy equilibria which would upset the result were they admissible. This is a serious drawback of this mechanism.

---

[10] The feasibility constraint can be enforced by the planner since he can verify feasbility.
[11] Again, feasibility is enforced by the planner.

Remark 2.2.2: No refinement beyond (pure strategy) subgame perfection is required. This is in contrast to the results to follow which impose stationarity as well.

Remark 2.2.3: The Proposition implies that the existence of a pure strategy SPE relies on the nonemptiness of the core of the game. Thus if (and only if) the game is balanced a pure strategy SPE exists. This is in contrast to the model of Perry and Reny (1994a) in which total balancedness is required.

Remark 2.2.4: Finding a simple, compelling game form that implements the core without excluding mixed strategies remains an open problem.

2.3 A Noncooperative Approach To The Core

2.3.1 A Canonical Discrete-Time Coalitional Bargaining Model

We now consider our second question "Are core outcomes inevitable?". Somewhat more precisely, must the outcome of a setting in which agents can interact in an unfettered manner be in the core? Note that the motivation here is quite different from that of the previous section. There is no planner here and the core is not viewed as a desirable goal that somehow must be attained. Rather, the focus is on identifying those settings in which the core *happens to be attained.* The significance of the results should then be judged in terms of the naturalness of the settings that are identified.

We shall focus on one particular setting whose underlying opportunities are summarized by the TU game (v, N). The following extensive form game with perfect information attempts to capture the idea that the players can interact in an "unfettered manner."

For each coalition $S \subseteq N$, let $\phi(S)$ denote an ordering of the members of S. Call ϕ a protocol. The game proceeds as follows. Calendar time, indexed by t, is initialized to zero. Players discount the future at the common discount rate $\delta \in (0,1]$. The first player according to $\phi(N)$ is chosen as the proposer. He makes a feasible proposal (x, S). The members of S respond (accept or reject) in order according to $\phi(S)$. If $i \in S$ is the first to reject the proposal, then calendar time moves ahead one unit, i becomes the proposer and the above process repeats. If all $i \in S$ accept (x, S), then each $i \in S$ receives payoff $\delta^t x_i$ and leaves the game. Calendar time does not move ahead. The game now proceeds as at the start, with the remaining players taking the role of N. Thus, if T denotes the set of remaining players, the first player according to $\phi(T)$ is chosen to be the proposer, etc. Call this the coalitional bargaining game.

This model (with $\delta = 1$) is due essentially to Selten (1981). A similar game is employed by Moldovanu and Winter (1991) and Chaterjee, Dutta, Ray and Sengupta (1993). The last paper assumes strict discounting (i.e. $\delta < 1$). Note that the discounting case of Rubinstein's (1982) bargaining model is a special case of the above game.

Our aim is to consider the subgame perfect equilibria of this game with a view to supporting only core (and all core) outcomes. A number of concerns are immediately apparent.

Stationarity: Without discounting many non core outcomes can be supported as SPE. Indeed, even with discounting there is a folk-theorem due to Shaked-like strategies (see Osborne and Rubinstein (1990) chapter 3). Hence, a refinement beyond subgame perfection is necessary. We'll therefore consider stationary SPE. These are subgame perfect equilibrium strategies in the usual sense that also satisfy the following stationarity property: players' actions depend only upon the set of remaining players, and the current proposal.[12]

Discounting: Since Rubinstein's (1982) model is a special case of the coalitional bargaining game, the latter cannot always support all core outcomes. For example, in Rubinstein's (1982) model, any division of the pie is in the core, yet only a single division is subgame perfect. Thus, if we insist on including discounting, there is no hope in supporting all core outcomes. On the other hand, we may be content so long as those outcomes that can be supported are in the core. The following example (due to Chaterjee et. al. (1993)) shows that the presence of discounting may preclude even this.

Example 1. $v(N) = 1 + \mu$, $v(\{1,2\}) = v(\{1,3\}) = 1$, $v(\{2,3\}) = \epsilon > 0$, where ϵ is small and $0 < \mu < 1/2$. For $\delta > \frac{\mu}{1-\mu}$, every stationary equilibrium has the common features described in the following table.

Players Remaining	Reservation Payoff
$\{1,2,3\}$	$\frac{\delta}{1+\delta}$
$\{1,2\}$ or $\{1,3\}$	$\frac{\delta}{1+\delta}$
$\{2,3\}$	$\frac{\epsilon\delta}{1+\delta}$
$\{1\}$, $\{2\}$, or $\{3\}$	0

The second row of the table, for instance, indicates that if the set of remaining players is $\{1,2\}$ (or $\{1,3\}$), then both players 1 and 2 (resp. 3) accept any proposal giving them a payoff of at least $\frac{\delta}{1+\delta}$. They reject all other proposals. Consequently, the grand coalition does not form. Although as we shall see below the protocol can play a role in determining whether or not the outcome is efficient, it plays no such role here. The inefficiency remains regardless of the protocol. It is the presence of discounting that accounts for the inefficiency here.

Player Order: The following examples indicate the important effect that the (essentially arbitrary) protocol, ϕ, has on the outcome of the game. The first is taken from Chaterjee et. al. (1993).

[12]This is a rather strong stationarity assumption. It is employed by Chaterjee et. al. (1993). Moldovanu and Winter (1991) also allow the accept/reject decision to depend upon the set of players who have so far accepted the current proposal. We adopt the stronger version here only so that we can provide a single analysis that is consistent with both papers.

Example 2. $v(N) = 1.3$, $v(\{1.2\}) = 1$, $v(\{1,3\}) = v(\{2,3\}) = .1$, and $v(i) = 0$ for $i = 1,2,3$. For $\delta \in (3/7, 1)$, and for each protocol, there is a unique stationary SPE. Although the equilibria may differ with different protocols, they each have the following in common. For $i \in \{1,2\}$, whenever it is $i's$ turn to make a proposal i proposes $((\frac{1}{1+\delta}, \frac{\delta}{1+\delta}), \{1,2\})$. Whenever it is player 3's turn to make a proposal, 3 proposes $((\frac{\delta}{1+\delta}, \frac{\delta}{1+\delta}, 1.3 - \frac{2\delta}{1+\delta}), \{1,2,3\})$. Consequently, if the protocol has player 1 or 2 going first, the outcome will be inefficient.

One might suspect that the protocol alone cannot account for inefficiency and that discounting is in fact always the culprit. The next example, in which there is no discounting at all shows this to be incorrect.

Example 3. $V(N) = 1$, $V(\{1,2\}) = v(\{1,3\}) = 1$, $v(\{2,3\}) = v(\{i\}) = 0$, for $i = 1,2,3$. The following table describes stationary strategies which constitute an SPE for $\delta = 1$.

Players Remaining	Reservation Payoff	Proposal
{1,2,3}	0 for $i = 1, 2$	((1,0,0),N) for $i = 1$
{1,2,3}	1 for $i = 3$	((0,0),{2,3}) for $i = 2, 3$
{1,2}	$\frac{1}{2}$	$((\frac{1}{2}, \frac{1}{2}), \{1,2\})$
{1,3}	$\frac{1}{2}$	$((\frac{1}{2}, \frac{1}{2}), \{1,3\})$
{2,3},{1},{2},{3}	0	offer each remaining player zero

The first row of the table indicates that if all players remain (first column), then players 1 and 2 accept all nonnegative offers (column 2), and when chosen as proposer player 1 would propose that the grand coalition form and that he get the entire surplus (column 3). It is straightforward to check that these strategies are stationary and that they form an SPE, regardless of the protocol. In addition it is clear that the outcome is (0,0,0) if the protocol is such that either player 2 or 3 is chosen as the first proposer. On the other hand, if player 3 happens to be the first proposer, then the outcome is the unique core point.

So, when the protocol affects the outcome, the outcome may not even be efficient let alone in the core. But when the protocol does not affect the outcome, it turns out that the outcome must be in the core as we now show. The following is based on Moldovanu and Winter (1991).

Definition 2.3.1: *Call a stationary SPE order independent if it remains an SPE and induces the same outcome for all protocols.*

Proposition 2.3.1 (Moldovanu and Winter (1991)): *Suppose that (v, N) is totally balanced. Then x is in the core of (v, N) iff x can be supported by an order independent SPE of the coalitional bargaining game with $\delta = 1$.*

Proof. Since (v, N) is totally balanced we may choose for each $S \subseteq N$, a vector $x^S \in \Re^{|S|}$ that is in the core of (v^S, S), where v^S is the restriction of v to S. If x is in the core, then consider the following strategies. If the set of remaining players is S :
(i) the proposer proposes (x^S, S),
(ii) a proposal (y, T) is accepted by $i \in T$ on his turn if and only if $y_i \geq x_i^S$.
It is straightforward to check that these strategies form a stationary, order independent SPE.

Conversely, suppose that x is supported by a stationary order independent SPE, σ. Assume by way of contradiction, that x is not in the core. Hence for some y and S with $y(S) \leq v(S)$, we have $y_i > x_i$ for all $i \in S$. Choose a particular $i \in S$ and consider the protocol in which i is the first proposer. By order independence, x remains a stationary SPE outcome under σ. Consider, however, the deviation in which i proposes (y, S). If some $j \in S$ rejects the proposal, then j becomes the proposer. But since σ is order independent, the resulting continuation outcome will also be x. Consequently, every member of S strictly prefers that (y, S) be accepted than that some member of S (including himself) reject it. Hence subgame perfection demands that (y, S) be accepted by every member of S. But this deviation by i is then profitable, a contradiction. ■

It is evident from the examples that the presence of discounting in the coalitional bargaining model can preclude the existence of stationary SPE outcomes that are in the core. Also evident is that the order in which players appear can be critical. In example 3 for instance, when player 1 is not the proposer, player 1 has no opportunity to respond to the proposal made by 2 or 3. In particular, player 1 does not have an opportunity to suggest a proposal that all players would prefer to (0,0,0). In short, player 1 is not given an opportunity to block the eventual non core outcome (0,0,0). But even our standard classroom story behind the core relies on the ability of all players to block a non core outcome. Thus, its no surprise that the coalitional bargaining model runs into trouble.

One might try to modify the coalitional bargaining model in the following way so that all players have an opportunity to block non core proposals: after a proposal is made, all remaining players must accept the proposal in order for it to go into effect, even those players who are outside the coalition whose payoffs are under consideration. Unfortunately, such a modification then allows all efficient outcomes to be supported, not just core outcomes.

There is however a way out of the "player order" difficulty. It must first be recognized that the problems we've encountered are artificial in the sense that they are artifacts of the discrete-time game theoretic model we have chosen to employ. If we were to imagine a real life setting in which people are gathered to trade and negotiate and there are "no rules" governing the negotiating procedure, then a priori there would be no prespecified ordering of the players. There would be proposals and counterproposals with players jumping in and out of the negotiations as they saw fit when they saw fit. This sort of free form bargaining can be captured by employing a continuous- (rather than discrete) time model.

As we shall see, giving the players the opportunity to strategically time their proposals adds just the kind of freedom needed to ensure core outcomes.

2.3.2 Continuous Time

The following is based on Perry and Reny (1994a). We will not present here the formal details of the continuous-time model. A number of (well known) subtleties must be addressed in order to ensure that such a model is well defined. Perry and Reny (1994a) contains the details. Rather, we shall be content to outline the model and discuss the results. The similarity to the coalitional bargaining model of the previous section should be noted. The continuous-time model has the following features:

1. Any player i can make a feasible proposal (x, S) at any time $t \geq 0$ (i need not be in S).
2. There is at most one active proposal (the most recent one) at any time.
3. An active proposal (x, S) can be accepted by members of S at any time (strictly) after it is made. Once all members of S accept it (including the proposer i if $i \in S$), we say that the proposal has been accepted.
4. Once accepted, (x, S) becomes a binding proposal. Members of S needn't leave and consume, although any member i of S may at any subsequent time choose to leave and receive payoff x_i (there is no discounting) forcing every member j of S to leave at the same time and receive payoff x_j.
5. A proposal is rendered inactive once it is accepted, or once it is replaced by a new proposal made by some remaining player.
6. If (x, S) is binding, then any proposal made to some member of S must be made to every member of S.

A strategy is a mapping from histories of play into actions, where an action is one of the folowing: make a proposal, accept a proposal, be quiet, or leave. The following restrictions are placed on strategies.

(i) Once a player has accepted a proposal he must remain quiet until the proposal is rendered inactive.
(ii) At any time t a player must be quiet a little before t and a little after t.

Restriction (i) eliminates the possibility of unwinnable races arising. For instance if i makes an irrational proposal to j, and i accepts it, then (if he could) i would like to render the proposal inactive before j accepts it and j would like to accept it before it is rendered inactive. In continuous time such a race to act first, yet after, a given time cannot be won. Thus without (i), after such a history no SPE would exist and consequently no SPE exists at all.

Restriction (ii) plays a dual role. It ensures that the continuous-time game is well defined and in this capacity its role is merely technical. In addition however, it serves a more significant purpose. The time during which all players are quiet although potentially arbitrarily small is time enough for one of them to make a proposal which blocks any non core outcome.

Definition 2.3.2: *A strategy is stationary if it depends only on the set of remaining players; the active proposal; those who have accepted it; the set of binding proposals; and the amount of time that the (currently) active proposal has been active, or the amount of time that has elapsed without an active proposal if currently there isn't one.*

Proposition 2.3.2: *If x is a stationary SPE outcome, then x is in the core of (v, N).*

Proof. Suppose not. Then there is a feasible proposal (y, S) such that $y_i > x_i$ for all $i \in S$. Let some particular $i \in S$ propose (y, S) at some time t such that according to their strategies all players are quiet at t and no one has accepted a proposal before time t. Such a time exists by restriction (ii) above. It suffices to show that all members of S must accept this proposal before it is rendered inactive, since this would then be a profitable deviation by i contradicting the equilibrium hypothesis. To see that all must accept, suppose that all but one member of S, player k say, have accepted. If according to the equilibrium strategies the proposal becomes inactive before k accepts it, it must be replaced by another proposal (z, T). Suppose that the continuation then leads to the payoff w_k for player k. Since k could have accepted (y, S) but did not in equilibrium, it must be the case that $w_k \geq y_k > x_k$. But player k could then propose (z, T) near enough to time zero to lead to a payoff for him (by stationarity) of $w_k > x_k$. But this contradicts the hypothesis that x is an equilibrium outcome. Consequently, if all but one member of S accepts (y, S) before it is rendered inactive, so will the last member. Using this, the same argument can be applied if all but two members of S have accepted (y, S). This can be carried all the way back to conclude that every member of S must accept (y, S) before it is rendered inactive. ■

Remark 2.3.1: Note how the line of proof mimics the usual classroom story motivating the core.

Remark 2..3.2: The Proposition implies that total balancedness of (v, N) is a necessary condition for the existence of a stationary SPE. For consider a subgame in which the set of remaining players is S. The Proposition applies equally well to this subgame considered as a game in its own right. Consequently, a stationary SPE must induce a stationary SPE on this subgame and by the Proposition the outcome on this subgame must be in the core of the TU game restricted to S. Thus, for all $S \subseteq N$, the core of the TU game restricted to S

must be nonempty. Therefore (v, N) must be totally balanced. The next result shows that total balancedness is sufficient for the existence of a stationary SPE as well.

Proposition 2.3.3: *If (v, N) is totally balanced, and x is in its core, then x can be supported as a stationary SPE.*

A proof can be found in Perry and Reny (1994a). We give here only a sketch.

Proof. (sketch) The main concern in constructing an equilibrium is to ensure that after any history there is an appropriate continuation with the following properties:

1. Those with binding proposals must obtain payoffs that are no lower than that associated with their binding proposal.[13]
2. If a proposal (x, S) is active, then (in equilibrium) acceptance of it does not reduce the payoff of those players not in S.
3. The subsequent outcome cannot be blocked (in the usual sense of blocking coalitions) by "admissible" (to be explained below) coalitions.

Property (1) is obviously necessary otherwise those with binding proposals would deviate by leaving. Property (2) is needed to ensure that unwinnable races do not arise. If members of S wish to accept (x, S) while members not in S wish to preclude this, an unwinnable race will arise and existence of an equilibrium will be thwarted. This brings us to property (3). Note that the discussion in Remark (2.3.2) establishes that the outcome in any subgame cannot be blocked by any proposal that can be made in the continuation. When there are no binding proposals among the players that remain, this means that the outcome must be in the core of the TU game restricted to the remaining players. However, when there are binding proposals present, the set of allowable proposals in the continuation is restricted by (6) of the description of the game. Thus a coalition is admissible if whenever it contains a member who is a part of a binding proposal (x, S), it contains all members of S.

With the above three properties in mind, we provide the essential ingredient for constructing strategies satisfying them. For each $T \subseteq N$, let x^T denote an element of the core of the TU game restricted to T. Consider any history of play. Let $\Pi = \{(y^1, S^1), ..., (y^m, S^m)\}$ denote the set of binding proposals, and let the set of remaining players be S. For each player $i \in S$, consider the following function:

---

[13] Note that by (6) of the description of the game, each player is a part of at most one binding proposal.

$$z_i(\Pi, S) = \begin{cases} y_i^k + \frac{\sum_{j\in S^k}(x_j^S - y_j^k)}{|S^k|} & \text{, if} \quad i \in S^k \\ x_i^S & \text{, if} \quad i \in S \setminus \cup_{k=1}^m S^k \end{cases}$$

An equilibrium continuation can now be constructed based on the vector-valued function $z(\Pi, S)$. The idea behind the construction is that $z(\Pi, S)$ is the equilibrium outcome after any history of play in which no remaining player (among S) has accepted the active proposal or in which there is no active proposal. We are content here to check that this ensures properties (1)-(3).

Property (1): It suffices to show that if $i \in S^k$ for some k then $\sum_{j\in S^k}(x_j^S - y_j^k) \geq 0$. Now since (y^k, S^k) is a feasible proposal, $\sum_{i\in S^k} y_i^k \leq v(S^k)$. In addition, because x^S is in the core of (v, S) and $S^k \subseteq S$, $v(S^k) \leq \sum_{i\in S^k} x_i^S$. These two inequalities yield the result.

Property (2): If there is no active proposal, then $i's$ payoff will be $z_i(\Pi, S)$. If there is an active proposal (y, T) and it is accepted, it becomes binding and is added to the list of binding proposals. Thus Π becomes Π' say. However, for all $i \notin T$, it is easy to check using the definition that $z_i(\Pi', S) = z_i(\Pi, S)$.

Property (3): It suffices to show that $z(\Pi, S)$ cannot be blocked by any coalition T such that for each k either T contains S^k or T is disjoint from S^k. So, consider such a T and suppose that (w, T) is a feasible proposal. Then $\sum_{i\in T} z_i(\Pi, S) = \sum_{i\in T} x_i^S \geq v(T) \geq \sum_{i\in T} w_i$, where the equality follows by definition of z_i, the first inequality follows since x^S is in the core of (v, S) and $T \subseteq S$, and the second inequality follows from the feasibility of (w, T). Consequently T cannot block $z(\Pi, S)$. ■

Remark 2.3.3: With an additional restriction on strategies Propositions 2.3.2 and 2.3.3 can be extended to the NTU case. The additional restriction is that only those to whom a proposal is directed can speak. All others must remain quiet until the proposal becomes inactive.

Remark 2.3.4: Although adding discounting to the discrete-time model gives the core no chance, discounting can be added to the continuous-time model without affecting the results in any significant manner. The only adjustment that must be made is that one must look instead at $\epsilon-$perfect equilibria rather than exact equilibria. Perry and Reny (1994a) contains the details.

So what have we gained from all of this? The hope is to have obtained some insights regarding those circumstances in which core outcomes are more likely to arise. The analysis of this section suggests that the players must employ relatively simple strategies that have a stationary structure, and that each player must have the opportunity to intervene (by making a proposal) before any player or group of players decides to leave. To the extent that these properties are present in settings in which the players can interact in an "unfettered" manner, we have then gone some way toward shedding light on the answer to Edgeworth's question: "Are core outcomes inevitable?".

References

Abreu, D. and H. Matsushima (1992a): "Virtual Implementation in Iteratively Undominated Strategies: Complete Information," *Econometrica*, 60, 993-1008.

Abreu, D. and H. Matsushima (1992b): "A Response to Glazer and Rosenthal," *Econometrica*,60, 1439-1442.

Abreu and Sen (1990): "Subgame Perfect Implementation: A Necessary and Almost Sufficient Condition," *Journal of Economic Theory,* 50, 285-299.

Chaterjee, K, B. Dutta, D. Ray, and K. Sengupta (1993): "A Non-Cooperative Theory of Coalitional Bargaining," *Review of Economic Studies,* 60, 463-477.

Edgeworth, F.Y. (1881): *Mathematical Psychics*, Reprinted 1967 by Augustus M. Kelley Publishers.

Glazer, J. and C-T. Ma (1989): "Efficient Allocation of a Prize— King Solomon's Dilemma," *Games and Economic Behavior* 1, 222-233.

Glazer, J. and R. W. Rosenthal (1992): "A Note on Abreu-Matsushima Mechanisms," *Econometrica,* 60, 1435-1438.

Jackson, M. (1992): "Implementation in Undominated Strategies: A Look at Bounded Mechanisms," *Review of Economic Studies,* 59, 757-775.

Maskin, E. (1977): "Nash Implementation and Welfare Optimality," mimeo Department of Economics, Massachusetts Institute of Technology.

Maskin, E. (1985): "The Theory of Implementation in Nash Equilibrium: A Survey," in L. Hurwicz, D. Schmeidler and H. Sonnenschein (eds.), *Social Goals and Social Organization: Essays in Memory of Elisha Pazner,* 173-204, Cambridge: Cambridge University Press.

Moldovanu, B. and E. Winter (1991): "Order Independent Equilibria, mimeo, Department of Economics, University of Bonn.

Moore, J. (1991): "Implementation in Environments with Complete Information," discussion paper #TE/91/235, London School of Economics.

Moore, J. and R. Repullo (1988): "Subgame Perfect Implemenation," *Econometrica* 46, 1191-1220.

Moore, J. and R. repullo (1990): "Nash Implementation: A Full Characterization," *Econometrica,* 58, 1083-1099.

Osborne, M. and A. Rubinstein (1990): *Bargaining and Markets.* New York: Academic Press.

Perry, M. and P. J. Reny (1994a): "A Noncooperative View of Coalition Formation and the Core," *Econometrica* 62, 795-817.

Perry, M. and P. J. Reny (1994b): "A General Solution to King Solomon's Dilemma," mimeo, Department of Economics, University of Pittsburgh, and forthcoming in *Games and Economic Behavior.*

Repullo, R. (1987): "A Simple Proof of Maskin's Theorem on Nash Implementation," *Social Choice and Welfare*,4, 39-41.

Rubinstein, A. (1982): "Perfect Equilibrium in a Bargaining Model," *Econometrica* 50, 97-109.

Serrano, R., and R. Vorha (1993): "Non-Cooperative Implementation of the Core," working paper #93-41, Department of Economics, Brown University.

Implementation Theory with Incomplete Information

Beth Allen

Department of Economics, University of Minnesota, Minneapolis, MN 55455 USA and Federal Reserve Bank of Minneapolis, 250 Marquette Avenue, Minneapolis, MN 55480 USA

Abstract. This paper surveys implementation theory when players have incomplete or asymmetric information, especially in economic environments. After the basic problem is introduced, the theory of implementation is summarized. Some coalitional considerations for implementation problems are discussed. For economies with asymmetric information, cooperative games based on incentive compatibility constraints or Bayesian incentive compatible mechanisms are derived and examined.

Keywords. Incentive compatibility, asymmetric information, incomplete information, implementation, Bayesian-Nash revelation principle, mechanisms, core, cooperative games, nontransferable utility

1 Introduction

In this paper, we examine part of the literature regarding implementation under incomplete or asymmetric information. Implementation includes not only the classical social choice problem of characterizing the functions or sets of allocations or outcomes that can be obtained as the result of a group decision—in this case, with truthful Bayes-Nash equilibrium in a direct mechanism—but also the extension of these group decisions to encompass cooperative game-theoretic solution concepts (rather than exclusively noncooperative equilibria) and the inclusion of incentive compatibility or mechanism considerations into cooperative games derived from economies with asymmetric or incomplete information.

After introducing the basic incentive compatibility problem, we proceed to examine implementation theory proper when there is incomplete information, with particular emphasis on Jackson's (1991) necessary and sufficient conditions for Bayes-Nash implementation. After some remarks concerning the possibilities for considering coalitional behavior in the implementation problem with incomplete information, we redirect our attention to economies with incentive compatibility constraints and the games they generate. Finally, we briefly consider a game-theoretic model of how agents could cooperatively select a Bayesian incentive compatible mechanism.

2 The Basic Problem

Consider a simple prototype implementation problem with asymmetric information. Suppose that the uncertainty is summarized by a set $S = \{s_1, s_2, s_3\}$ of states of the world or signals correlated with states. Let A be a set of possible actions. The problem is to pick a mapping from states to actions optimally.

Suppose further that there are two individuals, one of whom (distinguished by the subscript I for "informed") knows the state $s \in S$ while the other (denoted by the subscript U for "uninformed") does not know anything about which state $s \in S$ has occurred, although the set S, an objective probability on it, the utility functions, and the information structure are all common knowledge. The preferences of these two individuals are given by state-dependent utilities $u_I : A \times S \to I\!R$ and $u_U : A \times S \to I\!R$.

Incentive compatibility here means that $a : S \to A$ satisfies $u_I\big(a(s); s\big) \geq u_I\big(a(s'); s\big)$ for all $s, s' \in S$. This is also sometimes termed the self-selection constraint. It means that the informed agent is willing to reveal truthfully the state of the world, because in all states $s \in S$, the utility of $a(s)$ given s is never less than the utility the informed agent could obtain by stating $s' \in S$ and thus receiving $a(s')$ when the true state is $s \in S$. There is no incentive compatibility constraint for the second agent because his lack of information is common knowledge.

For the special case in which $A \subset I\!R$ and u_I is strictly monotone on A for each $s \in S$, incentive compatibility requires $a(s_1) = a(s_2) = a(s_3)$. The informed agent cannot be forced to reveal information truthfully if doing so would lead to this agent receiving less "money" than he or she could obtain in some other state of the world.

Changing the model to give the uninformed agent partial information so that he can distinguish $\{s_1\}$ from the event $\{s_2, s_3\}$ can alter the results. However, whether this partial information is verifiable—whether it can be confirmed by some third party who can act as a referee if necessary—matters greatly. If the information is verifiable, the partially uninformed agent can force an allocation in state s_1 which may be either better or worse for the fully informed agent than what the fully informed agent received in s_2 or s_3 [which still must satisfy $a(s_2) = a(s_3)$ in the strictly monotone one-dimensional example], and similarly for the informed agent. If $\{s_1\}$ versus $\{s_2, s_3\}$ is not verifiable, then we must add incentive compatibility constraints for the partially uninformed agent in order to force him to reveal correctly whether he believes that the true state lies in $\{s_1\}$ or $\{s_2, s_3\}$.

For convenience, one sometimes imposes excess incentive compatibility when there is asymmetric information. Doing so may decrease welfare, but it sometimes doesn't change the qualitative properties of the solution. Obviously, the advantage is to simplify notation by, for instance, treating informed and uninformed agents symmetrically by giving them all the same incentive compatibility constraint that properly applies only to the informed agents (when their identities are common knowledge). This procedure may be correct if the less informed agents cannot be distinguished and if all agents are permitted to announce the state of the world; in this case, the uninformed agents could always, for instance, announce the state $\bar{s}$ if $u_U(a(\bar{s}); s) \geq u_U(a(s'); s)$ for all $s, s' \in S$.

In the terminology of noncooperative game theory, incentive compatibility says that telling the truth is a Nash equilibrium in the game with strategies consisting of announcements about states of the world and payoffs defined by utilities evaluated at the proposed $a : S \to A$ mapping for the true state of the world. Implementation basically means that an allocation can be obtained as a truth-telling Nash equilibrium; this idea will be made more precise later.

An introduction to this literature can be found in d'Aspremont and Gérard-Varet (1979, 1982), Myerson (1991), and Postlewaite and Schmeidler (1987). Note, however, that I shall not attempt to give a complete reference list or even a historical summary of this topic.

3 Implementation with Asymmetric Information

A fundamental result in implementation theory is the revelation principle, which ***roughly*** states that anything which is incentive compatible (and hence implementable) can be implemented as a truth-telling (Nash) equilibrium of a direct mechanism, where a direct mechanism is a noncooperative game in which players' strategies consist of complete announcements of what they know about their "type" (i.e., preferences). The extension to incomplete information frameworks is due to Rosenthal (1978), Myerson (1979), and Harris and Townsend (1981); for a discussion, see the textbook by Myerson (1991, pp. 260-261) or the survey paper by Postlewaite and Schmeidler (1987).

Bayes-Nash Revelation Principle *With incomplete information, if an allocation function can be obtained as a Bayes-Nash equilibrium (of some mechanism or some communication game), then it can be implemented with truthful equilibrium strategies in a direct mechanism.*

An important insight is the importance of an informational condition, known as publicly predictable information (PPI) or nonexclusivity of information (NEI). The assumption states that no player has information which is not at least as coarse as the pooled information of all other players. In symbols, if we let the sub-σ-field $\mathcal{G}_i$ denote the information of player $i \in N$, where all of the $\mathcal{G}_i$ are sub-σ-fields of a given σ-field $\mathcal{T}$ of measurable events, publicly predictable information precisely requires that for all $i \in N$, $\mathcal{G}_i \subseteq \sigma\left(\bigcup_{j \neq i} \mathcal{G}_j\right)$. This means that all of the other players, acting together, can always detect lies by any individual. The significance of publicly predictable information is that it permits the use of "forcing contracts" or mechanisms in which an extremely bad outcome arises whenever a single player tells a lie. If the messages sent by players are inconsistent, the mechanism assigns the worst possible outcome so that any unilateral lie looks extremely risky; hence, truth must be a Nash equilibrium. The condition was discovered by—in alphabetical order—Blume and Easley (1990), Palfrey and Srivastava (1987), and Postlewaite and Schmeidler (1986); see also the discussion by Postlewaite and Schmeidler (1987).

An important research topic was the elucidation of necessary conditions and sufficient conditions for Bayes-Nash implementation. This work has resulted in a huge literature, including the articles by Blume and Easley (1990), Palfrey and Srivastava (1987), and Postlewaite and Schmeidler (1986) mentioned above. Contributions by Palfrey and Srivastava (1989) and Jackson (1991) are especially relevant here; Jackson's (1991) result for economic environments will be discussed in detail in the following section because he does obtain a set of conditions which are both necessary and sufficient. Further literature includes articles by Matsushima (1988, 1991) and Palfrey and Srivastava (1986, 1991). See also the recent survey by Palfrey and Srivastava (1992) and the background material on games with communication due to Forges (1986) and Myerson (1986).

Palfrey (1992) focuses on the problem of multiple equilibria for Bayes-Nash implementation. Mechanisms—even those direct mechanisms for which truth telling is a Nash equilibrium—typically exhibit many Nash equilibria. Therefore, the value of the revelation principle may be limited in the sense that some allocation could well be implementable as the unique equilibrium of some mechanism while being only one of a plethora of equilibria of direct mechanisms for which the given allocation arises as the truthful equilibrium. The notion of full implementation addresses this issue, as full implementability of an allocation means that it is implementable as the unique Bayes-Nash equilibrium of some suitable mechanism.

Ledyard (1986) expounds a critique of the concept of implementation. Using the mild hypotheses of strictly positive prior probabilities and monotonically increasing transformations of utilities, he points out that any undominated outcome can be rationalized as a Bayes-Nash equilibrium of some game. Of course, this means that Bayes-Nash implementation doesn't lead to interesting restrictions unless one either tightens the requirements of the definition of implementation (for instance, by requiring full implementation), restricts the class of allowable games, or insists on some refinement of Bayes-Nash equilibrium.

4 Jackson's Article

Jackson (1991) provides a set of necessary and sufficient conditions for Bayes-Nash implementation with or without the hypothesis of publicly predictable information. Previous work using the PPI assumption found two necessary conditions for Bayes-Nash implementability: incentive compatibility (also called self-selection) and a Bayesian analogue of Maskin's (1977) monotonicity condition. [See Postlewaite and Schmeidler (1986), Palfrey and Srivastava (1987), and Blume and Easley (1990).] In attempting to find a converse result, Jackson (1991) adds a closure condition [which was somewhat implicit in the Palfrey and Srivastava (1987) and Postlewaite and Schmeidler (1986) assumptions] that one can always "patch together" allocation functions at common points in players' information partitions. [The operation is reminiscent of Savage's (1954) treatment of personal probability and expected utility.] Subject to technicalities, incentive compatibility, Bayesian monotonicity, and closure are together necessary and sufficient for Bayes-Nash implementation.

For Jackson's (1991) theorem, we consider an exchange economy with at least three traders and strictly monotone utilities for every trader in every state of the world. The formulation could allow for public goods and externalities. To fix notation, let $N = \{1, \ldots, n\}$ be the set of economic agents and let $\preceq_i$ denote trader i's preference relation. I shall define and explain terminology after stating the result. To simplify, I restrict attention to economic environments; see Jackson (1991) for extensions to more general situations.

Theorem *A social choice set is implementable if and only if there exists a social choice set* $\widehat{F}$ *which is equivalent to* F *such that* $\widehat{F}$ *satisfies* (IC), (BM), *and* (C).

Definition 1 A *social choice set* F is a subset of the set of all social choice functions. In symbols, if $S = S_1 \times \ldots \times S_n$, where for $i \in N$, S_i is the finite information set of player i, and if A denotes the set of all feasible acts, which are assumed to be independent of elements in the set S (i.e., let A be the set of state-dependent allocations that are resource-feasible, given traders' initial endowments $e_i \in \mathbb{R}^\ell_+$ for $i \in N$, so that $A = \{(\widetilde{x}_i : S \to \mathbb{R}^\ell_+)_{i \in N} |$ for all $s \in S$, $\sum_{i \in N} \widetilde{x}_i(s) = \sum_{i \in N} e_i\})$, then F is a subset of $X = \{x | x : S \to A\}$.

We say that two social choice sets are *equivalent* if they are equal almost surely. Consequently, we need only work with those social choice sets that are defined on some convenient subset of S of full measure. If every $s = (s_1, \ldots, s_n) \in S$ occurs with strictly positive probability, no two distinct social choice sets can be equivalent; in this case, the theorem reduces to the statement that F is implementable if and only if it satisfies (IC), (BM), and (C).

Definition 2 A social choice set F satisfies *condition* (IC) if for all $i \in N$, all $x \in F$, all $s \in S$, and all $t_i \in S_i$, $x(s) \succeq_i (s_i)\ x(s)_{i(}, t_i)$, where $\succeq_i (s_i)$ denotes trader i's preference relation when his information set is $s_i \in S_i$.

Definition 3 A social choice set F satisfies *condition* (C) if for all common knowledge partitions $\{S', S''\}$ of S and all $x, y \in F$, there is $z \in F$ such that $z(s) = x(s)$ if $s \in S'$ and $z(s) = y(s)$ if $s \in S''$.

The closure condition is needed because equilibria of mechanisms can similarly be patched together based on common knowledge events. If a mechanism has two Bayes-Nash equilibria—call them x and y—then it must also have a third equilibrium, z, defined by doing x on part of S and y on the other part of S, providing that S can be divided into two or more pieces that are common knowledge.

Definition 4 F satisfies *condition* (BM) if, whenever $x \in F$ and $\alpha = (\alpha_1, \ldots, \alpha_n)$ is a *deception*, where $\alpha_i : S_i \to S_i$ for all $i \in N$ and $x \circ \alpha$ denotes the social choice function with outcomes $x\big(\alpha(s)\big) = x\big(\alpha_1(s_1), \ldots, \alpha_n(s_n)\big)$ for all $s = (s_1, \ldots, s_n) \in S$, and whenever there is no social choice function in F which is equivalent to $x \circ \alpha$, then there exists $i \in N$, $s_i \in S_i$ and $y \in X$ such that $(y \circ \alpha) \succ_i (s_i)\ (x \circ \alpha)$ and $x \succeq_i (t_i)\ y \circ \alpha_i(s_i)$ for all $t_i \in S_i$, where $\big(y \circ \big(\alpha_i(s_i)\big)\big)(s) = y\big(s)_{i(}, \alpha_i(s_i)\big)$ for all $s \in S$.

An interpretation of the Bayesian monotonicity condition is as follows (ignoring equivalence): If a mechanism implements F and if $x \in F$, then there is an equilibrium σ (of the game defined by the mechanism) which yields x. If agents use deception α, they obtain $x \circ \alpha$. If there does not exist a social welfare function in F which is equivalent to $x \circ \alpha$, then $\sigma \circ \alpha$ cannot be an equilibrium. Bayesian monotonicity ensures that, in fact, $\sigma \circ \alpha$ isn't an equilibrium. The idea is that agent i uses y to signal that α is being played; this makes trader i happier. The second condition says that player i cannot gain by falsely accusing others of deception.

5 Cooperative Implementation

By definition, implementation is a noncooperative concept; it requires allocations to arise as (truthful) Nash equilibria. Perhaps the most straightforward way to include the consideration of coalitional behavior is to replace Nash equilibrium with strong equilibrium. A disadvantage of this approach is that strong equilibria may not exist in general noncooperative games, whereas there are always Nash equilibria, at least in mixed strategies under fairly general technical conditions. A more radical strategy is to examine the possibilities for attaining outcomes as some cooperative solution in a game. In this case, the precise application of incentive compatibility constraints is unclear. Should one worry about incentives to lie within a coalition that is cheating?

Are blocking allocations required to be incentive compatible? Such considerations seem to have the flavor of bargaining sets (i.e., objections versus counterobjections) or coalition proofness.

Incentive compatibility can be incorporated into cooperative games on three levels. First, one can find some solution set and ask whether it satisfies incentive compatibility or, as a weaker alternative, at least contains some outcome satisfying incentive compatibility. This is the approach taken by Krasa and Yannelis (1994) and Koutsougeris and Yannelis (1992). Secondly, one can require incentive compatibility only in the definition of feasible actions for the grand coalition. This approach implicitly appears in the second-best efficiency considerations for the planner in the literature on incentives and mechanism design. Finally, one can consistently require incentive compatibility for the definition of feasible agreements for all coalitions. This strongest use of incentive compatibility treats all coalitions symmetrically but possesses the disadvantage of possibly leading to games that violate some of the standard properties one expects. This tack is followed in Allen (1991, 1992, 1993, 1994).

A further factor which complicates the analysis is that games without transferable utility are more appropriate when incentive considerations are present. To summarize the worth of a coalition by a single number—as is done in the definition of cooperative games with transferable utility (or *TU* games)—suggests that members of the coalition share a single objective function. Yet, if these players were indeed a team, they would necessarily be willing to share their information fully and honestly in order to better maximize the total payoff accruing to the coalition. This contradicts the spirit of incentive compatibility, which hypothesizes that players will hide information or will lie to further their own goals.

Finally, one can ask whether participation or individual rationality constraints should be imposed. Requiring that all players be willing to play the game is natural for some mechanism problems, as it is a weaker rationality requirement than Bayesian incentive compatibility. On the other hand, in a cooperative context, most solution concepts are automatically—by definition—individually rational, although out-of-equilibrium behaviors such as blocking and objecting may not always be individually rational compared to nonparticipation. Moreover, *ex ante* and *ex post* individual rationality are distinct concepts. The latter restricts risk sharing so that its imposition can prevent efficient outcomes such as those obtainable with fair insurance contracts.

6 Economies with Asymmetric Information

Consider a pure exchange economy with agent set $N = \{1, \dots, n\}$ in which Ω is a finite set. To simplify, assume that every state of the world occurs with strictly positive probability and that these probabilities are common knowledge. Agents' information either consists of partitions on Ω or can be specified by signals $\mathbf{s}_i : \Omega \to S_i$, where each S_i is also assumed to be a finite set. Write $S = \Pi_{i \in N} S_i$ and $\mathbf{s} = (\mathbf{s}_1, \dots, \mathbf{s}_n)$. Consumption sets are $\mathbb{R}^\ell_+$ and initial endowments are $e_i \in \mathbb{R}^\ell_+$ for $i \in N$. Endowments are assumed not to depend on Ω or S in order to guarantee that initial endowment vectors are incentive compatible and hence that there exist incentive compatible feasible allocations. Preferences are specified by state-dependent cardinal utilities $u_i : \mathbb{R}^\ell_+ \times \Omega \to \mathbb{R}$ where, for every $i \in N$ and every $\omega \in \Omega$, $u_i(\cdot; \omega) : \mathbb{R}^\ell_+ \to \mathbb{R}$ is continuous, strictly monotone, and strictly concave.

The classical incentive compatibility constraints are given by the restrictions that allocations $x_i : \Omega \to \mathbb{R}^\ell_+$ must satisfy, for all $i \in N$ and all $\omega \in \Omega$, $u_i\big(x(\omega); \omega\big) \geq u_i\big(x_i(\omega'); \omega\big)$ for all $\omega' \in \Omega$. Note that these constraints apply to every player regardless of the coalition to which he belongs. They are written in "overkill" fashion, as if each player were able to distinguish all states rather than in a form that reflects the player's individual information (which could depend on his coalition). Think of these incentive compatibility constraints as restrictions on the state-dependent consumption set of each agent.

Alternatively, for a framework in which traders receive signals about the state of the world, Bayesian incentive compatibility requires

$$\sum_{\omega \in \Omega} u_i\big(x_i(\mathbf{s}(\omega)); \omega\big) \mu_i(\omega | s_i) \geq$$

$$\sum_{\omega \in \Omega} u_i\big(x_i(s_i', \mathbf{s})_{i(}(\omega)); \omega\big) \mu_i(\omega | s_i)$$

for all $s_i \in S_i$, all $s_i' \in S_i$, and all $i \in N$, where the allocation $x_i : \Omega \to \mathbb{R}^\ell_+$ must be measurable with respect to the signals $\mathbf{s}(\cdot) = \big(\mathbf{s}_1(\cdot), \dots, \mathbf{s}_n(\cdot)\big)$, and $\mu_i(\omega | s_i)$ denotes player i's posterior probability of $\omega \in \Omega$, given that he or she has observed signal $s_i \in S_i$.

7 Incentives with Asymmetric Information

The study of cooperative solution concepts for economies with incentive considerations has focused primarily on the core, although the value has also been examined. One approach that has proved useful is to analyze the cooperative games with nontransferable utility that are generated by (exchange) economies with incentive compatibility constraints. Thus, one defines the cooperative games $V : 2^N \to \mathbb{R}^n$ with nontransferable utility (or *NTU* games) by $V(\emptyset) = \mathbb{R}^n$ and for $T \subseteq N$ with $T \neq \emptyset$, $V(T) = \{(w_1, \ldots, w_n) \in \mathbb{R}^n |$ for $i \in T$, there exists $x_i : \Omega \to \mathbb{R}^\ell_+$ such that, for all fully informed $i \in T$ and all $\omega, \omega' \in \Omega$, $u_i(x_i(\omega); \omega) \geq u_i(x_i(\omega'); \omega)$, where $\sum_{i \in T} x_i(\omega) = \sum_{i \in T} e_i$ for all $\omega \in \Omega$ and $w_i \leq \sum_{\omega \in \Omega} u_i(x_i(\omega); \omega)\mu(\omega)$ for all $i \in T\}$, where $\mu(\omega)$ is the probability of state ω and agents in N are assumed to be either fully informed (i.e., their information partitions on Ω precisely equal 2^Ω) or completely uninformed (i.e., their information partitions on Ω are the trivial partitions $\{\Omega\}$). One can modify the game to take careful account of players' partial information or to use the Bayesian version of incentive compatibility constraints.

The incentive compatible core was first introduced by Boyd and Prescott (1986) in a model of financial intermediation with risk neutrality. They demonstrate nonemptiness of the core by showing that certain systems of linear inequalities can be solved. Berliant (1992) and Marimon (1989) also examine the incentive compatible core for particular economic problems—those involving taxation and adverse selection. Allen (1991, 1994) follows the approach outlined above of deriving *NTU* games from economies with (classical or Bayesian) incentive compatibility constraints and finds that the game need not be balanced and can, in fact, have an empty core. For economies with asymmetric information, Koutsougeris and Yannelis (1992) define core allocations and check whether they are incentive compatible.

For the value, Allen (1992) derives the games from economies with (classical or Bayesian) incentive compatibility and shows that the value is well defined. Krasa and Yannelis (1994) focus on the private information value and ask whether the fine, coarse, and private information values satisfy incentive compatibility.

8 Mechanisms with Asymmetric Information

Instead of adding incentive compatibility constraints to the definition of the games derived from economies with asymmetric information, one can incorporate Bayesian incentive compatible mechanisms into the definition of these games. This approach builds on the work of Harsanyi (1967-68) on noncooperative games with incomplete information and its use by Myerson (1984) to model cooperative games with incomplete information.

Allen (1993) proposes a game containing both cooperative and noncooperative phases in which the feasible outcomes are taken to be Bayesian-Nash equilibrium outcomes of Bayesian incentive compatible direct mechanisms. Formally, the entire model is assumed to be common knowledge and, in the first strategic phase, players cooperatively pick a Bayesian incentive compatible mechanism. The choice of a mechanism is a binding agreement; the commitment is made *ex ante*. Then, after agents learn their types, the noncooperative game defined by the chosen mechanism is played. Traders send messages (about their types, since we can restrict ourselves to direct mechanisms by the Bayes-Nash revelation principle), which lead to an outcome according to the mechanism. The equilibrium concept used in the noncooperative (mechanism) game phase is Bayesian-Nash equilibrium, which (by the revelation principle) can be taken to be truthful. Somewhat more formally, the game given by $V(S) = \{(w_1, \ldots, w_n) \in \mathbb{R}^n \mid$ there exists a randomized direct Bayesian incentive compatible mechanism λ, and there is a (truthful) Bayesian-Nash noncooperative equilibrium σ for λ such that, if $i \in S$, i's payoff in λ under σ is at least as great as $w_i\}$. The use of incentive compatible mechanisms in cooperative economic contexts is also studied by Ichiishi and Idzik (1992), Page (1992), and Rosenmüller (1990).

Acknowledgements

This work was supported by the National Science Foundation (through grant SBR93-09854), the Federal Reserve Bank of Minneapolis, and the Curtis Carlson Chair in Economics at the University of Minnesota. The views expressed herein are those of the author and not necessarily those of the Federal Reserve Bank of Minneapolis or the Federal Reserve System. Mr. Xiaodong Huang of the University of Minnesota was the notetaker for the lecture on which this paper is based.

References

Allen, Beth (1991) Incentives in market games with asymmetric information: The core. Center for Analytic Research in Economics and the Social Sciences Working Paper #91-38, Department of Economics, University of Pennsylvania. Also distributed as CORE Discussion Paper No. 9221, Center for Operations Research and Econometrics, Université Catholique de Louvain, Louvain-la-Neuve, Belgium

Allen, Beth (1992) Incentives in market games with asymmetric information: The value. Center for Analytic Research in Economics and the Social Sciences Working Paper #92-11, Department of Economics, University of Pennsylvania

Allen, Beth (1993) Supermechanisms. Mimeo, Department of Economics, University of Minnesota

Allen, Beth (1994) Incentives in market games with asymmetric information: Approximate *NTU* cores in large economies. In: Barnett, W.A., Moulin, H., Salles, M., Schofield, N.J. (eds.) Social Choice, Welfare, and Ethics. Proceedings of the Eighth International Symposium in Economic Theory and Econometrics, Caen, 1992. International Symposia in Economic Theory and Econometrics, Cambridge University Press, 263-276

d'Aspremont, Claude and Gérard-Varet, Louis-André (1979) Incentives and incomplete information. Journal of Public Economics **11**, 25-45

d'Aspremont, C. and Gérard-Varet, L.-A. (1982) Bayesian incentive compatible beliefs. Journal of Mathematical Economics **10**, 83-103

Berliant, Marcus (1992) On income taxation and the core. Journal of Economic Theory **56**, 121-141

Blume, Lawrence and Easley, David (1990) Implementation of Walrasian expectations equilibria. Journal of Economic Theory **51**, 207-227

Boyd, John H. and Prescott, Edward C. (1986) Financial intermediary-coalitions. Journal of Economic Theory **38**, 211-232

Forges, Françoise M. (1986) An approach to communication equilibria. Econometrica **54**, 1375-1385

Harris, Milton and Townsend, Robert M. (1981) Resource allocation under asymmetric information. Econometrica **49**, 33-64

Harsanyi, John C. (1967-68) Games with incomplete information played by "Bayesian" players. Management Science **14**, 159-182, 320-334, 486-502

Ichiishi, Tatsuro and Idzik, Adam (1992) Bayesian cooperative choice of strategies. Mimeo, Department of Economics, Ohio State University

Jackson, Matthew O. (1991) Bayesian implementation. Econometrica **59**, 461-477

Koutsougeris, Leonidas and Yannelis, Nicholas C. (1992) Incentive compatibility and information superiority of the core of an economy with differential information. Mimeo, Department of Economics, University of Illinois at Champaign-Urbana

Krasa, Stefan and Yannelis, Nicholas C. (1994) The value allocation of an economy with differential information. Econometrica **62**, 881-900

Ledyard, John O. (1986) The scope of the hypothesis of Bayesian equilibrium. Journal of Economic Theory **39**, 59-82

Marimon, Ramon (1989) The core of private information economies. UAB/IAE Discussion Papers W.P. 131.90, Universitat Autònoma de Barcelona

Maskin, Eric (1977) Nash equilibria and welfare optimality. Mimeo, Department of Economics, Massachusetts Institute of Technology

Matsushima, Hitoshi (1988) A new approach to the implementation problem. Journal of Economic Theory **45**, 128-144

Matsushima, Hitoshi (1991) Incentive compatible mechanisms with full transferability. Journal of Economic Theory **54**, 198-205

Myerson, Roger B. (1979) Incentive compatibility and the bargaining problem. Econometrica **47**, 61-74

Myerson, Roger B. (1984) Cooperative games with imcomplete [*sic*] information. International Journal of Game Theory **13**, 69-96

Myerson, Roger B. (1986) Multistage games with communication. Econometrica **54**, 323-358

Myerson, Roger B. (1991) Game Theory: Analysis of Conflict. Harvard University Press

Page, Frank H. Jr. (1992) Bayesian incentive compatible mechanisms. Mimeo, Department of Finance, University of Alabama at Tuscaloosa

Palfrey, Thomas R. (1992) Implementation and Bayesian equilibrium: The multiple equilibrium problem in mechanism design. In: Laffont, Jean-Jacques (ed.) Advances in Economic Theory, Volume 1. Proceedings of the Sixth World Congress of the Econometric Society, Barcelona, 1990. Econometric Society Monograph Series, Cambridge University Press, 283-323

Palfrey, Thomas R. and Srivastava, Sanjay (1986) Private information in large economies. Journal of Economic Theory **39**, 34-58

Palfrey, Thomas R. and Srivastava, Sanjay (1987) On Bayesian implementable allocations. Review of Economic Studies **54**, 193-208

Palfrey, Thomas R. and Srivastava, Sanjay (1989) Implementation with incomplete information in exchange economies. Econometrica **57**, 115-134

Palfrey, Thomas R. and Srivastava, Sanjay (1991), Efficient trading mechanisms with pre-play communication. Journal of Economic Theory **55**, 17-40

Palfrey Thomas R. and Srivastava, Sanjay (1992) Bayesian Implementation. Fundamentals of Pure and Applied Economics, Harwood Academic Publishers

Postlewaite, Andrew and Schmeidler, David (1986) Implementation in differential information economies. Journal of Economic Theory **39**, 14-33

Postlewaite, Andrew and Schmeidler, David (1987) Differential information and strategic behavior in economic environments: A general equilibrium approach. In: Groves, T., Radner, R., and Reiter, S. (eds.) Information, Incentives, and Economic Mechanisms: Essays in Honor of Leonid Hurwicz. University of Minnesota Press, 330-348

Rosenmüller, Joachim (1990) Fee games: $(N)TU$-games with incomplete information. Institut für Mathematische Wirtschaftsforschung Working Paper Nr. 190, Universität Bielefeld

Rosenthal, Robert W. (1978) Arbitration of two-party disputes under uncertainty. Review of Economic Studies **45**, 595-604

Savage, L.J. (1954) The Foundations of Statistics. Wiley, New York

Coalitional Non-Cooperative Approaches to Cooperation

Rajiv Vohra

Department of Economics, Brown University, Providence, RI 02912, USA

1 Introduction

Cooperative and non-cooperative approaches to game theory represent two polar, and simplifying, extremes. In the former, it is assumed that players can make commitments that are binding, i.e., an agreement once made is enforceable. In contrast, non-cooperative game theory assumes that agreements cannot be enforced and equilibrium agreements or strategies, therefore, must be self-enforcing.

The present paper will explore two aspects of the inter-connections between these two polar extremes. First, we discuss the issue of consistency in coalitional deviations. This is, at least in spirit, a non-cooperative idea but one that can be applied even to cooperative equilibrium concepts in the characteristic function form. Second, we focus on the growing literature that studies the process through which equilibrium cooperative agreements are reached. Even in a setting in which binding agreements can, in principle, be written (and enforced), the negotiations that lead to a final 'cooperative' outcome are likely to depend on non-cooperative considerations. In particular, the ability to write binding agreements does not preclude a player(s) from choosing not to cooperate. The decision to cooperate and join a larger coalition will depend on the payoff corresponding to the non-cooperative equilibrium. This approach, therefore, directly incorporates many non-cooperative ideas into what is essentially a cooperative framework.

Cooperative game theory, with its emphasis on the possibilities for cooperation, has traditionally been associated with games in coalitional form, or characteristic function form. As a result, the primitive data of the problem being studied abstracts from the details of the negotiation process – details that may be completely specified in some underlying extensive form game[1]. One should not hasten, though, to the extensive form specification since often it then turns out that the outcomes begin to depend too critically on rather small details of these procedures. The characteristic function may be viewed as a reduced form of a more detailed specification of a model. Its value lies in bringing to the fore

[1] We will turn presently to the normal form specification of a game, which can also be seen as a primitive basis from which a characteristic function is derived.

the cooperative abilities of various coalitions and from that to an analysis of cooperation in the grand coalition.

We shall begin, in the next Section, by considering games in characteristic function form to discuss the influence of non-cooperative ideas on cooperative game theory or, more precisely, to discuss the notion of consistency in coalitional deviations. As a starting point, consider the notion of the core. An outcome is said to be in the core if no coalition has an objection to it, i.e., if no coalition can, on its own, improve the utility of its members.[2] One source of dissatisfaction with the core relates to the notion of objections. It seems reasonable to argue that the same considerations that qualify an objection as a means to discredit a status-quo should (for the same reasons) be used to check if the objection itself can be similarly objected to. In other words, objections should be tested against counterobjections. The notion of a bargaining set incorporates this idea. An imputation belongs to the bargaining set if to every possible objection there exists a counterobjection. Variants of this concept differ in their precise formulation of objections and counterobjections; the one that we define in Section 2 is due to Mas-Colell (1989). While these concepts are very much part of the cooperative theory, the idea of checking the credibility of an objection has some hint of non-cooperative considerations too. This becomes even more explicit if one incorporates fully, or consistently, the ideas of objections and counterobjections. One could take the objection-counterobjection idea to its logical conclusion by subjecting counterobjections to a similar test, and so on. This was formalized as the consistent bargaining set in Dutta *et al* (1989). While this is not a non-cooperative notion, it is reminiscent of subgame perfection and backwards induction; the idea that if you do such and such, the very same considerations that underlie the solution concept will permit a further reaction – a new subgame.

The coalitional form does not readily lend itself to a study of the formal connections between cooperation and non-cooperation. It is the normal form specification of a game that seems more appropriate for this purpose. And we will argue that this leads to a cooperative theory in which equilibria are very critically dependent on the non-cooperative equilibria. We begin in Section 3 by a brief review of several core-like solution concepts for normal form games that were introduced in Aumann (1961). The essential idea is to construct for each coalition, the set of feasible utility profiles and then appeal to the notion of the core as the equilibrium concept for the derived game in coalitional form. The immediate issue that arises is that of defining what the feasible utility sets for the various coalitions are. Since the primitive model is specified as a normal form, and the utilities of players depend on the complete strategy profile, the feasible utility set of a coalition is not independent of the strategies of the complementary coalition. One option is to assume, *a la* Nash, that the complementary coalition keeps its strategy fixed according to the status-quo. For any given status-quo

[2]The formal definitions will appear in the next Section.

then, we can derive the corresponding coalitional form. The corresponding core-like, cooperative solution is the strong equilibrium. A strong equilibrium can be viewed as a strategy profile with the property that the corresponding utility profile belongs to the core of the associated coalitional form with this strategy profile as the status-quo. Another possibility is to take the view that the feasible utility set of a coalition should reflect the utilities that it can guarantee itself, i.e., a coalition is pessimistic in its outlook and fears the worst from its complement. This approach makes it possible to define the feasible sets independently of a status-quo. The cooperative solution concept it leads to is the α-core, and the related notion of the β-core. Analogous to the critique of the core and the bargaining set in the previous paragraph, it can be argued that these solution concepts too, suffer from not consistently taking account of the repercussions that follow from an initial objection. This argument is only strengthened by the fact the strong equilibrium, the α-core and the β-core, despite having a common connection to the core, lead to very diverse outcomes. Indeed, in the context of the normal form, even a purely non-cooperative but coalitional approach that consistently takes account of the reactions stemming from an initial objection can bring new insights to the issue at hand.[3]

To see how consistency would apply to cooperative theory in the normal form consider what happens in a two-player game when one of the players deviates from the grand coalition. This corresponds, of course, to each player acting independently, or non-cooperatively, and the resulting outcome is then a non-cooperative equilibrium, say a Nash equilibrium. This is far more convincing a prediction of the result of a deviation than what is prescribed by the strong equilibrium, the α-core or the β-core. It should also be clear that the equilibrium cooperative outcome will then depend very critically on the non-cooperative equilibrium. A general formulation of this idea was recently developed in Ray and Vohra (1994), and will be the subject of Section 3.2.

The present paper does not deal with non-cooperative coalitional approaches in the extensive form, an omission that is amply justified by the papers by Reny and Mas-Colell in this volume.

2 Games in Characteristic Function Form

Let $N = \{1, \ldots, n\}$ denote the set of players and let $\mathcal{N} = 2^N \setminus \{\emptyset\}$ denote the set of all non-empty subsets of N. An element of $\mathcal{N}$ is referred to as a coalition. For any coalition $S \in \mathcal{N}$, let R^S denote the $|S|$ dimensional Euclidean space with coordinates indexed by the elements of S. For $u \in R^N$, u_S will denote its projection on R^S. We shall use the convention $\gg, >, \geq$ to order vectors in R^N. A game in characteristic function form is defined as $(N, V(S)_{S \in \mathcal{N}})$, where $V(S) \subseteq R^S$ refers to the feasible set of payoffs or utilities of coalition S.

[3]The reference here is to the notion of a coalition proof Nash equilibrium developed by Bernheim, Peleg and Whinston (1987); see Section 3.1 below.

The set of *imputations* is defined as $V^*(N) = \{x \in V(N) \mid \nexists y \in V(N)$ such that $y > x\}$.

A pair (S, y), where $S \in \mathcal{N}$ and $y \in V(S)$ is said to be an *objection* to an imputation x if $y > x_S$.

The *core* of is defined as

$$C(V) = \{x \in V(N) \mid \text{ there does not exist an objection to } x\}.$$

To check the credibility of an objection one can test it against a counterobjection, which refers to the ability of some other coalition to make an improvement in the utility of its members; for players common to both coalitions the comparison is to their utility in the objection, and for others the comparison is to the status-quo.

Let (S, y) be an objection to x. (T, z), where $T \in \mathcal{N}$ and $z \in V(T)$ is said to be a *counterobjection* to (S, y) if $z > (y_{S \cap T}, x_{T \setminus S})$.

An objection (S, y) to x is said to be a *justified objection* if there does not exist any counterobjection to (S, y).

The *Bargaining Set* is defined as

$$B(V) = \{x \in V^*(N) \mid \text{ there does not exist a justified objection to } x\}.$$

This definition of the bargaining set is based on Mas-Colell (1989), and is the one that we shall discuss in the present paper.[4] For a comprehensive survey of various other notions of the bargaining set the reader is referred to Maschler (1992). What is common to this family of solution concepts is the idea that an imputation is not excluded simply because it has an objection. Objections are taken seriously, and termed justified, only if they do not admit counterobjections. If one accepts this objection-counterobjection logic, one is immediately faced with the following question: Does a counterobjection itself pass the test that objections are now being subjected to? It is compelling then, to put counterobjections to the same test that objections are subjected to. This idea was formalized by Dutta *et al* (1989), who defined a notion of a consistent bargaining set in which the credibility of counterobjections, and of further objections to them, and so on, is consistently evaluated. To define the consistent bargaining set we need some additional notation.

---

[4]Zhou (1993) argues that a counterobjecting coalition should also be required to have a proper intersection with the objecting coalition. See also Shimomura (1994).

Consider an imputation x. Define a *collection* $\mathcal{A}$ as $\{x; (S^i, x^i)_{i=0}^m\}$, where x is an imputation and, for each $i = 0, \ldots, m$, x^i is an imputation[5] for S^i. Define $b(\mathcal{A}) \in R^N$ by

$$b_j(\mathcal{A}) = \max\{x_j, (x_j^i)_{\{i|j \in S^i\}}\}.$$

A pair $(\hat{S}, \hat{x})$, where $\hat{S} \in \mathcal{N}$ and $\hat{x} \in V^*(\hat{S})$, is an *objection to the collection* $\mathcal{A}$ if

$$\hat{x} > b_{\hat{S}}(\mathcal{A})$$

A collection $\{x; (S^i, x^i)_{i=0}^m\}$ is a *chain* if (S^0, x^0) is an objection to x, and for each $i = 1, \ldots, m$, (S^i, x^i) is an objection to the collection $\mathcal{A}^{i-1} = \{x; (S^j, x^j)_{j=0}^{i-1}\}$.

A pair $(\hat{S}, \hat{x})$ is a *terminating objection to the chain* $\mathcal{A} = \{x; (S^i, x^i)_{i=0}^m\}$ if it is an objection to $\mathcal{A}$ such that there is no objection to the chain $\{x; (S^i, x^i)_{i=0}^m, (\hat{S}, \hat{x})\}$.

An objection $(\hat{S}, \hat{x})$ to $\mathcal{A}$ is *valid* if there is no valid objection to $\{\mathcal{A}, (\hat{S}, \hat{x})\}$. It is *invalid* if there exists a valid objection to $\{\mathcal{A}, (\hat{S}, \hat{x})\}$.

Given that the number of coalitions is finite, and that an objection is drawn from the set of imputations of the dissenting coalition, it follows that all chains must be of finite length. Since a terminating objection to a chain is necessarily valid, we can work backwards from the valid terminating objections to uniquely determine the "label" of each objection.

The *Consistent Bargaining Set* is defined as

$$CB(V) = \{x \in V^*(N) \mid \nexists \text{ a valid objection to } x\}.$$

To see, heuristically, how the consistent bargaining set differs from the bargaining set consider Figure 1. It shows some of the chains that emanate from an imputation x. The objections in boxes denote terminating objections to a chain. Since the objections (S^1, y^1) and (S^2, y^2) both have counterobjections, neither one of them can be used to rule out x from being in the bargaining set. In fact, (S^1, y^1) is not only not justified, it is also invalid (since it is terminated by (T^2, z^2)). However, (S^2, y^2) is a valid objection if *all* objections to it are, like (T^3, z^3) and (T^4, z^4), terminated by some other objection.[6] Moreover, if *all*[7] objections to x have counterobjections, then, in this example, $x \in B(V) \setminus CB(V)$.

[5]The set of imputations for S is defined as $V^*(S) = \{x \in V(S) \mid \nexists\, y \in V(S) \text{ such that } y > x\}$.

[6]Note that figure 1 is extremely incomplete. To establish the validity of (S^2, y^2) it is necessary to consider all possible objections to the chain $\{x, (S^2, y^2)\}$, including the objections (W^2, w^2) and (W^3, w^3).

[7]Including $(T^1, z^1), \ldots, (T^4, z^4), (W^1, w^1), \ldots (W^3, w^3)$ etc.

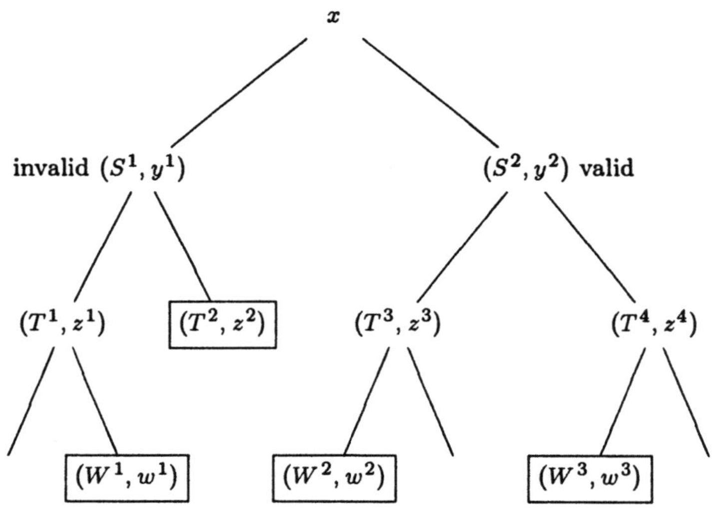

Figure 1: Valid and Invalid objections.

It is, of course, easy to see that

$$C(V) \subseteq CB(V) \subseteq B(V).$$

There are interesting, special cases in which $CB(V) = B(V)$. In such cases, the objection/counterobjection logic is itself enough to check the validity of an objection.

Proposition 2.1 *(Dutta et al (1989)). In three-player games, and in all strictly comprehensive, superadditive, ordinally convex games $CB(V) = B(V)$.*

Proposition 2.2 *(Mas-Colell (1989)). In exchange economies with an atomless measure space of consumers $B(V) = C(V)$.*

It is also worth remarking that if all objecting coalitions to a chain are required to be subsets of the last coalition in the chain, then again consistency has no additional bite. Defining internal consistency by restricting attention to 'internal chains', we can define the internally consistent bargaining set as the set of imputations to which there does not exist an internally valid objection. Certainly, this set contains the core. Moreover, if there exists an objection which is not internally valid, it is possible to find one which is. This implies that the internally consistent bargaining set is identical to the core, or that the core is internally consistent; see Ray (1989).

Notice that the von-Neumann Morgenstern stable set solution is also designed to look consistently, beyond a single step deviation. In fact, as Greenberg (1992) shows, the consistent bargaining set can alternatively be obtained by formulating the notion of objections and counterobjections in the context of an abstract stable set. He also demonstrates the fact in such solution concepts, the results are very sensitive to whether or not objections are defined with strict inequalities or semi-strict inequalities.

Since the bargaining set contains the core, its existence can be established under weaker conditions. In transferable utility games, the bargaining set typically exists even in games that are not balanced; see Maschler (1992), Vohra (1991) and Zhou (1993). Dutta *et al* (1989) provide an example of a four player superadditive, TU game in which the consistent bargaining set is empty. Simple, sufficient conditions (weaker than balancedness) under which the consistent bargaining set is non-empty are not yet known.

3 Games in Normal Form

In this section we shall see how the idea of consistency discussed in the previous Section can be applied to non-cooperative and cooperative solution concepts in the context of games in normal form. We begin by reviewing some well-known coalitional solutions for normal form games introduced by Aumann (1961).

Let $N = \{1, \ldots, n\}$ denote the set of players. Each player has a strategy set $X_i \subseteq R^l$ and a utility function $u_i : X \mapsto R$, where $X = \Pi_{i=1}^n X_i$. A game in normal form is defined as (N, X_i, u_i). We shall use $u_S(x)$ to denote $(u_i(x))_{i \in S}$ and $-S$ to denote the coalition $N \setminus S$.

A strategy profile $x \in X$ is said to be a *Strong Equilibrium* (SE) of a game (N, X_i, u_i) if there does not exist a coalition S and $\bar{x}_S \in X_S$ such that

$$u_S(\bar{x}_s, x_{-S}) > u_S(x).$$

Given $x \in X$, we can define for each coalition a set of feasible utilities, conditional on the complementary coalition's strategies as fixed. Let

$$V(S; x) = \{u \in R^S \mid \text{ there exists } \bar{x}_S \in X_S \text{ such that } u \leq u_S(\bar{x}_S, x_{-S})\}.$$

Thus for every $x \in X$ we can define a game in characteristic function form $(N, V(.; x))$ and x is a strong equilibrium for the original game if and only if $u(x)$ belongs to the core of $(N, V(.; x))$.

A strategy profile $x \in X$ is said to be in the *α-core* of a game (N, X_i, u_i) if there does not exist a coalition S and $\bar{x}_S \in X_S$ such that for all $y_{-S} \in X_{-S}$,

$$u_S(\bar{x}_s, y_{-S}) > u_S(x).$$

A strategy profile $x \in X$ is said to be in the *β-core* of a game (N, X_i, u_i) if there does not exist a coalition S such that for every $y_{-S} \in X_{-S}$ there exists

$\bar{x}_S \in X_S$ such that

$$u_S(\bar{x}_s, y_{-S}) > u_S(x).$$

Observe that

$$SE \subseteq \beta - core \subseteq \alpha - core.$$

While a strong equilibrium exists only in very restrictive cases, Scarf (1971) showed that in an exchange economy with externalities, the α-core is non-empty if all utility functions are quasi-concave.

All these solution concepts are designed to focus on strategy profiles that are immune to objections. However, in contrast to games in characteristic function form, one must now also specify what the complementary coalition is assumed to do in the face of an objection. The three alternatives above consider the case in which the complementary coalition is assumed to keep its actions unchanged or to act in a way that minimizes the benefits to the dissenting coalition. In light of the discussion of the previous Sections we need not dwell on the fact that each of these assumptions is somewhat ad-hoc. Take the notion of the strong equilibrium, and consider the idea of checking the credibility of an objection. In particular, we can ask if a subcoalition from the objecting coalition can make a similar objection (counterobjection). In fact, one can develop this idea in a purely non-cooperative vein as has been proposed by Bernheim, Peleg and Whinston (1987).

3.1 Consistency in a Non-Cooperative Coalitional Theory

Consider a model in which the players can communicate freely but cannot make binding commitments. In a two-player game this would allow refining Nash equilibria to those that are Pareto undominated by other Nash equilibria. In games with more than two-players, this idea can be applied recursively provided one restricts attention to internal deviations, i.e., assume that dissenting coalitions must be subcoalitions of previously objecting coalitions.

Given $y_{-S} \in X_{-S}$, we define the utility function of $i \in S$ as $\bar{u}_i : X_S \to R$ as $\bar{u}_i(x_S) = u_i(x_S, y_{-S})$. We can now define the game $\Gamma(y_{-S})$ induced on S by the actions y_{-S} of the coalition $-S$ as

$$\Gamma(y_{-S}) = (S, (X_i, \bar{u}_i)_{i \in S}).$$

In a game with a single player i, $x_i \in X_i$ is a *coalition-proof Nash equilibrium* if and only if $x_i = \operatorname{Argmax}_{X_i} u_i(.)$.

Let $n > 1$ and suppose coalition-proof Nash equilibrium has been defined for games with fewer than n players. Then

(a) For a game Γ with n players, x is *self-enforcing* if for all coalitions S such that $|S| < n$, x_S is a coalition-proof Nash equilibrium in the game $\Gamma(x_{-S})$.

(b) For a game Γ with n players, x is a *coalition-proof Nash equilibrium* if it is self-enforcing and there does not exist another self-enforcing y such that $u_i(y) > u_i(x)$ for all $i = 1, \ldots, n$.

The assumption that a deviating coalition only takes into account further dissension from within, and not from coalitions that contain members from its complement, is a strong one. Its power lies in being able to develop the concept recursively. Recall that in defining the consistent bargaining set we could rely on the finiteness of chains to work backwards and label objections as valid or invalid. In a normal form game, the analogous notion of chains would lead, in general, to chains that are not of finite length – unless we confine ourselves to internal deviations. By concentrating on internal objections, it is possible to develop coalition proof Nash equilibrium by defining chains of objections in the normal form and following the steps in the definition of the consistent bargaining set. It can also be developed by appealing to the notion of abstract stable sets, as shown by Greenberg (1990).[8] While it is important to extend these ideas beyond the confines of internal objections, this issue has not yet been completely resolved; see, however, Chakravorti and Kahn (1991).

3.2 Consistency in a Cooperative Theory

We now turn to an application of consistency to a purely cooperative theory in the normal form. At the outset, it is important to keep in mind a simple rule of thumb; for a solution concept to qualify as a cooperative solution it ought to have the property that in the prisoner's dilemma it picks out the 'cooperative outcome' and rules out the 'non-cooperative outcome'. This the α-core and the β-core satisfy; the strong equilibrium is empty in the prisoner's dilemma. However, these solutions are not, in general, consistent. For a concrete motivation, consider a cooperative solution to the Cournot duopoly with a linear demand curve and constant average costs. The strategy set of each firm is the real line – representing its output level. In this example too, there exists no strong equilibrium. The α-core, which in this example is identical to the β-core, consists of *all* strategies that are individually rational and Pareto optimal. In particular, one firm producing zero and the other producing the monopoly output is a strategy profile in the α-core. The reason is simply that neither firm can object to getting zero profit *if* it assumes (as the α-core prescribes) the very worst, i.e., if it assumes that its rival will flood the market. It would be reasonable to argue that when a coalition deviates it should not take as given the strategies of its complement, nor should it fear the worst. It should look ahead to a resulting 'equilibrium' that its actions induce. The real issue then, is to describe what the equilibrium is that emerges when one firm breaks away

[8]In both these alternative approaches, however, there is a complication that creeps in if the set of self-enforcing strategies is not compact; see Claim 7.2.5 in Greenberg (1990) and Kahn and Mookherjee (1992).

from the grand coalition, i.e., the outcome that corresponds to the break-up of cooperation.

We take the view that a break down of cooperation induces a non-cooperative situation. Taking Nash equilibrium as the 'equilibrium' corresponding to the state of non-cooperation then, no firm should fear that by breaking off negotiations it will receive less than the Nash equilibrium profit. The equilibrium cooperative outcome would then be the set of all Pareto optimal outcomes in which each firm receives at least its Nash payoff. Somewhat more generally, this line of argument leads to the conclusion that in a two-player game with a unique Nash equilibrium, a solution concept we seek should predict as equilibria all Pareto optimal outcomes that weakly dominate the Nash equilibrium. The problem, of course is more complex in games with three or more players, but this basic idea has recently been formalized by Ray and Vohra (1994) in the notion of equilibrium binding agreements. The problem also becomes more interesting in games with more than two players because then full cooperation in the grand coalition and pure non-cooperation are not the only possibilities; partial break down of cooperation in the form of some intermediate coalition structure also becomes a possibility; see example 3.2.1 below.

There are three basic ingredients in the concept of equilibrium binding agreements:

1. Equilibrium binding agreements are meant to capture the idea that any coalition can, in principle, write a binding agreement among its members but this agreement must be independent of the actions of players outside this coalition. Non-cooperative play across coalitions is modeled *a la* Nash. Thus, one feature of this equilibrium concept is that if in equilibrium the coalition structure that emerges is $\mathcal{P}$, then the equilibrium strategy profile x must satisfy the best response property with respect to $\mathcal{P}$. Formally, this is the requirement that for every $S \in \mathcal{P}$ there is no $y_S \in X_S$ such that $u_S(y_S, x_{-S}) \gg u_S(x)$. Let $\beta(\mathcal{P})$ denote the set of best response strategy profiles for $\mathcal{P}$. A necessary condition for x to constitute a binding agreement for $\mathcal{P}$ is that $x \in \beta(\mathcal{P})$.

2. It is assumed that agreements can only be written between members of an existing coalition. Thus, deviations can only make an existing coalition structure finer – mergers are ruled out. Recall that this is also the assumption on which the notion of a coalition proof Nash equilibrium is based. And here again the strength of this assumption lies in allowing for a recursive definition.

3. Deviations from an existing coalition structure must be based on a consistent prediction of any further reorganization of the coalition structure that may follow from the initial deviation.

Notice that these conditions immediately yield the conclusion that for the finest coalition structure, $\mathcal{P}^*$, the set of equilibrium binding agreements, denoted

$\mathcal{B}(\mathcal{P})$, is precisely the set $\beta(\mathcal{P})$, which is the set of Nash equilibria of the game. In other words, if all players are in singleton coalitions, the equilibrium binding agreements are precisely the Nash equilibria of the game.

To describe binding agreements for an arbitrary coalition structure we will consider the simple case in which there are only three players; for the more general case the reader is referred to Ray and Vohra (1994). It will be useful to begin by describing the corresponding story. Consider the three players negotiating in this grand coalition and discussing the possibility of agreeing upon a weakly Pareto optimal strategy profile x. Suppose player i contemplates leaving the grand coalition expecting to do better than x. After such a deviation there are only two kinds of coalition structures that can eventually emerge. One possibility is that this coalition structure, $(\{i\}, \{j, k\})$, is 'stable'. The other possibility is that this coalition structure breaks up into $\mathcal{P}^*$. In the first case, player i should deviate if there exists an 'equilibrium binding agreement' x' for the coalition structure $(\{i\}, \{j, k\})$ such that $u_i(x') > u_i(x)$. In the second case, i should deviate if there exists $y \in \beta(\mathcal{P}^*)$ such that $u_i(y) > u_i(x)$. This provides a consistent set of rules for player i to decide whether or not to sign the agreement x. The only concept left undefined so far is the notion of 'an equilibrium binding agreement' for the coalition structure $(\{i\}, \{j, k\})$. But this is easy to do, using similar arguments, given that the only possible change in this coalition structure is to $\mathcal{P}^*$, which will arise if and only if there exists $y \in \mathcal{B}(\mathcal{P}^*)$ such that either $u_j(y) > u_j(x')$ or $u_k(y) > u_k(x')$. Similar considerations can be used to determine whether or not a two-player coalition would agree to sign the agreement x.

We can now provide the formal definitions based on this discussion. If $N = \{1, 2, 3\}$, there are only five different coalition structures, $N = \{1, 2, 3\}$, $\mathcal{P}^* = (\{1\}, \{2\}, \{3\})$ and $\mathcal{P}^i = (\{i\}, N_{-i})$, $i = 1, 2, 3$. Consider the coalition structure $\mathcal{P}^i$. Suppose $x \in \beta(\mathcal{P}^i)$. By assumption the only subcoalitions that can deviate for this coalition structure are $\{j\}$ and $\{k\}$, where $j, k \neq i$. And in the coalition structure that they induce, $\mathcal{P}^*$, the equilibrium outcomes are $\mathcal{B}(\mathcal{P}^*) = \beta(\mathcal{P}^*)$. Given $x \in \beta(\mathcal{P}^i)$, we say that $(\mathcal{P}^*, x')$ blocks $(\mathcal{P}^i, x)$ if $x' \in \mathcal{B}(\mathcal{P}^*)$ and there exists $j \neq i$ such that $u_j(x') > u_j(x)$. We can now define $\mathcal{B}(\mathcal{P}^i)$ as the set of $x \in \beta(\mathcal{P}^i)$ that are not blocked. Having defined the equilibrium binding agreements for the intermediate coalition structures, we can now consider the blocking possibilities starting from the grand coalition. Note that $\beta(N)$ is the set of all weakly Pareto optimal strategies. There are two kinds of coalition structures that can be used to block (N, x), where $x \in \beta(N)$. We say that $(\mathcal{P}^i, x')$ blocks (N, x) if $x' \in \mathcal{B}(\mathcal{P}^i)$ and either $u_i(x') > u_i(x)$ or $u_j(x') > u_j(x)$ for $j \neq i$, i.e., either a single player can do better by deviating or a coalition of two players can do better by deviating. Note that in each case, the new coalition structure $\mathcal{P}^i$ is immune to further deviations, since $x' \in \mathcal{B}(\mathcal{P}^i)$. It is also possible that $(\mathcal{P}^*, x')$, where $x' \in \mathcal{B}(\mathcal{P}^*)$, blocks (N, x). This can happen in two ways, either there exists i such that $u_i(x') > u_i(x)$ and $\mathcal{B}(\mathcal{P}^i) = \emptyset$, or there exist i, j such that $u_i(x') > u_i(x)$ and $u_j(x') > u_j(x)$. And $\mathcal{B}(N)$ is the set of $x \in \beta(N)$ such that it cannot be blocked.

		x_{2a}	x_{2b}	x_{2c}
	x_{1a}	2.6, 2.6, 2.6	3.2, 2.2, 3.2	3.7, 1.7, 3.7
x_{3a}	x_{1b}	2.2, 3.2, 3.2	2.7, 2.7, 3.7	3.1, 2.1, 4.1
	x_{1c}	1.7, 3.7, 3.7	2.1, 3.1, 4.1	2.6, 2.6, 4.6

		x_{2a}	x_{2b}	x_{2c}
	x_{1a}	3.2, 3.2, 2.2	3.7, 2.7, 2.7	4.1, 2.1, 3.1
x_{3b}	x_{1b}	2.7, 3.7, 2.7	3.1, 3.1, 3.1	3.6, 2.6, 3.6
	x_{1c}	2.1, 4.1, 3.1	2.6, 3.6, 3.6	2.9, 2.9, 3.9

		x_{2a}	x_{2b}	x_{2c}
	x_{1a}	3.7, 3.7, 1.7	4.1, 3.1, 2.1	4.6, 2.6, 2.6
x_{3c}	x_{1b}	3.1, 4.1, 2.1	3.6, 3.6, 2.6	3.9, 2.9, 2.9
	x_{1c}	2.6, 4.6, 2.6	2.9, 3.9, 2.9	3.3, 3.3, 3.3

Figure 2: An Example of Inefficiency.

The concept of equilibrium binding agreements can be used to address an important issue concerning the efficiency of equilibrium outcomes when binding agreements are feasible. It is widely believed that if there are no informational imperfections, then the ability to make binding agreements must result in all gains from cooperation being exploited. As we have already indicated, this conclusion is valid in two-player games. More precisely, in a two-player game with a unique Nash equilibrium the outcome will be efficient; the set of binding agreements correspond to all weakly Pareto optimal strategy profiles that weakly dominate the Nash equilibrium. But, surprisingly, in games with three or more players, the theory does not bear out the conclusion that equilibrium binding agreements are always efficient. It is possible that every Pareto optimal outcome that dominates the Nash equilibrium is blocked by a coalition, and leads to an inefficient outcome with an equilibrium coalition structure that is neither N nor $\mathcal{P}^*$. This is shown by the following example, which will also serve to illustrate the notion of an equilibrium binding agreement.

Example 3.2.1 *(Ray and Vohra (1994)).*

Consider a three-consumer economy with one private good and one public good. Let x_{ia}, x_{ib} and x_{ic} denote, for consumer i, successively higher levels of contribution towards the public good. It is possible to specify well behaved, quasi-linear utility functions such that they result in the normal form game depicted in Figure 2, where player 1 chooses rows, player 2 chooses columns and player 3 chooses matrices.

We claim that in this example there is no efficient equilibrium binding agreement, and that the grand coalition breaks up into an intermediate coalition structure. Notice first that every player i has a dominant strategy, x_{ia}. Thus the unique Nash equilibrium, and the only equilibrium binding agreement for $\mathcal{P}^*$, is (x_{1a}, x_{2a}, x_{3a}), which is Pareto dominated by (x_{1c}, x_{2c}, x_{3c}).

Next, we examine the equilibrium binding agreements for an intermediate coalition structure $\mathcal{P} = (\{i\}, \{j, k\})$. Since the game is symmetric, there is no loss of generality in considering the coalition structure $\mathcal{P} = (\{3\}, \{1, 2\})$. Since player 3's dominant strategy is x_{3a}, any $z \in \beta(\mathcal{P})$ must be such that $z_3 = x_{3a}$. Thus we need only look at the first matrix. Clearly, both (x_{1a}, x_{2a}) and (x_{1c}, x_{2c}) are dominated by (x_{1b}, x_{2b}). In fact, it is easy to see that $(x_{1b}, x_{2b}, x_{3a}) \in \beta(\mathcal{P})$. Moreover, this strategy cannot be blocked by a deviation to $\mathcal{P}^*$. It is, therefore, an equilibrium binding agreement. Indeed, this is the only one for this coalition structure. To see this, notice that in all other best response equilibria, either player 1 or player 2 receives less than 2.6, the unique Nash payoff. Since the game is symmetric, we can now claim that for every intermediate coalition structure $(\{i\}, \{j, k\})$, the only equilibrium strategy profile is (x_{ia}, x_{jb}, x_{kb}). The payoffs to i, j and k are 3.7, 2.7 and 2.7 respectively. But this outcome is not efficient. It is Pareto dominated by (x_{ib}, x_{jc}, x_{kc}).

Finally, consider the grand coalition. For any strategy profile it must be the case that there exists a player, i, who gets less than 3.7. This player can then block this proposal by deviating to $(\{i\}, \{j, k\})$ and earning 3.7. This in fact, is the *only* coalition structure that i can induce by deviating from the grand coalition. Thus, the grand coalition breaks up into some intermediate coalition structure with an inefficient equilibrium. And the only equilibrium in the finest coalition structure too is inefficient. Since all the best response equilibria are strict, it follows that this example is robust.

A more detailed analysis of the public goods model is contained in Ray and Vohra (1994). The Cournot oligopoly is another interesting model to which these ideas have been applied; see Bloch (1992), Ray and Vohra (1994) and Yi (1993).

A drawback of the approach we have described is that it is restricted to internal deviations – it is assumed that deviating coalitions cannot merge with others. While it is possible to define a more general solution concept, it seems difficult to do so while retaining the relative transparency and tractability that comes with a recursive definition. For example, consider Greenberg's (1990) framework of abstract stable sets. This approach does incorporate consistency, and Greenberg suggests two new notions for normal form games, contingent threats equilibrium and coalitional commitments equilibrium, that appear to be related to the idea of binding agreements. However, the former does not exist even in the prisoner's dilemma and the latter does not always predict cooperation even in two-player games.[9] It is also possible that further progress

[9]It is also worth keeping in mind Chwe's (1994) critique of the abstract stable set solution,

on this issue might come from studying an extensive form model of coalitional bargaining. And one can expect that progress here would also help in extending the notion of a coalition proof Nash equilibrium beyond the case of internal deviations.

Acknowledgments

I am indebted to Debraj Ray for many discussions, over the last several years, on the subject of this chapter. All opinions expressed here and any errors, of fact or interpretation, may be attributed to us jointly. Support from NSF grant SBR 9414142 is gratefully acknowledged.

References

Aumann, R. (1961), The Core of a Cooperative Game without Sidepayments, *Transactions of the American Mathematical Society*, 98, 539-552.

Bernheim, D., B. Peleg and M. Whinston (1987), Coalition-Proof Nash Equilibria. I. Concepts, *Journal of Economic Theory*, 42, 1-12.

Bloch, F. (1992), Sequential Formation of Coalitions with Fixed Payoff Division, Brown University Working Paper No. 92-5.

Chakravorti, B. and C. Kahn (1991), Universal Coalition Proof Equilibrium, mimeo, University of Illinois, Champaign.

Chwe, M. (1994), Farsighted Coalitional Stability, *Journal of Economic Theory*, 63, 299-325.

Dutta, B., D. Ray, K. Sengupta and R. Vohra (1989), A Consistent Bargaining Set, *Journal of Economic Theory*, 49, 93-112.

Greenberg, J. (1990), *The Theory of Social Situations*, Cambridge University Press, Cambridge.

Greenberg, J. (1992), On the Sensitivity of von Neumann and Morgenstern Abstract Stable Sets: The Stable and Individually Stable Bargaining Set, *International Journal of Game Theory*, 21, 41-55.

Kahn, C. and D. Mookherjee (1992), The Good, the Bad, and the Ugly: Coalition Proof Equilibrium in Infinite Games, *Games and Economic Behavior*, 4, 101-121.

Mas-Colell, A. (1989), An Equivalence Theorem for a Bargaining Set, *Journal of Mathematical Economics*, 18, 129-138.

Maschler, M. (1992), The Bargaining Set, Kernel and Nucleolus, in *Handbook of Game Theory*, Eds. A. Aumann and S. Hart, North-Holland, Amsterdam.

namely that far-sightedness is not automatically incorporated in this notion. Notice that the notion of equilibrium binding agreements, albeit in a different framework, does explicitly incorporate 'far-sightedness'.

Ray, D. (1989), Credible Coalitions and the Core, *International Journal of Game Theory*, 18, 185-187.

Ray, D. and R. Vohra (1994), Equilibrium Binding Agreements, mimeo, Boston University.

Scarf, H. (1971), On the Existence of a Cooperative Solution for a General Class of n-Person Games, *Journal of Economic Theory*, 3, 169-181.

Shimomura, K-I, (1994), The Bargaining Set and Coalition Formation, mimeo, Brown University.

Vohra, R. (1991) An Existence Theorem for a Bargaining Set, *Journal of Mathematical Economics*, 20, 19-34.

Yi, S-S. (1993), Stable Coalition Structure with "Forward-Looking" Players, mimeo, Dartmouth College.

Zhou, L. (1993), A New Bargaining Set of an N-Person Game and Endogenous Coalition Formation, *Games and Economic Behavior*, 6, 512-526.

Situation Approach to Cooperation

Joseph Greenberg

Department of Economics, McGill University, 855 Sherbrooke St. West, Montreal, Quebec, CANADA H3A 2T7

The *Theory of Social Situations* (TOSS) is an integrative approach to the study of formal models in social and behavioral sciences. TOSS unifies the representation of "cooperative" and "non-cooperative" social environments, allowing for diverse coalitional interactions. It does so using the notion of a (social) "situation". TOSS disassociates the solution concept from the representation of social environments. The unified solution concept in TOSS — "stable standard of behavior" — employs stability as the sole criterion. One of the important merits of TOSS is that by representing a social environment as a situation, it specifies the exact negotiation process and the way in which players and coalitions use the set of outcomes (actions, alternatives) available to them. Moreover, the flexibility of TOSS enables the analysis of social environments that cannot be studied within the classical paradigm of game theory. This lecture is divided into three parts: (1) Motivation for the notion of a social situation, (2) Formal definitions of a situation and of a stable standard of behavior, and (3) Some applications of TOSS to cooperation. [1]

1. Motivation

In this section I shall show that none of the three types of games provides an adequate representation of a social environment. In particular, the negotiation process and the behavioral assumptions are not specified when a social environment is described as either a cooperative game, or a normal form game, or an extensive form game. Thus, for example, within the context that is interest to us here, no answer is provided to questions such as:

> What, in fact, is the meaning of forming a coalition — is it a binding commitment of the players to remain together, or to never re-negotiate with nonmembers, or is it merely a "declaration of intentions", which can

[1] Unless otherwise referenced, results in this lecture can be found in Greenberg (1990).

be revised? Do players first form a coalition and only then discuss their payoffs there, or are the two decisions made simultaneously?

As a result, many different negotiation processes can be associated with each of these three types of games. [2] I shall now illustrate the validity of this claim for each of the three types of games.

1.1 Cooperative Games

A cooperative game is a pair (N, v), where N is the nonempty set of players and v is the characteristic function which assigns to every coalition $S \subset N$ a nonempty subset of $\mathcal{R}^S$ denoted $v(S)$. For $S \subset N$, the subgame (S, v_s) is given by $v_S(T) \equiv v(T)$ for every $T \subset S$.

Cooperative games were the main object of investigation by von Neumann and Morgenstern in their pioneering work (1944), and have been extensively studied since then. However, as I shall argue, the description of a social environment provided by cooperative games is incomplete. While it specifies the set of payoffs available to coalitions if and when they form, it is totally silent on the crucial issues of how exactly this set can be used in the negotiation process and what it means to form a coalition. As a result, different negotiation processes and behavioral and institutional assumptions can be associated with a given cooperative game. The following is a sample of the many different scenarios in which the game can be played.

(1) A coalition first forms and only then its members decide on the distribution of payoffs within the coalition. Moreover, once a coalition forms the game is over (at least for its members). Schematically,

$$x \in v(N) \overset{S \subset N}{\longrightarrow} (S, v(S)).$$

That is, when x is offered, coalition S can form and then choose the payoff it will adopt. No further modifications are thereafter possible.

(2) A coalition forms and decides on the payoff at the same time. Moreover, once a coalition forms the game is over (at least for its members). Schematically,

$$x \in v(N) \overset{S \subset N}{\longrightarrow} \{(S, y) \mid y \in v(S)\}.$$

That is, when x is offered, coalition S can form and adopt a payoff $y \in v(S)$. No further modifications are thereafter possible.

---

[2] It is the disparate solution concepts for these three types of games that often involve behavioral and institutional assumptions that should be part of the description of the social environments.

(3) A coalition first forms and then decides on the payoff distribution. Moreover, once a coalition forms its members will never again approach nonmembers. However, a subset of the members of the coalition can further deviate, and then decide on its payoff. Schematically,

$$x \in v(N) \overset{S\subset N}{\longrightarrow} (S, v_S); y \in v(S) \overset{T\subset S}{\longrightarrow} (T, v_T); z \in v(T) \ldots$$

That is, when x is offered, coalition S can form. Once S forms, its members decide upon the payoff vector $y \in v(S)$ that it intends to adopt. At this time, another coalition, T, which is a subset S, can in turn form, and then decide on the payoff vector $z \in v(T)$ that it intends to adopt, and so on.

(4) A coalition forms and decides on the payoff distribution at the same time. Moreover, once a coalition forms its members will never again approach nonmembers. However, a subset of the members of the coalition can further deviate in the same manner. Schematically,

$$x \in v(N) \overset{S\subset N}{\longrightarrow} \{(S,y) \mid y \in v(S)\} \overset{T\subset S}{\longrightarrow} \{(T,z) \mid z \in v(T)\} \longrightarrow \ldots$$

That is, when x is offered, coalition S can form and adopt $y \in v(S)$. Another coalition $T \subset S$ can, in turn, form and adopt $z \in v(T)$, and so on.

The distinctive feature of all the above negotiation processes can be viewed as follows: once a coalition S objects to a proposed outcome $x \in v(N)$, its members "leave the room" and will never negotiate again with players in $N \setminus S$. Although the above negotiation processes are different, the "stable standard of behavior" for each of them leads to the an important solution concept in cooperative games – the *core*. This is also true for the following procedure:

(5) A coalition forms and proposes a payoff for the *entire society*. Moreover, once a coalition forms the game is over. Schematically,

$$x \in v(N) \overset{S\subset N}{\longrightarrow} \{(N,y) \mid y \in v(N), y^S \in v(S)\}.$$

That is, when x is offered, S can propose a payoff vector y that is feasible both for the grand coalition and for S itself.

There are, however, many other negotiation processes that can be associated with a cooperative game. In particular, individuals can engage in "open negotiations", i.e., every offer or counter offer has to include *the members of the entire society*. And, no coalition is excluded from making counter proposals to the one which is currently offered. In particular, members of $N \setminus S$ remain active throughout the negotiation process. The following two negotiation processes belong to this category:

(6) Open negotiation among all players: When a coalition $S \subset N$ offers a payoff (to the entire society), any coalition $T \subset N$ (not necessarily a subset of S) can propose another payoff (again, to the entire society). Moreover, every offer and counter-offer has to be feasible for the proposing coalition as well as for the entire society. Schematically,

$$x \in v(N) \xrightarrow{S} \{(N, y) \mid y \in v(N), y^S \in v(S)\} \xrightarrow{T}$$
$$\{(T, z) \mid z \in v(N), z^T \in v(T)\} \longrightarrow \ldots$$

The stable standard of behavior for this negotiation process yields the von Neumann and Morgenstern solution.

(7) Updating "reservation prices" according to the last "tender offer": Assume that a payoff x is offered. A coalition S can object to x if there is an S-feasible payoff $y^S \in v(S)$ which makes each member strictly better off, that is, $y^S \gg x^S$. The new modified offer then becomes $y \equiv (y^S, x^{N \setminus S})$. Now, another coalition, T, may object to y if there is a payoff $z^T \in v(T)$ such that $z^T \gg y^T$. The resulting new modified offer is then $z \equiv (z^T, y^{N \setminus T})$, and the bargaining process continues in this manner. Schematically,

$$x \in v(N) \xrightarrow{S} \{(N, y) \mid y^S \in v(S), y^{N \setminus S} = x^{N \setminus S}\} \xrightarrow{T}$$
$$\{(T, z) \mid z^T \in v(T), z^{N \setminus T} = y^{N \setminus T}\} \longrightarrow \ldots$$

Observe that in contrast to the negotiation process (6) where each tender offer has to be feasible for the entire society (i.e., belong to $v(N)$), the bargaining procedure described here is such that a coalition S does not have to offer a payoff that is feasible for the entire society, but only a payoff that is S-feasible. The stable standard of behavior for this negotiation process leads to *the stable bargaining set* (see Greenberg 1990 and Greenberg 1992a).

1.2 Normal Form Games

As is the case with cooperative games, the description of a social environment as a normal form game is also inadequate. Again, different negotiation processes can be associated with a normal form game. Recall that a normal form game is a triple $G = (N, \{Z^i\}_{i \in N}, \{u^i\}_{i \in N})$, where N is the set of players, Z^i is a nonempty set of strategies of player i, and u^i is player i's payoff function, $u^i : Z^N \to \Re$. For $S \subset N$, let Z^S denote the Cartesian product of Z^i over $i \in S$, i.e., $Z^S = \prod_{i \in S} Z^i$.

In order to know how the players can use their strategy sets, we must answer the question: given $x \in Z^N$, what can a coalition $S \subset N$ do? Clearly, the answer to this question depends on the negotiation process that is applied. The following is a sample of the many different scenarios in which the game can be played.

(1) When a coalition forms it assumes that nonmembers will adhere to the proposed recommendation. Moreover, once a coalition forms the game is over (at least for its members). Schematically,

$$x \in Z^N \xrightarrow{S \subseteq N} \{(x^{N \setminus S}, y^S) \mid y^S \in Z^S\}.$$

That is, when $x \in Z^N$ is offered, members of a coalition S may decide to deviate from $x^S \in Z^S$, assuming that members of $N \setminus S$ will continue to follow $x^{N \setminus S}$. Moreover, once S is formed, it will never dissolve, that is, forming a coalition is a binding agreement. The stable standard of behavior for this negotiation process yields the strong Nash equilibrium.

(2) When a coalition forms it assumes that nonmembers will adhere to the proposed recommendation. Moreover, once a coalition forms its members will never again approach nonmembers. However, a subset of the members of the coalition can further deviate. Schematically,

$$x \in Z^N \xrightarrow{S \subseteq N} \{(S; x^{N \setminus S}, y^S) \mid y^S \in Z^S\} \xrightarrow{T \subseteq S}$$
$$\{(T; (x^{N \setminus S}, y^{S \setminus T}, z^T) \mid z^T \in Z^T\} \ldots$$

That is, when $x \in Z^N$ is offered, members of a coalition S may decide to deviate from $x^S \in Z^S$, assuming that members of $N \setminus S$ will continue to follow $x^{N \setminus S}$. However, forming a coalition is not binding; after S deviates, a coalition T, which is a *subset* of S can then decide to further deviate. The stable standard of behavior for this negotiation process yields the coalition proof Nash equilibrium.

(3) Forming a coalition is transitory, and is used to openly negotiate with the entire society. Every coalition can counter a proposal. Schematically,

$$x \in Z^N \xrightarrow{S} \{(N; x^{N \setminus S}, y^S) \mid y^S \in Z^S\} \xrightarrow{T}$$
$$\{(T; (x^{N \setminus \{S \cup T\}}, y^{S \setminus T}, z^T) \mid z^T \in Z^T\} \ldots$$

Thus, in contrast to (2) where only subsets of a deviating coalition can further deviate, this negotiation process does not restrict further deviations to subsets of the deviating coalition.

(4) Forming a coalition allows its members to (irrevocably) commit to a (correlated) choice of strategies. Schematically,

$$x \in Z^N \xrightarrow{S \subseteq N} \{(N \setminus S; x^{N \setminus S}, y^S) \mid y^S \in Z^S\} \xrightarrow{T \subseteq N \setminus S}$$
$$\{(N \setminus \{S \cup T\}; (x^{N \setminus \{S \cup T\}}, y^S, z^T) \mid z^T \in Z^T\} \ldots$$

That is, when $x \in Z^N$ is under consideration, a coalition $S \subset N$ can object to x and choose an S-tuple of strategies $y^S \in Z^S$, to which it henceforth commits. Another coalition T, a subset of $N \setminus S$, can in turn further commit to some $z^T \in Z^T$, and so on.

1.3 Extensive Form Games

The representation of a social environment as an extensive form game is also not satisfactory. On one hand, it requires the rigid and exact specification of the precise order of moves. On the other hand, each node belongs to a single individual, thereby considerably limiting the analysis of coalition formation within this framework. Moreover, despite the excessive rigidity concerning the order of moves, an extensive form game reveals little about the precise negotiation process [for example, who can (effectively) recommend a path or a strategy profile, and how can players communicate?]. As a result, a game in extensive form can be associated with different social situations that represent different negotiation processes. [3]

Current analysis of extensive form games employs the notions of strategy and Nash equilibrium. A strategy is a complete plan that specifies an action in every eventuality (even in those that might not arise). Hence it is complex and artificial. Moreover, an equilibrium strategy profile forces commonality of beliefs on the part of the players. That is, all players have exactly the same beliefs concerning other players' choices of actions, including those "off the equilibrium path".

One advantage of TOSS is that it allows to analyze extensive form games using the more natural notion of paths which specify only the relevant actions of the players. (See Greenberg, Monderer, and Shitovitz 1993, and Greenberg 1994a,b.) Consider, for example, the environment in which before playing the game, individuals discuss/suggest how to play the game, that is, which *path* is to be followed. The negotiation process depends, then, on the answer to the question: who can, and at what juncture, effectively recommend a path? The following are three (of many) ways in which this question can be answered.

(1) Upon reaching his [4] decision node, player i can move, induce a subtree, and then he, and he alone, can recommend a course of actions (a path) to be followed thereafter. Consider, for example, the game in Figure 1.3.1.

At the root of the tree, player 1 can either induce the subtree T^u and suggest one of the two paths in that subtree (for example a) to be followed, or induce the subtree T^v and suggest one of the paths in this subtree (for

[3] Another deficiency of extensive form games stems from the "tree structure": there is a unique path from the root of the game tree to a node. It is, therefore impossible to use such games to analyze situations in which players have "human rationality" and thus often ignore (perhaps even relevant) information.

[4] Gender-neutrality of all the masculine nouns/pronouns/adjectives is, of course, assumed throughout.

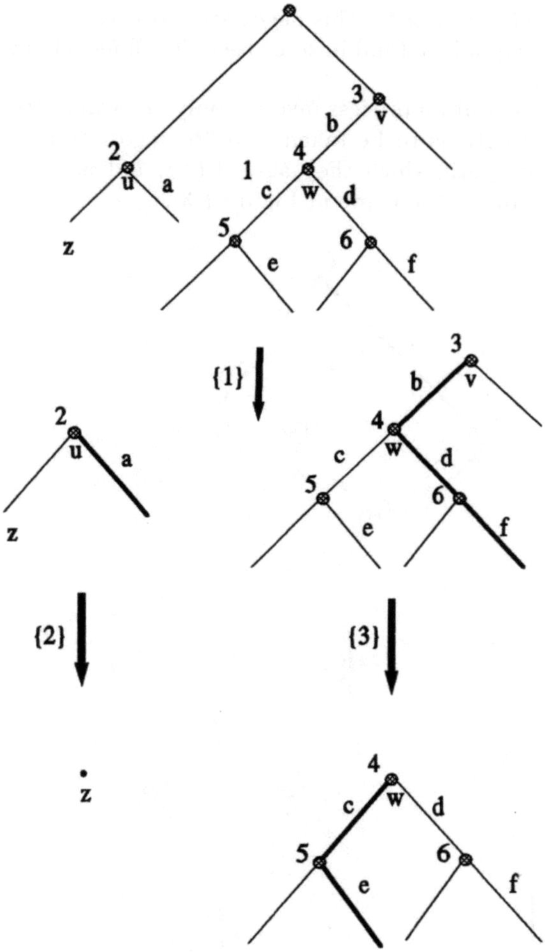

Figure 1.3.1

example, *bdf*). If player 1 induces T^u, then it is player 2's turn to move. Player 2 is under no obligation to follow 1's recommendation. He is free to induce any of the terminal nodes in that subtree, including z, even if player 1 suggested the path a for T^u. If player 1 induces T^v then player 3 is the next to move. Player 3 can induce any subtree of T^v and then recommend a path in the induced subtree. For example, he can induce T^w and recommend the path *ce*. This process continues till a terminal node is reached. The stable standard of behavior for this negotiation process yields a refinement of subgame perfect equilibria (and in some cases it refines all other refinements).

(2) A different negotiation process involves only one single discussion concerning the course of actions to be taken. No "re-negotiation" occurs if players deviated from the path which they agreed (but did not commit) to follow. Consider, for example, the game in Figure 1.3.2.

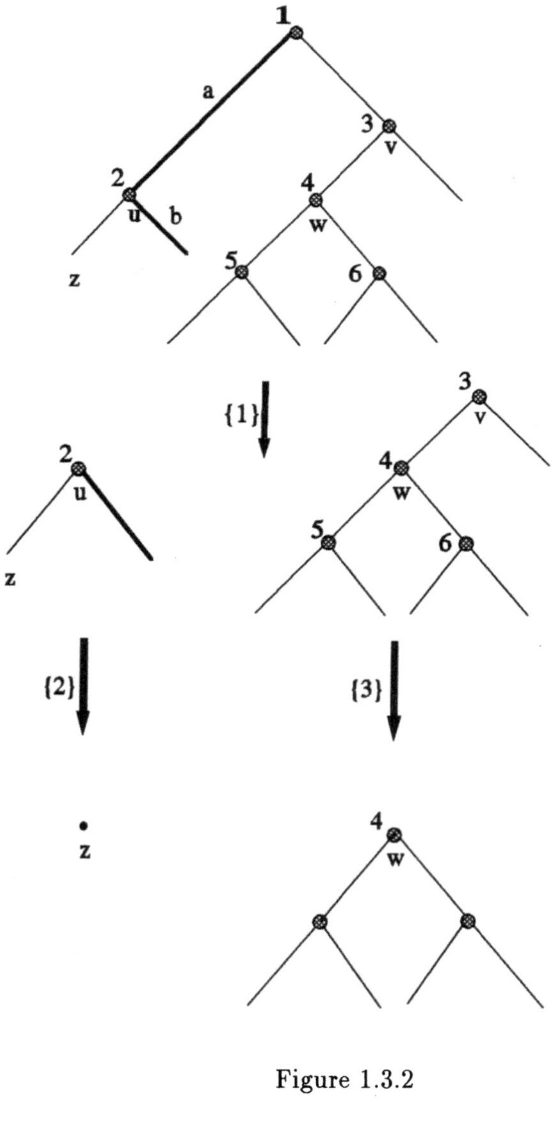

Figure 1.3.2

At the beginning of the game players consider following the path ab. As in the previous negotiation process, player 1 can induce either the subtree T^u or the subtree T^v. He can, however, make no new recommendations concerning the paths to be followed. Thus, if he induces T^u it is common knowledge that the path a is the one that is supposed to be followed. If, on the other hand, he induces T^v, no such path exists. This negotiation process is particularly appealing in view of the fact that most agreements are incomplete; they fail to specify what might happen in every contingency that might arise should deviations take place. Moreover, players often know what is the existing social norm, or have information concerning how the game has been played in the past (by other players). This knowledge serves as focal point or the status quo. The stable standard of behavior for this negotiation process yields the new solution concept, that of "stable paths" (Greenberg 1994a). A path is stable if it would be followed by rational players were it recommended at the beginning of the game.

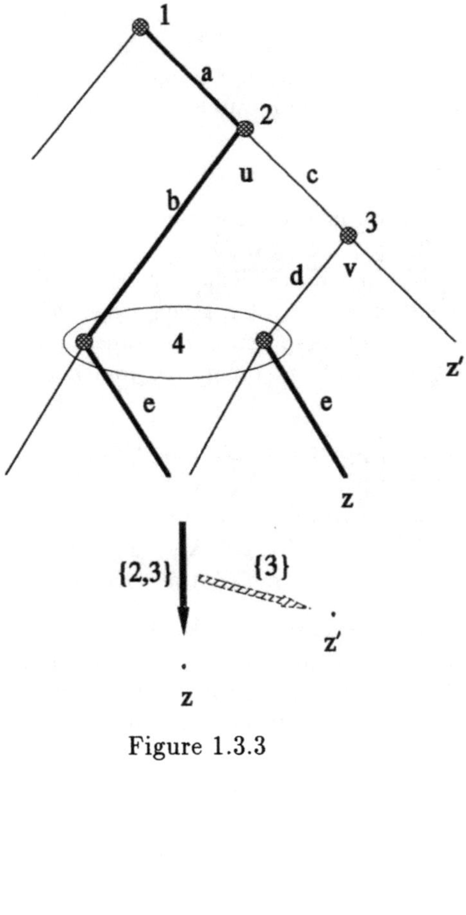

Figure 1.3.3

(3) TOSS easily accommodates coalition formation. For example, we can modify (1) and (2) to allow not only single individuals, but also coalitions, to jointly deviate from a proposed path. To illustrate how this is accomplished, consider the game in Figure 1.3.3 and the negotiation process in (2) above except that now coalitions can form. Suppose that path *abe* is recommended at the root of the game. When u is reached, players 2 and 3 can form a coalition and coordinate their actions. In particular, player 2 can choose action c and player 3 can choose d. Since player 4 is not part of this coalition, (players 2 and 3 expect that) he will continue to follow the original recommendation, namely, choose action e. Thus, coalition $\{2,3\}$ can, in this case, induce the terminal node z. If forming a coalition is not binding, then player 3, upon reaching v, might wish to "double-cross" and induce the terminal node z' instead.

2. The Theory of Social Situations (TOSS)

I now turn to the formal definitions of the two main concepts in TOSS: the notion of a "situation" that provides a complete description of the social environment, and the notion of "stability" of a "standard of behavior" that provides a unified criterion for recommendations that are likely to be acceptable to rational, free individuals.

2.1 Situations

The concept of situation involves two elements: a *position* that describes the "current state of affairs", and an *inducement correspondence* that specifies the alternatives available to a player if he decides to reject a proposed outcome in the present position. Thus, at a given stage, a position specifies the set of individuals, the set of all possible outcomes, and the preferences of the individuals over this set of outcomes. Formally, a "position" is defined as follows:

Definition 2.1.1. *A* **position**, *G, is a triple $G = (N_G, X_G, \{u_G^i\}_{i \in N_G})$, where N_G is the set of* **players**, *X_G is the set of all feasible* **outcomes**, *and u_G^i is the* **utility function** *of player i in position G over the outcomes, that is, $u_G^i : X_G \to \Re$. Thus, for all $x, y \in X_G$, and for all $i \in N_G$, $u_G^i(x) > u_G^i(y)$ if and only if i prefers, in position G, the outcome x over the outcome y.*

The set of outcomes describes the feasibility constraints at the particular stage, and not the choices that are likely to, or should, be made. The only requirement is that the domain of the utility functions of the players be the set of outcomes; players in position G are able to evaluate every outcome in X_G.

Now, consider a position G and suppose that an outcome $x^* \in X_G$ is proposed. For a player $i \in N_G$ to be able to decide whether to accept or reject

x^*, he must know the alternatives available to him. Such a specification is provided in TOSS by means of the *set of positions*, denoted by $\gamma(\{i\} \mid G, x^*)$, that a player i can, then, "induce". [5] Moreover, he must also be able to anticipate what might happen once such a position $H \in \gamma(\{i\} \mid G, x^*)$ is induced — in particular, what positions can, in turn, be induced from (the induced) position H. This reasoning is applicable to individuals as well as coalitions. Denote, therefore, by $\gamma(S \mid G, x^*)$ the set of positions a coalition $S \subset N_G$ can induce when an outcome x^* is proposed in position G.

A complete description of a social environment is, therefore, given by the following concept of a "situation".

Definition 2.1.2. *A* **situation** *is a pair* (γ, Γ), *where* Γ *is a set of* **positions**, *and the mapping* γ, *called the* **inducement correspondence**, *which satisfies the condition that for all* $G \in \Gamma$, $S \subset N_G$, *and* $x \in X_G$, $\gamma(S \mid G, x) \subset \Gamma$.

The requirement that Γ be closed under γ guarantees a complete specification of what might happen from any admissible position $G \in \Gamma$, when a feasible outcome $x \in X_G$ is proposed. Note that γ may assign the empty set to some (or even all) coalitions, that is, some coalitions may be unable to induce any position at all. [6]

2.2 Standards of Behavior

Consider a situation (γ, Γ), a position $G \in \Gamma$, an outcome $x^* \in X_G$, and a position $H \in \gamma(S \mid G, x^*)$. For members of S to be able to decide whether to reject x^* and induce H, it is not sufficient for the players to know the set X_H of potentially feasible outcomes; they must also know the outcomes that are expected to result once position H is induced. Such a set will be called a "solution for H", which can most generally be defined as a subset of X_H, denoted by $\sigma(H)$. That is, the acceptability of an outcome x^* in X_G depends on the outcomes that are (predicted to be) accepted in the positions that can be induced from G when x^* is proposed. It is for this important reason that the concept of a standard of behavior is introduced. Given the description of the social environment as a situation, a standard of behavior specifies a solution to *every* position in Γ.

Definition 2.2.1. *Let* Γ *be a set of positions. A mapping* σ *that assigns to each position* $G \in \Gamma$ *a solution,* $\sigma(G) \subset X_G$, *is called a* **standard of behavior (SB)** *for* Γ.

---

[5] Observe that, in contrast to classical game theory, the actions available to a player are not explicitly modeled. What is important, and modeled by the inducement correspondence, is what impact can such actions have.

[6] It is convenient to impose the mild restriction that the set of players in each position that a coalition S can induce, includes, but need not coincide with, the players in S, that is, for all $G \in \Gamma$, $S \subset N_G$, and $x \in X_G$, if $H \in \gamma(S \mid G, x)$, then $S \subset N_H$.

An SB, σ, for Γ is any *arbitrary* mapping. However, rational players cannot be expected to follow a "senseless" standard of behavior. Therefore, for σ to be adopted, some restrictions on σ seem to be necessary. These restrictions are manifested by the requirement that the standard of behavior be "stable".

2.3 Stability

Consider a situation (γ, Γ), and let σ be an SB for Γ. If all players adopt σ, it seems reasonable to stipulate that a group of players, S, will reject an outcome, $x \in X_G$, if it can induce a position $H \in \gamma(S \mid G, x)$, whose solution, $\sigma(H)$, benefits all members of S. It is important to note that the rejection of x must depend only on those outcomes that belong to $\sigma(H)$, not on the entire set of feasible outcomes, X_H. Therefore, we shall say that the SB, σ, is **internally stable** for (γ, Γ) if for all $G \in \Gamma$, $x \in \sigma(G)$ implies that there exist no coalition $S \subset N_G$ and position $H \in \gamma(S \mid G, x)$, such that S benefits by rejecting x and inducing H, realizing that the solution to H is given by $\sigma(H)$. That is, if σ is internally stable, then accepting σ as the SB implies the willingness of all players to follow its recommendations in *every* position.

Internal stability requires the consistency of outcomes recommended by σ. TOSS requires that, in addition, the decision to exclude certain outcomes from σ be not arbitrary. (Note that by never recommending any outcome, no inner contradictions might arise. That is, a standard of behavior σ such that $\sigma(G) = \emptyset$ for all $G \in \Gamma$ is always internally stable.) That is, for every position $G \in \Gamma$, the SB σ must account not only for elements in $\sigma(G)$ but also for those in $X_G \setminus \sigma(G)$. TOSS maintains that the only reason for excluding an outcome $x \in X_G$ is that, were it included in $\sigma(G)$, it would have been rejected by players who adopt the SB σ. Formally, TOSS insists that the SB σ be **externally stable**: for all $G \in \Gamma$, $x \in X_G \setminus \sigma(G)$ implies that there exist a coalition $S \subset N_G$ and a position $H \in \gamma(S \mid G, x)$, such that S benefits by rejecting x and inducing H, realizing that the solution to H is given by $\sigma(H)$.

The overall single consistency requirement imposed on an SB is that it be **stable** — both internally and externally. Thus, if an SB is stable, then the solution it assigns to each position contains those and only those outcomes that are not rejected by any coalition, whose members are aware of and believe in the specification of the SB.

While intuitively appealing, if not compelling, there is, however, a technical difficulty in formally defining a stable SB. The expression "members of S prefer the set $\sigma(H)$ over the outcome $x \in X_G$" involves comparisons of *a single outcome* with *a set of outcomes*. This is a general difficulty which confronts any analysis of social environments in which players do not have a (subjective or an objective) probability distribution over the set of outcomes. It is important to note that this issue concerns players' preferences and thus ought to be part of the description of the social environment.

Most of the analysis that employs TOSS has, to date, considered one of two extreme forms of players' preferences: optimistic and conservative. [7] The assumption of optimistic behavior entails that members of S prefer $\sigma(H)$ over $x \in X_G$ *if there exists* an outcome $y \in \sigma(H)$ that all members of S prefer to x. The other extreme assumption is that players behave "conservatively", and S will reject a proposed outcome x in position G, and induce the position $H \in \gamma(S \mid G, x)$, only if *all* outcomes in $\sigma(H)$ make all members of S better off. Formally, this is expressed by the following definitions. Note that it is the individuals that are optimistic or conservative not the concept of stability.

Definition 2.3.1. *Let σ be an SB for the situation (γ, Γ). We shall say that σ is*

(i) **optimistic internally stable** *for (γ, Γ) if for all $G \in \Gamma$, $x \in \sigma(G)$ implies that there do not exist a coalition $S \subset N_G$, a position $H \in \gamma(S \mid G, x)$, and an outcome $y \in \sigma(H)$ such that for all $i \in S$, $u^i_H(y) > u^i_G(x)$.*

(ii) **optimistic externally stable** *if for all $G \in \Gamma$, $x \in X_G \setminus \sigma(G)$ implies that there exist $S \subset N_G$, $H \in \gamma(S \mid G, x)$, and $y \in \sigma(H)$ such that $u^i_H(y) > u^i_G(x)$ for all $i \in S$.*

(iii) **optimistic stable standard of behavior** *(OSSB) if it is both optimistic internally and externally stable.*

Definition 2.3.2. *Let σ be an SB for the situation (γ, Γ). The SB σ is:*

(i) **conservative internally stable** *for (γ, Γ) if for all $G \in \Gamma$, $x \in \sigma(G)$ implies that there exist no $S \subset N_G$ and $H \in \gamma(S \mid G, x)$ such that $\sigma(H) \neq \emptyset$, and for all $y \in \sigma(H)$, $u^i_H(y) > u^i_G(x)$ for all $i \in S$.*

(ii) **conservative externally stable** *if for all $G \in \Gamma$, $x \in X_G \setminus \sigma(G)$ implies that there exist $S \subset N_G$ and $H \in \gamma(S \mid G, x)$ such that $\sigma(H) \neq \emptyset$, and for all $y \in \sigma(H)$, $u^i_H(y) > u^i_G(x)$ for all $i \in S$.*

(iii) **conservative stable standard of behavior** *(CSSB) if it is both conservative internally and externally stable.*

Remark 2.3.3. The terms "standard of behavior" and "stability" have been borrowed from von Neumann and Morgenstern's (1947) seminal work. However, the formalism, motivation, and rationale of these two approaches are quite different. A situation provides the relevant details about the negotiation process. It turns out, as was pointed out by Shitovitz (Greenberg, 1990, Theorem 4.5), that the OSSB can be formally derived from a *von Neumann and Morgenstern* (vN&M) *abstract stable set* of an "associated abstract system". This result should not be misinterpreted to imply that the OSSB, or more generally TOSS, can be identified with vN&M's notion because of the following two reasons:

[7] Though these assumptions yield, as we shall see, many interesting results, the investigation of alternative (and perhaps, more plausible) behavioral assumptions may yet prove to be much more valuable.

(i) The standard of behavior that results from *any* behavioral assumptions other than optimism cannot be formally derived from a vN&M abstract stable set. (For example, CSSB's may be nested in contrast to the fact that OSSB or vN&M abstract stable set can never be nested.)

(ii) Even if individuals are optimistic, different situations (with different negotiation processes, beliefs, and institutions) might yield the same abstract system. That is, the solutions (captured by the OSSB or the vN&M abstract stable set) for these situations coincide, but not the underlying data.

2.4 A Few General Results

I shall now mention some general properties of OSSB and CSSB. [8] First, there exist situations that: (i) admit both an OSSB and a CSSB, (ii) admit neither an OSSB nor a CSSB, (iii) admit either an OSSB or a CSSB but not both. Second, in many situations the OSSB and the CSSB coincide. These situations include all those that represent a social environment where the negotiation process is such that at any stage a specific outcome ("bill / proposal") is being considered. This "status quo" can, in turn, be replaced ("amended / countered") by another outcome. Third, a large class of situations, called ***strictly hierarchical situations***, admit a unique OSSB as well as a unique CSSB, for which explicit formulae can be provided. Loosely speaking, strictly hierarchical situations are those situations that can be represented by a finite acyclic directed graph. Many situations that represent games or economic models of particular interest are strictly hierarchical.

Observe that, by external stability, if σ is either an OSSB or a CSSB for a situation (γ, Γ), then there exists at least one position $G \in \Gamma$ with $\sigma(G) \neq \emptyset$. We shall say that the SB σ is ***nonempty-valued*** if $\sigma(G) \neq \emptyset$ *for all* $G \in \Gamma$. If σ and τ are two SBs for a situation (γ, Γ), we say that σ *includes* τ if $\tau(G) \subset \sigma(G)$ for all $G \in \Gamma$. For CSSB, Greenberg, Monderer, and Shitovitz (1993) proved the following general result.

Theorem 2.4.1. *Let (γ, Γ) be a situation and let Σ be the set of all conservative internally stable nonempty-valued SBs. If Σ is nonempty, then it admits a largest element, σ^L, with respect to the inclusion order. Moreover, σ^L is the largest nonempty-valued CSSB.*

Corollary 2.4.2. *If σ is a nonempty-valued OSSB, then there exists a CSSB that includes it.*

As was discussed in Remark 2.3.3, an OSSB can be formally derived from a vN&M abstract stable set of an "associated abstract system". This result enables us to apply results from works on abstract stable sets to TOSS. Using one of these results, Shitovitz proved the following theorem.

---

[8] As stated in the introduction, for precise statements and proofs of results quoted in this lecture, see Greenberg (1990a), unless otherwise specified.

Theorem 2.4.3. *Let* (γ, Γ) *be a situation such that* Γ *contains a finite number of positions, and each position* $G \in \Gamma$ *contains a finite number of outcomes. Assume, in addition, that the inducement correspondence is such that for all* $G, H \in \Gamma$, $x \in X_G$ *and* $S \subset N_G$, *if* $H \in \gamma(S \mid G, x)$ *then* $N_H = S$. *Then,* (γ, Γ) *admits a unique OSSB.*

Another implication of the formal relationship between OSSB and abstract stable set is that, in contrast to CSSB, it is impossible for one OSSB to include another.

3. TOSS and Cooperation

3.1 Cooperative Games

As was mentioned in subsection 1.1, the (optimistic or conservative) stable standards of behavior for situations that represent a cooperative game yield some of the better-known solution concepts (such as the core and the vN&M solution) as well as offer new concepts (such as the stable bargaining set). In addition to "bridging" classical game theory and TOSS these results also highlight the underlying negotiation processes, thereby enhancing our understanding of these solution concepts, and perhaps making us more critical of them.

For example, it can be shown that the only candidate for an OSSB for the situation that represents the negotiation process whereby a coalition S can induce the subgame (S, v_S) [outlined in (3) of subsection 1.1], is the core mapping. However, not every game admits an OSSB. This result points out the following deficiency in the definition of the core. The core of a cooperative game (N, v) contains all those payoff vectors in $v(N)$ that are not blocked by any coalition S, using *any* payoff vector in $v(S)$, including a payoff that can, in turn, itself be blocked. But if we are to consider only payoffs that cannot be blocked, (i.e., that belong to the core), then the same property should be required of the blocking payoffs. Using the terminology of TOSS, the definition of the core lacks "external stability": outcomes that are not recommended (in this case, payoffs that do not belong to the core) must be those (and only those) that if offered would be rejected *given the recommendation for the "induced positions"* (in this case, the core of the subgames).

The external stability of the core is interesting on its own right. Moreover, as mentioned above, external stability of the core implies the existence of OSSB. The following are two important classes of games in which the core mapping is externally stable.

(i) In every game where the set of players is finite, the core mapping is an OSSB (see Ray 1989 and Greenberg 1990).

(ii) In a variety of mixed economies with "large" (atoms) and "small" (atomless) traders, the core mapping is an OSSB (see Mas-Colell 1989, Greenberg and Shitovitz 1994, Einy and Shitovitz 1994, and Shitovitz and Weber 1994).

These results naturally raise the following **open question**:

Let (N, v) be a cooperative game (with an infinite number of players) with the property that each of its subgames admits a nonempty core. Is the core correspondence an OSSB for the core situation? That is, is it true that every payoff outside the core is blocked by a payoff in the core of some subgame?

The fact that the OSSB for the situation representing the negotiation process (6) in subsection 1.1 yields the vN&M solution lead to another interesting open question: *in which games does the core coincide with the* (*unique*) *vN&M solution?* Here is a partial answer to this question. We shall say that a game has the **"extension property"** if for every payoff x in the core of a subgame (S, v_S) there exists a core payoff y whose restriction to S coincides with x, i.e., $x = y^S$. It is easy to verify that if the extension property holds, then the core coincides with the (unique) vN-M solution if and only if the core mapping is optimistic (externally) stable. Convex games is one class of games that has this property. Another class was recently discovered by Einy, Holtzman and Shitovitz (1994), who proved the following result: Let $\mu_1, \mu_2, ..., \mu_n$ be non-atomic probability measures on (T, Σ) which are absolutely continuous with respect to μ, where $T = [0, 1]$ is the set of agents, and Σ is the set of Borel subsets of T, representing the set of admissible coalitions. For each $S \in \Sigma$ let $v(S) = \min\{\mu_1(S), \mu_2(S), ..., \mu_n(S)\}$. Then, the core of (N, v) is the unique vN-M stable set.

Another open question is: *Does the stable bargaining set* [*see negotiation process* (7) *in subsection 1.1.*] *contain* (*only*) *Pareto optimal payoffs?*

In addition to these specific open questions, it is also interesting to investigate the (optimistic, conservative, or any other type of) stable standard of behavior for different negotiation processes, as well as to find a "plausible" negotiation processes that yield known solution concepts such as the Shapley value.

3.2 Cooperation in "Noncooperative Games"

When TOSS is applied to either normal or extensive form games, the OSSB often yields "cooperative outcomes". The following are a few examples that demonstrate this phenomenon.

(1) The OSSB in the situation associated with a game tree often yields appealing refinements of subgame perfect equilibria that involve cooperation. This is clearly illustrated by the following example of "retrospective voting".

There is a single voter, player 1, who can vote for either candidate 2 or candidate 3. The candidate that player 1 elects becomes the incumbent, and must then choose a policy from the set $\{a, b\}$. After the incumbent chooses a policy, the voter must vote again for the candidate who will be the incumbent in the next period. The voter's preferences are a function only of the policy selected, and he gets one unit of utility every time a is chosen, and 0 every time b is chosen. The candidates get utility only from being elected, obtaining one unit of utility whenever they are elected. This game proceeds a finite number of periods. An important feature of this model is that candidates cannot commit themselves before the election to adopting particular policies.

Inherent to this game is that the decision made by the voter affects the utility of the candidate, but not the utility of the voter, and similarly, when the candidate adopts a policy, it affects the utility of the voter, but not the utility of the candidate. It is for this reason that, as is easily verified, there are many subgame perfect equilibria in this game [as well as many "trembling hand" (Selten 1975) and "stable" (Kohlberg and Mertens 1986) equilibria]. In fact, every path in the game tree is supported by a subgame perfect equilibrium. Clearly, the "plausible" equilibrium is where the voter "rewards" the candidate who chooses the policy a, and "punishes" the candidate who implements policy b, thereby inducing the candidate to select the preferred policy, in anticipation of being reelected. These equilibria result in paths which always give the voter maximum utility, and the initial incumbent always gets reelected. Quite remarkable, these are precisely the paths assigned by the unique OSSB for the associated situation. Winer (1989) shows that this characterization holds also for the infinite case.

The appeal of the refinement provided by the OSSB for game trees is also demonstrated by Tadelis (1994) who shows that for a large class of games, that include all games with common interest, the "non-discriminating" OSSB of the infinitely repeated extensive form game yields only Pareto optimal outcomes.

An interesting **open question** is *the existence of an OSSB in "continuous game trees"*. A partial answer to this question was recently provided by Shitovitz (1994).

(2) TOSS is particularly useful when players are allowed to negotiate or discuss the course of actions they wish to take. This feature of TOSS has already been extensively applied. For example, in normal form games, the OSSB for the situation that represents the negotiation process given in (2) of subsection 1.2 yields the notion of "coalition proof Nash equilibrium" (CPNE). This characterization enables us to extend the definition of CPNE to games with an infinite number of players. [The original definition, due to Bernheim, Peleg, and Whinston (1987) is recursive.] This extension was recently used by Alesina and Rosenthal (1993) who study a model with a continuum of

voters and use the CPNE to explain the phenomena of "mid-term electoral cycle", "split-ticket voting", and "divided government". Khan and Mookherjee (1993) use our characterization to study the CPNE in insurance economies with incomplete information.

Among the many **open questions** in this area is finding *general conditions for the existence and uniqueness of OSSB in the "CPNE situation"* [described in (2) of subsection 1.2].

Another situation that can be associated with a normal form game is the "contingent threat" situation, described in (3) of subsection 1.2. It captures "open negotiations" among players who can make "tender threats". Muto and Okada (1992) study the OSSB for this situation in a duopoly market with quantity-competition. They show that if the two firms can form a coalition (but cannot sign binding agreements) then outputs supported by the OSSB yield higher profits than those of the Cournot-Nash level. Another application of the OSSB for the open negotiation process was explored by Muto and Nakayama (1991) in the context of "resale-proof" trade of information when communication and non-binding agreements between players, as well as resale of information, are allowed. Resale-proof trade is obtained as the unique stable SB of the situation that describes an open negotiation process.

(3) TOSS also provides new "cooperative concepts" within the context of "noncooperative games". For example, Asheim (1988, 1990) derived the notion of *renegotiation proofness* from the OSSB of the following situation: in every subgame of the original (infinite) game tree, in addition to deviations by single players, the grand coalition can "reconsider" its course of actions. That is, the entire set of players, N, can induce ("recommend") any path in any subgame (position).

Another concept that can naturally be defined using TOSS is that of "farsighted behavior". The way in which individuals view their alternatives and the consequences of their actions is captured by the set of alternatives at each position and the inducement correspondence. Chwe (1992) and Xue (1993) study coalition formation of farsighted individuals in social environments with diverse coalitional interactions.

(4) TOSS allows also the analysis of social environments that cannot be represented as "games". For example, there are social environments in which players may be forced (because of, e.g., legal, historical, social, or ethical considerations) to restrict their actions so that, for example, the resulting outcomes be Pareto optimal. That is, only Pareto optimal outcomes can be considered, and similarly objections to a proposed outcome cannot be based on deviations to non-Pareto outcomes. But the set of Pareto optimal strategy profiles is, in general, a strict subset of the set of all a-priori possible profiles, and, moreover, it cannot be represented as a Cartesian product of the individual strategy sets, and therefore, cannot be represented as a (normal

form) game. Greenberg, Monderer, and Shitovitz (1993) have analyzed such environments using the notion of "multistage situations". Interestingly, they show that the unique CSSB in the repeated prisoner dilemma when only Pareto optimal paths are "admissible", consists only the two extreme non-cooperative paths.

(5) Using TOSS Greenberg, Monderer, and Shitovitz (1993) defined the notion of *"k bounded rationality"*. This notion captures the fact that players "look ahead" at most k periods, (where k is less then the length of the game). Thus, players will not deviate from a path agreed upon if and only if they cannot benefit from a deviation in the *next k periods.* Of course, when deviating, players consider only paths that are "k-rational" in the induced subgame. Greenberg, Monderer, and Shitovitz (1993) show that if players are "conservative", "k boundedly rational", and their discount factor is close to 1, then it is possible to have (*full*) cooperation in the finitely repeated Prisoner's Dilemma game.

(6) My last example in this lecture concerns negotiations in extensive form games with imperfect information. Consider the "peace-negotiation" example in Figure 3.2.1.

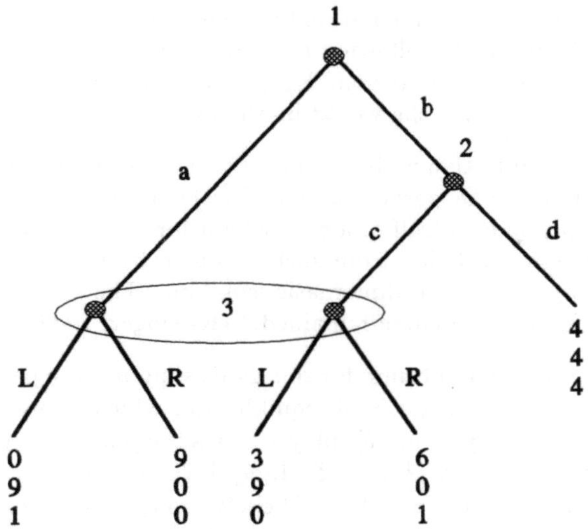

Figure 3.2.1

Each of the two warring countries, 1 and 2, has to decide whether or not to reach a peace agreement, represented by the path (bd). Failing to reach an agreement, country 3 would "re-evaluate" its policy – a decision that will affect both countries 1 and 2. Assume that country 3 has no way to know which of the two countries caused the break-down of the negotiations (otherwise, it could threaten to retaliate against that country). All it knows is whether or not the negotiations are successful. As the payoffs indicate, it is in the best interest of country 3 that the two warring countries sign the peace agreement. Both countries 1 and 2 may (correctly) anticipate the set of "plausible/rational" re-evaluated policies country 3 can take if no agreement is reached. As country 3 cannot know who is responsible for the break-down of the peace negotiations, both policies L and R are "rational". [L is the best policy for 3 if it is country 1 that refused the treaty, and R if it is country 2.] Therefore, unless country 3 *pre-determines* (or *reveals in advance*) the policy it would adopt should the peace treaty not be reached, countries 1 and 2 have no way to *know* (*even probabilistically*) which policy would be adopted by country 3. It is, then, conceivable that each country will follow the path (bd), *each because of different reasons*: country 1 for fearing that policy L is more likely to be adopted than policy R, and country 2 for fearing that policy R is more likely to be adopted than policy L. It is important to observe that if both countries held the same beliefs on the likelihood of the adoption of policies L and R, at least one of these two countries would find it in its best interest to jeopardize the peace talks.

The success of the peace mediation between Israel and Egypt (players 1 and 2) by the U.S. (player 3) following the 1973 war, may be, at least partially, attributed to such a phenomenon. Egypt and Israel were each afraid that if negotiations broke down, she would be the looser.

"And once a negotiation is thus reduced to details, it has a high probability of success - unless one party has consciously decided to make a show of flexibility simply to put itself in a better light for a deliberate breakup of the talks. Egypt was precluded from such a course by the plight of the Third Army, Israel by the fear of diplomatic isolation. The odds favored success, even though major differences remained." (Kissinger, 1982, p. 802.)

However, no Nash equilibrium for this game supports the path (bd). In fact, this game possesses a single Nash equilibrium, which is given by: Player 1 uses the mixed strategy ($\frac{1}{2}\ a, \frac{1}{2}\ b$), player 2 uses the pure strategy c, and player 3 uses the mixed strategy ($\frac{1}{2}\ L, \frac{1}{2}\ R$). Indeed, the notion of Nash equilibrium in *strategies* implies "commonality of beliefs": players have exactly the *same* beliefs concerning other players' actions, including those "off the equilibrium path" (which are, therefore, unobservable). In contrast, TOSS enables us to use of the notion of paths rather than strategies, thereby allowing for the possibility that off the ("stable path") players may have different beliefs concerning the actions other players might take. Thus, players agree to follow a course of actions, each for his own reasons. *"Stable paths"* (Greenberg

1994a) are those that would be followed if recommended to rational players (at the beginning of the game). In the "peace negotiation" example above, (*bd*) is a stable path.

The above analysis and the notion of "stable paths" suggests a new avenue of research for stable outcomes in situations where there is a "single public recommendation". (This recommendation may represent the existing social norm, past observations, incomplete contracts, etc.) Recall that the definition of stability entails that if σ is a stable standard of behavior, then a (publicly recommended) outcome $x \in X_G$ does not belong to $\sigma(G)$ if and only if there exist a coalition $S \subset N_G$ and a position $H \in \gamma(S \mid G, x)$ such that S "prefers $\sigma(H)$ to x" [9]. Thus, it is implicitly assumed that once position H is induced, a new recommendation [from $\sigma(H)$] will be *publicly* [10] made. While this may well be the case in many social environments, such a procedure does not prevail in general. It is, therefore, interesting to investigate stable outcomes in social environments where a public recommendation is made only in one ("the initial") position, and no new recommendations are publicly made in subsequent induced positions.

4. Concluding Remarks

I argued that none of the three types games provides complete description of social environments. It is for this reason that the disparate solution concepts involve (often implicit) assumptions which should be part of the description of social environments. More importantly, there are social environments that cannot be represented by any of the three types of games. The theory of social situations amends these deficiencies. In particular, it offers an integrative approach to the study of social environments with diverse coalitional interaction. A *situation* provides a complete and unified description of a social environment while the solution concept uses *stability* as the sole criterion. I have shown that the flexibility and merits of the theory through several applications. There are many interesting open questions concerning both applications and theoretical issues. Some of the possible directions, in addition to those mentioned in the previous sections, along which the theory could be extended are: analyzing situations with behavioral assumptions other than optimistic or conservative; designing more pertinent institutions and negotiation processes; studying persisting social norms (or incomplete agreements) where players do not share the same beliefs "off the equilibrium path"; and exploring new variants of bounded rationality that constrain not only on the computational but also the perceptual abilities of the players.

---

[9] We ignore the difficulty of formalizing this clause and assume that it is well-defined, either by optimistic or conservative behavior, or by assigning some probability distribution over outcomes in $\sigma(H)$.

[10] In the sense that every player in H is aware of this recommendation.

Acknowledgements

Financial support from the Research Councils of Canada [both the Natural Sciences and Engineering (NSERC) and the Social Sciences and Humanities (SSHRC)] and from Quebec's Fonds (FCAR) is gratefully acknowledged.

References

Alesina, A., and Rosenthal, H. (1993), "A theory of divided government", mimeo, Harvard University.

Arce M., D.G. (1993a), "Standards of behavior that lead to undominated strategies", mimeo, University of Alabama.

Arce M., D.G. (1993b), "Coalition dynamics, social norms, and the persistence of inflation", to appear in *Quarterly Review of Economics and Finance.*

Asheim, G.B. (1988), "Rawlsian inter-generational justice as a Markov-perfect equilibrium in a resource technology", *Review of Economic Studies* 55, 469–484.

Asheim, G.B. (1988), "Renegotiation-proofness in finite and infinite stage games through the theory of social situations", mimeo, Norwegian School of Economics and Business Administration.

Asheim, G.B. (1990), "Extending renegotiation-proofness to infinite horizon games", *Games and Economic Behavior* 3,278–294.

Asheim, G. B. (1991), " Individual and collective time consistency", mimeo, Norwegian School of Economics and Business Administration.

Asheim, G.B. (1992), "A unique solution to the n-person sequential bargaining game", *Games and Economic Behavior* 4, 169–181.

Asheim, G. B. (1993), " Defining rationalizability in 2-player extensive form fames", mimeo, University of Oslo.

Bernheim, D., and Whinston, M. (1987), "Coalition-proof Nash equilibria. II. Applications", *Journal of Economic Theory* 42, 13–29.

Bernheim, D., Peleg, B., and Whinston, M. (1987), "Coalition-proof Nash equilibria. I. Concepts", *Journal of Economic Theory* 42, 1–12.

Chakravorti, B., (1993) "Dynamic Public Goods Provision with Coalitional Manipulation", to appear in *Journal of Public Economics.*

Chakravorti, B., (1993) "Universal Coalition-proof Equilibrium: Concepts and Applications", mimeo, Bellcore.

Chwe, M. S. (1993), "Farsighted Coalitional Stability", to appear in *Journal of Economic Theory.*

Dutta, B., Ray, D., Sengupta, K., and Vohra, R. (1989), "A consistent bargaining set", *Journal of Economic Theory* 49, 93–112.

Einy, E., Holtzman, R., and Shitovitz, B. (1994), "Core and stable sets of large games arising in economies", to appear in *Journal of Economic Theory.*

Einy, E., and Shitovitz, B. (1994), "The optimistic stability of the core mapping in public goods production economies", to appear in *Economic Theory.*

Fudenberg, D. and Levine, D. (1993), "Self-confirming equilibrium", *Econometrica* 61, 523-545.

Greenberg, J. (1989a), "Deriving strong and coalition-proof Nash equilibria from an abstract system", *Journal of Economic Theory* 49, 195–202.

Greenberg, J. (1990), *The Theory of Social Situations: An Alternative Game Theoretic Approach.* Cambridge: Cambridge University Press.

Greenberg, J. (1992a), "On the sensitivity of von Neumann and Morgenstern abstract stable sets: The stable and individual stable bargaining set", *International Journal of Game Theory* 21, 41–55.

Greenberg, J. (1992b), "The stable value", mimeo, McGill University.

Greenberg, J. (1994a), "Social situations without commonality of beliefs — Worlds apart, but acting together", mimeo, McGill University.

Greenberg, J. (1994b), "The Nash equilibrium trap", mimeo, McGill University.

Greenberg, J. (1994c) "Implementing strong and lower strong positive association social choice rules by social procedures", in *Imperfections and Behavior in Economic Organization*, R. Gilles and P. Ruys, eds. Kluwer Academic Publishers.

Greenberg, J., Monderer, D., and Shitovitz, B. (1993), "Multistage situations", mimeo, McGill University.

Greenberg, J., and Shitovitz, B. (1994), "The optimistic stability of the core of mixed markets", *Journal of Mathematical Economics* 23, 379–386.

Khan, C. M., and Mookherjee, D. (1993), "Coalition-proof equilibrium in an adverse selection insurance economy", mimeo, University of Illinois.

Kissinger, H. (1982), *Years of Upheaval*, Little Brown and Company.

Mariotti, M., (1994) "A theory of agreements in normal form games", mimeo, University of London.

Mas-Colell, A. (1989), " An equivalent theorem for a bargaining set", *Journal of Mathematical Economics* 18, 129-138.

Muto, S., and Nakayama, M. (1991), "The resale-proof trade of information as a stable standard of behavior — An application of the theory of social situations", mimeo, Tohoku University.

Muto, S., and Okada, D. (1992), "von Neumann-Morgenstern Stable Sets in Cournot Competition", mimeo, Tohoku University.

Ray, D. (1989), "Credible coalitions and the core", *International Journal of Game Theory* 18, 185-187.

Ray, I. (1993), "Deriving correlated equilibrium as the stable standard of behavior of a social situation: A note", mimeo, CORE, Belgium.

Roemer, J. (1989), "Rawlsian justice as a stable standard of behavior", mimeo, University of California at Davis.

Shitovitz, B. (1994) "Optimistic stability in games of perfect information", to appear in *Mathematical Social Sciences.*

Shitovitz, B., and Weber, S. (1994), "Internal and external stability of the set of Lindahl equilibria", mimeo, SMU.

Tadelis, S. (1994), "Pareto optimality and optimistic stability in repeated extensive form games", mimeo, Harvard University.

van den Brink, R. (1992) "Hierarchical Organization", mimeo, Free University of Amsterdam.

Vannucci, S. (1992) "On stable standards of behavior and capacities of simple games", mimeo, Universita' di Siena.

Vannucci, S. (1993) "Generalized Situation and Admissible Implementation", mimeo, Universita' di Siena.

von Neumann, J., and Morgenstern, O. (1947), *Theory of Games and Economic Behavior.* Princeton NJ: Princeton University Press.

Winer, Z. (1989), M.Sc. thesis, University of Haifa, Israel.

Xue, L. (1993) "Farsighted optimistic and conservative coalitional stability", mimeo, McGill University.

PART C

DYNAMIC MODELS

Cooperation Through Repetition : Complete Information

Sylvain Sorin

Université Paris X-Nanterre and Ecole Normale Supérieure, Paris, France [1]

Keywords: repeated games, standard signalling, general signalling, folk theorems, discounted game, finitely repeated games, infinitely repeated games.

Introduction

Cooperative game theory deals with feasible outcomes, whereas non cooperative game theory is concerned with strategic equilibrium. Repeated games provide a bridge between these two theories: folk-theorem-like results deal with the relation between feasible payoffs in a one shot game and equilibrium payoffs in the corresponding repeated game.
Moreover, in repeated games new strategic objects appear (cooperative plans, threats...) that are closely related to cooperation. More precisely, through repetition all individually rational Pareto efficient allocations can be achieved as equilibrium payoffs.
For instance, the infinitely repeated prisoner's dilemma has equilibrium strategies that lead to the equilibrium path where both players behave friendly.
This lecture will be basically divided in two main sections: the first one focusing on the standard signalling case, where players know all past history, and the second one dealing with general signalling, where players get only private signals about the past. All the games we consider are with complete information.
This area is currently very active. We would like to refer to the following surveys for further results and points of view: [4], [9], [24], [26], [28], [33], [34].

1 Standard signalling

1.1 The model

We deal only with finite games.
$G = (N,\ (S^i)_{i \in N},\ g)$ is a normal form game defined by:
- N, a finite set of players ($N = \{1, ..., m\}$).
- For any i in N, S^i a finite set of pure moves. Let S denote $\prod_{i \in N} S^i$.

[1] Notes written by Dinah Rosenberg and revised by the author (December 1994).

- g, the payoff function : $g^i(s)$ is i's payoff if the players' independent moves result in the profile $s = (s^1, \ldots, s^m)$.
We extend the definition of g to mixed and correlated strategies. If σ is a probability distribution on S, $g(\sigma)$ is defined by: $g(\sigma) = \int_S g(s)d\sigma(s)$.

The repeated game associated to G is played in stages:

Stage 1: Each player i chooses a move s^i in S^i. $s = (s^1, \ldots, s^m)$ is announced.

Stage n: Let s_k stand for the profile of moves at stage k, $(s_k^1, \ldots, s_k^m)$. (Subscripts stand for time and superscripts for players). Denote by $h_n = (s_1, \ldots, s_{n-1})$ the history available at stage n. Knowing h_n each player i chooses s_n^i in S^i.
A play is an infinite history, we denote by H the corresponding set and by H_n the set of histories available at stage n.
Given a play $s = (s_j)_{j=1,2,\ldots}$, we have a well defined sequence of payoffs $g_j = g(s_j)_{j=1,2,\ldots}$.

Payoffs
We can define different games according to the way we evaluate the stream of payoffs:

a) the finite game G_n: the payoff is the arithmetic average of the payoffs of the n first stages: $\gamma_n = (g_1 + \ldots + g_n)/n$.

b) the discounted game G_λ: the payoff is the geometric average of the sequence of payoffs: $\sum_{m=1}^{\infty} \lambda$, for $\lambda \in (0, 1]$.
Since we are interested in long games, we will consider asymptotic properties: $n \to \infty$, $\lambda \to 0$.

c) the infinitely repeated game G_∞: the payoff is some limit of the sequence γ_n; one may choose liminf, or limsup, or some Banach limit. The choice of the limit we consider is arbitrary, so there is no well defined intrinsic payoff.

Strategies
We define the same set of strategies for the three games in order to compare them. In fact we consider that there is only one repeated game form, and that G_n, G_λ and G_∞ are three evaluations of outcomes in that game form.
A pure strategy for player i is a sequence $(\theta_1^i, \ldots, \theta_n^i, \ldots)$ with $\theta_n^i : H_n \to S^i$.
A mixed strategy is a probability distribution on pure strategies.
The games are with perfect recall, so we can use Kuhn's theorem (and its extension to the infinite case by Aumann), that allows us to consider only behavioral strategies, where a behavioral strategy for player i is a sequence $\sigma^i = (\sigma_1^i, \ldots, \sigma_n^i, \ldots)$, with σ_n^i being a function from H_n to $\Delta(S^i)$.(Δ(X) is the set of probability distributions over X, for any finite set X).

Remarks:
- Since strategies depend on histories, we have used the standard signalling hypothesis to define them.

- Altogether, G_n is a finite game, i.e. for any n the set of pure strategies is finite: there is no problem of existence of equilibria.
G_λ has compact sets of pure (hence mixed) strategies for each player, and continuous payoff functions: there is no problem of existence of equilibria either.
In order to avoid the problem of definition of the payoffs in G_∞, one will deal with *uniform equilibrium* in the following sense:

Definition: $x \in I\!R^n$ is a *uniform equilibrium payoff* if $\forall \epsilon > 0$, $\exists K$, $\exists \sigma$ such that $\forall n \geq K$ σ is an ϵ-equilibrium in G_n with payoff within ϵ of x.

We have to introduce a few more definitions and notations before we state the results.

Definition: σ is a *subgame perfect equilibrium* of the game Γ if it is a strategy profile such that for every history h, $\sigma[h]$ is an equilibrium of the subgame of Γ starting after h, where $\sigma[h]$ is defined for all history h' by $\sigma[h](h') = \sigma(hh')$, and hh' stands for h followed by h'.

Notations:
- D, the set of *feasible payoffs*, is the convex hull of $g(\mathrm{S})$.
- E_n (resp. E_λ, E_∞) is the set of *equilibrium payoffs* in G_n (resp. G_λ, G_∞).
- E'_n (resp. E'_λ, E'_∞) is the *set of subgame perfect equilibrium payoffs* in G_n (resp. G_λ, G_∞).
- The *individually rational level* for player i is defined by:
$v^i = \min_{\sigma^{-i}} \max_{\sigma^i} g_i(\sigma^{-i}, \sigma^i)$,
where the maximum and the minimum are taken respectively over the set of mixed strategies of player i and of his opponents.
A payoff x is *individually rational* if for all i in N, $x^i \geq v^i$.
$v = (v^1, \ldots, v^m)$ is the *threat point*. We denote by IR the set of *individually rational payoffs*.
- F is the set of individually rational payoffs that are in D.

Example 1: The battle of the sexes.
The game is represented by the following matrix:

$$\begin{pmatrix} (2,1) & (0,0) \\ (0,0) & (1,2) \end{pmatrix}$$

The threat point v is $(2/3,\ 2/3)$.
D is the set: $\{(x,y)\ /\ (2y-x) \geq 0,\ (y-2x) \leq 0,\ x+y-3 \leq 0\}$.
The following picture shows the point v, the sets D, IR, F.

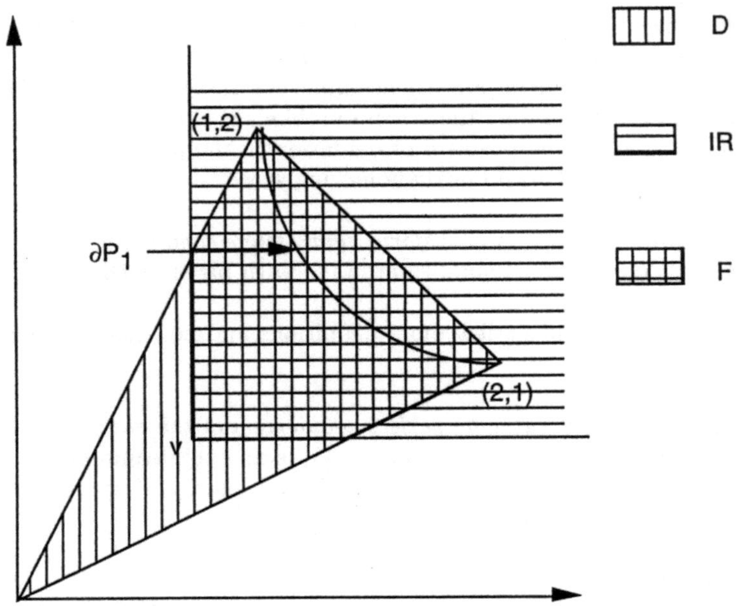

The set of payoffs achievable in G_1, P_1, is also represented, and one can see that it is strictly smaller than D.
In fact $P_1 = \{(x, y) \ / \ x^2 + y^2 - 2xy - 2x/3 - 2y/3 + 1 \geq 0\} \cap D$. It is easy to check that (3/2, 3/2) is not in P_1.

1.2 The Results

What kind of results are we looking for? A "folk theorem" like result is a statement of the kind "$\lim E_b^a = F$" with $b \in \{n, \lambda, \infty\}$, and with E^a standing for E or E', expressing the fact that asymptotically, all individually rational and feasible payoffs are sustainable by equilibrium strategies. This kind of results is quite striking because F depends only on the one shot game. In fact, both the set of feasible payoffs, D, and the individually rational level are parameters of the one shot game.

The meaning of this result in terms of cooperation is the following: as all points in F are achievable as equilibria, this theorem relates a cooperative notion (feasibility) to a non cooperative one. The theorem implies in particular that all Pareto optimal payoffs in IR are achievable as equilibrium payoffs, hence that there are Pareto optimal equilibrium payoffs.
The main results are summed up in the following table; each row concerns the folk theorem in one case (G_n, G_λ, or G_∞), one column concerns the theorems of the form $E = F$, and the other one the theorems of the form $E' = F$. If the result is true under some condition, it is mentioned in the corresponding cell.

	Nash Equilibrium	Subgame perfect equilibrium
G_∞	folk theorem (~1960)	Aumann and Shapley [5] Rubinstein [27]
G_λ	Sorin [32] under condition A	Fudenberg and Maskin [13] under condition B
G_n	Benoît and Krishna [7] under condition C	Benoît and Krishna [6] (pure strategies) Gossner [14] (mixed strategies) under condition D

The conditions are the following:
Condition A: *there is an element* f *in* F *such that for all* i *in* N, $f^i > v^i$.
Condition B: F *has a non empty interior.*
Condition C: *For all* i *in* N *there exists* $e(i)$ *in* E_1 *such that* $e(i)^i > v^i$.
Condition D: *for all* i *in* N *there exists* $e(i)$ *and* $f(i)$ *in* E_1 *such that* $e(i)^i > f(i)^i$, *and* F *has a non empty interior.*

As E' is included in E, condition B is obviously stronger than condition A, and condition D than condition C.
Moreover it is clear that condition D is stronger than condition B and condition C than condition A (because any Nash equilibrium belongs to F). This was not obvious a priori.

In the following we will try to give the main ideas involved in the theorems and the sketch of some proofs.
In all the proofs there are two inclusions. One of them is easy :

Lemma 1.1: *If* a *stands for* ∞, n, *or* λ,
E'_a *is included in* E_a.
E_a *is included in* F.

Proof:
Each subgame perfect equilibrium is a Nash equilibrium, which proves the first statement.

The random one stage payoff takes its value in $g(S)$ which is included in D and hence, as D is convex and closed, expectation, average, and limits are in D also. So E_a is included in D.

To show that any equilibrium is in IR, we use the standard signalling hypothesis.
If x is such that $x^i < v^i$, we construct a deviation for player i from a strategy σ^i (the profile σ leading to the payoff x), that gives him at least v^i at each stage. This will contradict the fact that x may be an equilibrium payoff.
If $s^{-i}(1)$ is the vector of mixed moves of the players other than i at stage one, by definition of v^i, i has a best reply, $s^i(1)$ which satisfies: $g_i(s^i(1), s^{-i}(1)) \geq v^i$. At each stage, i knows the previous history (here we use standard signalling). So, knowing the strategy of the other players, i can deduce their mixed moves at that stage: in fact the other players' strategies take into account their information, say h_n at stage n, to choose their moves, $\sigma_n^{-i}(h_n)$; as i knows this information, he can compute the moves if he knows the strategies. So he can play a one stage best reply that gives him at least v^i. ∎

To show the reverse inclusion, two main ideas are used: plan and threat.
A *plan* is a play associated to f in F, i.e. that leads to this payoff. It is seen as a cooperative device, from which players can eventually deviate. Note that the play is a sequence of *pure* moves so that potential deviations are observable under standard signalling.
A *threat* is a profile of strategies that gives to one player a "bad" payoff, whatever he does. The maximal threat against player i leads to a payoff of at most v^i for him.
To prove that a payoff is an equilibrium payoff, it will be enough to find a plan that gives this payoff, and a threat that is sufficient to prevent deviation from this plan. It is conceived as a punishment that is applied to a deviating player.
In the case of subgame perfect equilibria, the threat must be an equilibrium strategy in the subgame starting after the deviation. Hence it is more difficult in that case to find good threats. You have to reward a non deviating player who has punished a deviator.
In discounted games, the present has more impact than the future, so some threats may not be effective. Punishment is more difficult, but as $\lambda \to 0$, it becomes possible to sustain as equilibrium payoffs all vectors in F. As time counts, postponing threats and rewards is more difficult compared to the undiscounted case.
In finite games, some backward induction effects may appear. This means that at the last stage, the players play a one shot game. This is due to the fact that the end of the game is the same for everybody, and that at that stage it is public knowledge that it is the end of the game. This effect would not appear if the players did not know the last stage of the game or if the end of the game was not the same for everybody, see the chapter of professor Neyman about "Cooperation through repetition" in these proceedings.

We will give a more detailled proof of $E'_\infty = F$, and some ideas of how conditions A, B, C, D are used to have the corresponding results.

Result 1: $E_\infty = F$ is the so-called "Folk Theorem".
It is obviously implied by the following:

Result 2: $E'_\infty = F$.(see Aumann and Shapley [5], and Rubinstein [27])

This result was proved in the 70's, for more details, see Meggido, [23] p.0.

Sketch of the proof of $F \subset E'_\infty$:

Let f be in F, we want to show that f is a subgame perfect equilibrium payoff. f is a convex combination of $m+1$ elements of $g(S)$ (by Caratheodory's theorem), let us write $f = \sum_\alpha \mu_\alpha f_\alpha$, with $\mu = (\mu_\alpha)$ in the simplex of $\mathbb{R}^{m+1}$ and f_α in $g(S)$.

If the coefficients are rational, there is a plan p where the players play the pure moves leading to f_α with frequency μ_α, that gives payoff f. If people are supposed to follow such a plan, they should play pure moves at each stage, and so, deviations are observable. One can find such a play in a uniform way, i.e. such that after any history, the frequency of each move in the future remains the same.
Let the strategies be the following : play as indicated by the plan p, and if you see a deviation by player i punish him, during a determined finite number of stages, such that any one stage gain by deviating at stage n is inferior to the loss due to the punishment up to $1/n$. The payoff of a deviating player being the limit of the average payoff over n stages, for a deviation to be profitable, a player has to deviate at infinitely many stages. There is no gain by deviating if the gain per deviation becomes smaller and smaller when the deviation is made later. Finally a deviation during the punishment phase is ignored, but it is never profitable since it lasts for finitely many stages.
If many players deviate choose the first one in some order, and punish him.
So, there is a plan giving f as its payoff, and a threat that makes any deviation unprofitable. As the punishment is of finite length, applying the threat is an equilibrium strategy in the subgame starting after a deviation (it doesn't affect the punisher's payoff). Hence f is a subgame perfect equilibrium payoff.

If the coefficients are not rational, approximate them by rationals: if $f = \sum_\alpha \mu_\alpha f_\alpha$, where some μ_α are irrational, define a sequence of rationals $\mu^n = (\mu^n_\alpha)$ converging to μ. The sequence $(f^n = \sum_\alpha \mu^n_\alpha f_\alpha)_n$ converges to f. For any n let q^n be the common denominator of the μ^n_α. Define the following plan: play so as to get f_α with frequency (μ^1_α) during q^1 stages, then f_α with frequency (μ^2_α) during q^2 stages and so on... The threat is the same, these strategies lead to a subgame perfect equilibrium. ■

We will now comment on the conditions A, B, C, D. We are going to sketch the proofs showing why the folk theorems are true under these conditions and to explain how they are used.
The convergence results are always considered to be under the Haussdorff topology. We want to prove that any vector in F is the limit as $\lambda \to 0$ or $n \to \infty$ of vectors in E_b^a (with the notation we already used above for a and b).

Result 3: $\lim_{\lambda \to 0} E_\lambda = F$. (see Sorin [32])

Condition A: *there is an element f in F such that for all i, $f^i > v^i$.*
This condition implies that for any f in F there is an f' in F close to f such that $f'^i > v^i$ for all i.
So one can assume that one starts with an f in F such that for all i, $f^i > v^i$.
Let us choose a plan that achieves f as a payoff in E_λ for λ small enough, in such a way that after any history, the future payoff is within $\epsilon/2$ of f. (The future payoff may not be equal to f, but for λ small enough it can be as close as possible to f). The strategies are to follow this plan and to punish a deviator i by decreasing his payoff to the level v^i. If $\min_{i \in N}(f^i - v^i)$ is bigger than ϵ, then one can punish any deviation by decreasing the future payoff of the deviating player by at least $(1-\lambda)\epsilon/2$. As soon as $\lambda a < (1-\lambda)\epsilon/2$ with a being the maximum one shot gain from deviating, this threat is enough to prevent deviation, because it makes it unprofitable.
So, there is a λ^* such that for all $\lambda \leq \lambda^*$, f is an equilibrium payoff of E_λ, hence the result. ■

An example due to Forges, Mertens, and Neyman ([8]) showing that the folk theorem may not be true if condition A is not satisfied, is of the following kind: Consider a three player game, where player 3 is a dummy and players 1 and 2 are active players; the payoffs are given by the following matrix:

$$\begin{pmatrix} (1,0,0) & (0,1,0) \\ (0,1,0) & (1,0,1) \end{pmatrix}$$

There is a unique equilibrium payoff of the one shot game: (1/2, 1/2, 1/4) (players 1 and 2 both playing (1/2, 1/2)).
The point (1/2, 1/2, 1/2) is in F. But on any path that leads to this payoff, one of the active players has a profitable one shot deviation (a one shot deviation is valuable because the future is discounted), and that no threat can prevent him from deviating because the maximum threat gives him 1/2, which is the original equilibrium payoff. In fact, E_λ is reduced to (1/2, 1/2, 1/4), and (1/2, 1/2, 1/2) cannot belong to the limit of E_λ .

Result 4: $\lim_{\lambda \to 0} E'_\lambda = F$. (see Fudenberg and Maskin [13])

Condition B: *F has a nonempty interior.*
Condition B tells us that for all f and all δ, there is an element f' of F such that $\|f - f'\| < \delta$, and f' is in the interior of F. So it is enough to prove the result for f in the interior of F.
The idea is that it may be costly to punish somebody, so you must get a reward if you do so, because we are looking for perfect equilibria. But the reward the punisher gets must not also be a reward for the deviator. That is why a dimension condition is needed.
Moreover, the players must not be tempted to deviate during the punishment phase. But during that phase, the players play mixed moves, hence deviations may not be detectable: the other players don't see the randomizing distribution. In order to prevent deviation in that phase, you have to manage to give the punishing player the same payoff whatever moves he chooses during the punishment phase (at least in the support of the punishing strategy); hence he has no incentive to deviate. This is possible because the set of feasible payoffs is convex in the discounted game for λ small, it would not be possible in a finitely repeated game.

Definition: Let f be a vector in F and s a plan inducing f as its payoff, let h be any finite truncation of s of length r, then the *continuation payoff* after h leading to f is a vector $c(h)$ in $I\!R^m$ such that

$$f = \lambda \sum_{k=1}^{r} (1-\lambda)^{k-1} g(s_k) + (1-\lambda)^r c(h).$$

Note that the continuation payoff after a history is in general different from f. For any f in F there is a plan s that leads to payoff f, for λ small enough. Choose a smooth plan in the following sense: given any finite history, the continuation payoff has to be close to f. We are going to use the following lemma:

Lemma 1.2: *For any f in F and positive δ, there exists a λ_1 such that for all $\lambda \leq \lambda_1$, there is a plan inducing f as its payoff, such that the continuation payoff after any history h is at a distance of at most $\delta/4$ of f.*
Let us assume that f is at a distance δ of the boundary of F and try to play the plan defined in the lemma.

Remark: If you want to achieve a payoff f as subgame perfect equilibrium payoff, you have to prove that after any history, there is no profitable deviation. The lemma says that after any history, the continuation payoff for player i is at least $f^i - \delta/4$. Hence you have to prove that there is a threat that leads to a payoff of less than $f^i - \delta/4$. Then you can punish a deviating player in the following way: first you give him his individually rational level for a sufficiently long time, in order to cancel his gain from deviating, and then you switch to a plan that gives him less than $f^i - \delta/4$.

If there is a deviation by a player i, the other players punish him. We focus on punishments of the following kind: first the deviator's gain by deviating is compensated, and then a new continuation payoff is fixed.
More precisely, after a deviation of player i, the other players punish him (at his individually rational level) during R periods. R is a fixed integer such that any one shot deviation of any player j is made unprofitable: by getting v^j for R periods, the average for these $R+1$ stages will be at most $v^j + \delta/2$ for any λ smaller than some λ_2. If the strategies specify the moves after these R periods such that the continuation payoff of the deviator i is less than or equal to the payoff under the initial plan, this one being at least $v^j + \delta/2$, then a deviation is unprofitable.

In order to get a subgame perfect equilibrium, one has to check that a punisher has no incentive not to punish, and that there is no profitable deviation for any player after these R periods.
For every history of length R, h, and for all i, define the vector payoff $d[i](h)$ such that playing h during R periods and then receiving $d[i](h)$ in the future gives, as a whole payoff $v^j + \delta$ to player $j \neq i$, and $d[i](h)^i = v^i + \delta/2$ to player i. Note that for λ small enough, one has also $d[i](h)^j \geq v^j + \delta - \delta/4$. If $d[i](h)$ can be implemented as a perfect equilibrium, no one has an incentive not to punish a deviator since whatever he does during the R periods of punishment he gets the same payoff. So it remains to prove that $d[i](h)$ can be achieved as a subgame perfect equilibrium payoff for any h.

Let $s[i](h)$ be a plan inducing the payoff $d[i](h)$ and satisfying Lemma 1.2. In particular we want to prevent a new deviation of the same player i, after a history hh' (histories begin here after the first deviation). The problem is the following: after h the future payoff of i is $v^i+\delta/2$, and after hh' it may be $v^i+\delta/4$ because of the remark we made above. We hence have to justify $v^i + \delta/4$ as a perfect equilibrium payoff. But after some history, according to the same remark the future payoff of i may be v^i. After this history, no threat can prevent i from deviating. That is why one has to achieve $v^i + \delta/2$ through a plan that gives to i his "bad"payoffs first, i.e. $s[i](h)$ is such that the continuation payoff for player i is always bigger than $v^i + \delta/2$. Such a plan exists.
We can now check that no deviation is profitable for any player at any time:
- We have already seen that a deviation of player i is profitable, only if he deviates also after h. If not, his one shot gain by deviating is compensated during the first R stages after the deviation, and the continuation payoff afterwards is $v^i + \delta/2 < f^i - \delta/4$.
Let us assume now that a deviation of player i has already occurred.
In fact, we forget the plan leading to the payoff f, we switch to a situation where the players are supposed to follow the plan $s[i](h)$.

- During the first R periods after the deviation, i plays a best reply and gets at most v^i, hence there is no profitable deviation for him.
- If i deviates after hh', the other players will switch to the plan $s[i](h)$. If he had followed the plan, his continuation payoff would have been greater than $v^i + \delta/2$. After the deviation, his continuation payoff is $v^i + \delta/2$. The threat is effective.
- A player $j \neq i$ has no profitable deviation during the first R stages after i's deviation, because whatever he does, his overall payoff is $v^j + \delta$.
- If player j deviates after hh', the players punish him for R periods and then they switch to the plan $s[j](h')$. His continuation payoff is then $v^j + \delta/2$, and this threat is effective because if he did not deviate his continuation payoff would be at least $d[i](h)^j - \delta/4$, which is greater or equal than $v^j + \delta \ - \delta/4 \ - \delta/4$. ∎

Remark: If there is a public extensive form correlation device, i.e. if the players receive a public signal before each stage of the game, then, they can achieve f through a plan such that for all history h, $c(h) = f$, hence the previous construction is much easier.

There is an example of a game where this condition is not satisfied in [13], and the result is false, so a condition is indeed necessary.
The example is the following game G:
Player 1 chooses a row, player 2 a column and player 3 a matrix. The payoffs are given by the following matrices:

$$\begin{pmatrix} (1,1,1) & (0,0,0) \\ (0,0,0) & (0,0,0) \end{pmatrix} \quad \begin{pmatrix} (0,0,0) & (0,0,0) \\ (0,0,0) & (1,1,1) \end{pmatrix}$$

The threat point is (0,0,0). The players always have the same payoffs. So one can define w as the minimal payoff achievable through a subgame perfect equilibrium in G_λ.
Let σ be a subgame perfect equilibrium. It is easy to see that at least one player can (eventually by deviating) get 1/4 at stage one. Since his future payoff is at least w, this implies that w is bigger than 1/4.
Hence (0,0,0) cannot be approached by E'_λ.

Result 5: $\lim_{n\to\infty} E_n = F$. (see Benoît and Krishna [7])

Condition C: *for all i there exists $e(i)$ in E_1 with $e(i)^i > v^i$.*
Some condition is needed: there are games where some feasible individually rational payoffs cannot be achieved as equilibrium payoffs. In fact, in the following prisoner's dilemma, $E_n = \{(1,1)\}$ for all n so the result does not hold.

Example 2: The prisoner's dilemma is the two player game defined by:

$$\begin{pmatrix} (3,3) & (0,4) \\ (0,4) & (1,1) \end{pmatrix}$$

This above property is due to the fact that E_1 is reduced to the threat point. More precisely we have the general result [32]:

Lemma 1.3: *If* $E_1 = \{v\}$, *then* $E_n = \{v\}$ *for all* n.
Proof: As the game is finite, at stage n each player knows it is the last stage. So on the equilibrium path, at that stage the payoff will be in E_1, hence it will be v. Let us consider the longest history h (let p denote its length) of positive probability on the equilibrium path after which the one stage payoff is not in E_1. So one player, say i, has a one shot profitable deviation after h. Hence, since we are on an equilibrium path, there is a punishment after stage $p+1$, that prevents player i from deviating. But, the payoff after stage $p+1$ on the equilibrium path is already the worst possible fot him: there can be no threat. Hence the payoff will be v at any stage along the equilibrium path. ■

Notice that the failure of the folk theorem in this case is not due to the existence of a unique equilibrium in strictly dominating strategies. In the following game there is a unique Nash equilibrium in strictly dominating strategies, but condition C holds and the result is true.

$$\begin{pmatrix} (3,1) & (0,0) \\ (4,2) & (1,0) \end{pmatrix}$$

In this case the sequence of moves (Top, Left) (Bottom, Left) is an equilibrium path in G_2. The strategies of player 1 (resp. 2) are the following: play as recommended above, and if there is a deviation at stage 1, player 2 plays Right. Hence one can exhibit in this case an equilibrium payoff in E_2, $(\frac{3+4}{2}, \frac{1+2}{2})$ that is different from the payoff in E_1, (4,2), in spite of the existence of a unique Nash equilibrium in strictly dominating strategies.
So one needs a condition to avoid that $E_n = \{v\}$, and we will show how to get the result under the sufficient condition C.

The spirit of the proof is the following:
- find a play of length T that leads to a payoff that approaches f in F within ϵ.
- the strategies are:
first follow the plan that consists in playing the above play cyclically up to Rm stages before the end of the game. (Recall that m is the number of players). Then, play R cycles leading to $(e(1), \ldots, e(m))$. This is crucial: playing equilibrium strategies at the end of the game excludes the possibility of profitable deviations at the end. It avoids the "backward induction" effect we have seen in the prisoner's dilemma case.

If there is a deviation of player i during the first phase of the strategy, switch to strategies punishing him to v^i for the remaining stages. The gain of deviating will never exceed $R(e(i) - v^i)$ if R is large enough. It means that if a deviation from the equilibrium strategies occurs far enough from the end of the game, you have time to punish a deviator.
R is a constant, so as n goes to infinity, the presence of these last stages payoffs does not influence the limit payoff. ∎

Result 6: $\lim_{n\to\infty} E'_n = F$. (see Benoît and Krishna [6] and Gossner [14])

Condition D: *for all i there exist $e(i)$ and $f(i)$ in E_1 such that $e(i)^i > f(i)^i$, and F has a non empty interior.*
The proof uses both the ideas of the proof of Result 5 (add terminal stages where you play an equilibrium strategy, to avoid backward induction effects), and the ideas of the proof of Result 4 (rewarding the punishers, that is why we need the interior of F not to be empty).
At the end of the game play cycles of $(e(1), \ldots, e(m))$; in that part of the game there will be no profitable deviations. Before, play cycles of a path that leads to a payoff close to f. If there is a deviation punish during a finite number of stages, and then reward the punishers (as in the theorem about E'_λ). If the deviation occurs too late to do all this, then, at the end of the game switch from cycles leading to $(e(1), \ldots, e(m))$, to cycles leading to $(e(1), \ldots, e(i-1), f(i), e(i+1), \ldots, e(m))$ if i was the deviator. If the lengths of the cycles are chosen appropriately this is a perfect equilibrium.
The above procedure works if you can identify the deviator, i.e. if you restrict yourself to pure strategies (even in the punishment phase). If not, the proof is much more intricate, and requires a statistical test and a family of late adapted payoffs (see [14]). ∎

Comments

a) In two player zero sum games, D is a segment and F is reduced to a single point, $(\bar{v}, -\bar{v})$, where $\bar{v}$ is the value of the game.

b) For the two player case, the results hold under weaker conditions. The reason for this is that both players can minmax each other at the same time (see [13], [6], [7]).

c) It is clear that if the players could correlate their moves, they could not achieve more payoffs since the set of correlated payoffs is the convex hull of the set of feasible payoffs. However, if the players could correlate secretely to punish somebody, the individually rational level could change, but this is not the case if the correlation signal is public.

d) If only pure strategies are allowed, v^i is higher, so D may be smaller, even empty. It is the case in the following zero sum game:

$$\begin{pmatrix} 1 & 0 \\ 0 & 1 \end{pmatrix}$$

With mixed strategies, $v = (1/2, -1/2)$, $D = \{(x, -x)\ ,\quad x \in [-1, 1]\ \}$, hence $F\ =\ \{\ v\ \}$.
With pure strategies, D is the same but $v = (1, 0)$, hence $F\ =\ \emptyset$.

e) If mixed moves are observable, and not only their realizations, the game is similar to a game with continuous strategy sets, and pure observable strategies. It is no longer a finite game: the set of strategies of i are $\Delta(S^i)$ and the histories are $h_n = (s_1, \ldots, s_n)$, with $s_k \in \Delta(S^1) \times \ldots \times \Delta(S^m)$.

f) If the players have different discount factors, we may get feasible payoffs and even equilibrium payoffs outside D. For example, in the battle of sexes (cf. Example 1), if player 1 is much more patient than player 2, it may be an equilibrium strategy to play first (Bottom, Right), and next (Top, Left). So we may have an equilibrium payoff close to (2,2) in the repeated game with different discount factors for the two players.

g) In overlapping generation models people play during a finite number K of stages. Generations follow each other; each generation lives for K periods, and is then replaced by a new generation, starting their lives.
In this case, a folk theorem holds with simpler conditions (see [15], [16], [30], [31]).
For instance, assume the people live two periods, one in which they are young, and one in which they are old. Assume that at each period one young and one old are faced and that the game is a prisoner's dilemma. The following strategies are in equilibrium: play cooperatively when you are young and non cooperatively when you are old; if a young person deviates, he will be punished when old by a new young.

h) Another question is whether this result can be extended to extensive form games. In extensive form games, the information of the players at a terminal node is not precisely defined. But one usually assumes that people know the node they have reached. Take the following game with player 1 and 2:

Example 3:

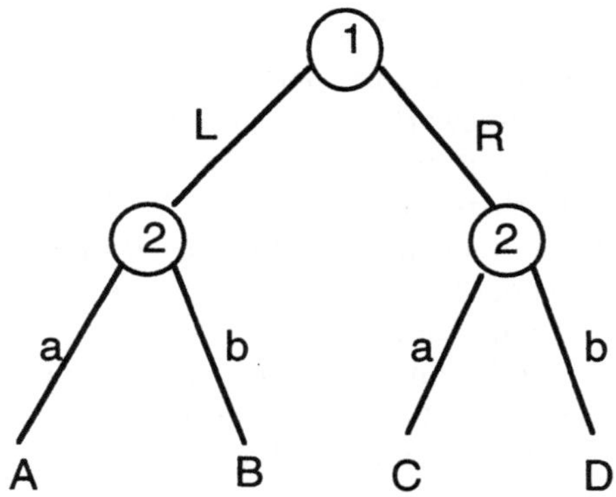

In the normal form of this game, the set of strategies of player 1 is $\{L, R\}$, and the set of strategies of player 2 is $\{aa, ab, ba, bb\}$. In the extensive form repeated game, if player 1 played L in one period, and if he saw A, he knows that the strategy of player 2 was aa or ab, but he does not know which of them. In the normal form repeated game under standard signalling, he can distinguish aa from ab. Hence repeating an extensive form game is equivalent to repeating a normal form game with non standard signalling (see [35]).

i) Another point of view would be the following: instead of looking for conditions under which, asymptotically, the set of equilibrium payoffs is equal to the set of feasible individually rational payoffs, one can try to characterize the set of equilibrium payoffs in general. This problem is studied in a paper by Wen [36].

2 General signalling

We are going to consider here the same kind of models as in the previous section, but now the players only get signals about the past. They do not know the whole history. In many cases this kind of models is more realistic than the previous one, but there is no straightforward generalization of the results of the previous section. The general signalling framework is much more difficult to study.

2.1 The Model

The model is basically the same as in the previous section, except that the players are no longer informed of the moves and only get a private signal on them.

More precisely, A^i is the finite set of private signals for player i and h^i is the signalling function of player i; it is a map from S to A^i. More generally, one can define a function h from S to $\Delta(A^1 \times \ldots \times A^m)$; in that case the law of the signal received by player i is the i'th component of $h(s)$; this allows for random signals. This more general framework can be useful when dealing with moral hazard situations. But here we will restrict ourselves to the case of deterministic signals i.e. to the case where there exist functions $h^i : S \to A^i$.
As before, the repeated game associated to G is played in stages :

Stage 1: each player i chooses a move s^i in S^i. $s = (s^1, \ldots, s^m)$ is not announced. Each player i is now informed of $h^i(s) = a^i$.

Stage n: if $s_k = (s_k^1, \ldots, s_k^m)$, and $a_k = (a_k^1, \ldots, a_k^m)$ denote the profile of moves and signals at stage k, then at stage n, knowing $k_n^i = (a_1^i, \ldots, a_{n-1}^i)$ each player i chooses s_n^i in S^i, and is informed of $h^i(s_n) = a_n^i$.

So the players know only a function of the previous moves, for instance, in a Cournot framework, they observe the realized price and not the vector of quantities.

In the extensive form game presented in example 3, $h^1(K,\ xy) = x$ (K stands for L or R and x and y for a or b).

In this framework, standard signalling corresponds to the case where h^i is the identity function: the signal each agent gets is the profile of moves.

In the two player finite case, one represent each function h^i by a matrix. For example, if there are two strategies and two players, let h^1 be represented by :

$$\begin{pmatrix} a & a \\ a & b \end{pmatrix}$$

This means that if 1 plays top, he cannot observe if 2 played left or right, whereas he can observe it if he plays bottom.

Note that we always assume that each player knows the moves he played and recalls all his previous information. Hence we usually define through signalling matrices only the incremental information at each stage, for each player, in addition to his own move.

Remark: public information (all the players have the same information) and perfect recall (each player always remembers all he did and all he knew) imply standard signalling.

Consequences

You cannot rely on histories, and you may not be able to check deviations. So it may be more difficult to define a cooperative plan.

Moreover, it may be the case that some players get signals allowing them to correlate their moves, without another player, say i, knowing about it. In that case, to evalutate the individually rational level for player i, one must take the minimum over $\Delta(\prod_{j \neq i} S^j)$ and not over $\prod_{j \neq i} \Delta(S^j)$, i.e. player i must take into account the fact that his opponents may correlate their moves. So the threat point may be different from the previous value v.

Moreover, the individually rational value may change during the game. In fact the correlation between the players depends only on the signals they get hence on the actions taken.
This general problem is an open question. Remark nevertheless that in the two player case v can be defined as before because there is no problem of correlation.

2.2 Recursive Structure and Public Signals

In this section we concentrate on discounted games, subgame perfect equilibria, and public signals. Some more results will first be needed in order to characterize the set of subgame perfect equilibria in the discounted game with public signals. The tools we introduce here are inspired from the work of Abreu [1], Abreu, Pearce and Stacchetti [2] and Mertens [24].

2.2.1 Optimality Principle

In this paragraph we will moreover restrict ourselves to standard signalling. We show a few useful results that we will extend later to public equilibria of games with general signalling.
The first result will be used to characterize the set E'_λ.

Definition: A *one shot deviation* of player i from a behavioral strategy t is a behavioral strategy s such that:
- there is a history h after which the mixed move induced by s differs from the one induced by t.
- after any other history, s and t induce the same mixed moves.

Lemma 2.1: (*Optimality Principle*) *In a repeated game with continuous payoffs, a strategy profile σ is a subgame perfect equilibrium iff there is no one shot profitable deviation, for any player i.*

Sketch of the proof :
The condition is obviously necessary.
Let σ be a profile of strategies, that is not a subgame perfect equilibrium. Hence there is a player, i and a pure strategy τ^i that is a profitable deviation for i from σ^i after some history h.
Let $\theta(n)$ be the following strategy: play σ as long as h has not occurred; when h has occurred play τ^i for n stages and then return to play σ^i. As the payoff is continuous, there is an integer $N^\star$ such that for all $n \geq N^\star$, $\theta(n)$ is a profitable deviation from σ^i.
Consider the smallest n, say $\bar{N}$ such that $\theta(n)$ is a profitable deviation from σ^i. Thus $\theta(\bar{N})$ is a profitable deviation from σ^i, and $\theta(\bar{N}-1)$ is not. Hence there

is a history h' such that playing $\theta(\bar{N}-1)$ as long as hh' has not occurred, and after history hh', playing τ^i instead of σ^i once, is a profitable deviation from $\theta(\bar{N}-1)$.
Consider now the following strategy: play σ^i as long as hh' has not occurred; after hh' play τ^i for one period and then play σ^i. It is a one stage profitable deviation from σ^i. ∎

This lemma will help us to characterize E'_λ in the standard signalling case. The idea is the following: in a discounted game, the game starting today and the game starting tomorrow are of the same structure. There is a kind of stationarity along the play. Denote by f tomorrow's evaluation of the future payoff, as a function of today's moves. The overall payoff is $E_\sigma[\lambda g(s_1)+(1-\lambda)f(s_1)]$, where s_1 is the first stage profile of moves, and σ is the profile of strategies. If σ is a subgame perfect equilibrium, the range of f is included in the set of subgame perfect equilibrium payoffs (even for s_1 outside the equilibrium path, because σ is a subgame perfect equilibrium).

Notation: Let $F(W)$ be the set of functions $u: S \to W$, where W is a fixed bounded subset of $\mathbb{R}^m$. For $f \in F(W)$, let $G(f,\lambda)$ be the one shot game with set of players N, pure moves S^i and payoff function $\lambda g + (1-\lambda)f$. We want to characterize E'_λ using $E_f(\lambda)$, the set of equilibrium payoffs of $G(f,\lambda)$.
Let T_λ be the following operator : $T_\lambda(W) = \bigcup_{f\in F(W)} E_f(\lambda)$.

Theorem 2.2: *E'_λ is the largest fixed point of T_λ.*

Sketch of the proof:
We show first that E'_λ is a fixed point of T_λ, and then that any fixed point of T_λ is included in E'_λ.

1- E'_λ is a fixed point of T_λ.
Let σ be a subgame perfect equilibrium of G_λ, and u the associated payoff; denote by $f(s_1)$ the payoff for the future after history s_1. As the game is discounted, and as the equilibrium is perfect, $f(s)$ is in E'_λ for all s in S. By definition of σ, σ_1 (the mixed move induced by σ at stage 1) is a Nash equilibrium of $G(f,\lambda)$, and it leads to the payoff u. Thus u is in $E_f(\lambda)$. Hence for all u in E'_λ, there is a f in $F(E'_\lambda)$ such that u is in $E_f(\lambda)$: $E'_\lambda \subset T_\lambda(E'_\lambda)$.
Let now u be in $T_\lambda(E'_\lambda)$. There is an f in $F(E'_\lambda)$, such that u is in $E_f(\lambda)$. As f is in $F(E'_\lambda)$, there is a subgame perfect equilibrium $\tau(s)$ leading to payoff $f(s)$ for all s in S; let σ_1 be the equilibrium strategy in $G(f,\lambda)$, that leads to u. Clearly, playing $(\sigma_1, \tau(s))$ is a subgame perfect equilibrium of G_λ that leads to payoff u; so, u is in E'_λ, and $T_\lambda(E'_\lambda) \subset E'_\lambda$.
Hence, E'_λ is a fixed point of T_λ.

2- If a set A is included in $T_\lambda(A)$, A is included in E'_λ.
If f_1 is in A, there exists an f_2 in $F(A)$ such that f_1 is in $E_{f_2}(\lambda)$, with the strategy profile σ_1. $f_2(s)$ is in A, for all s, so we can define $f_3(s)$, and $\sigma_2(s)$, and so on....We thus define a sequence of strategies and payoffs. We now want to show that $(\sigma_1, \ldots, \sigma_n, \ldots)$ is a subgame perfect equilibrium (σ_n is a function of all the previous history). As the game is discounted, the payoff is continuous, so the optimality principle applies. It is clear that σ induces the payoff f_1, hence it is enough to show that there is no one shot profitable deviation. But this follows from the definition of σ_n since $(\sigma_{n+1}, \cdots)$ induces the payoff f_{n+1}. So f_1 is in E'_λ, and hence, $A \subset E'_\lambda$. ∎

Comments:
- The result is not true for Nash equilibrium payoffs; in fact, there is no optimality principle for Nash equilibria.
- The fact that the game is discounted is also crucial because it implies continuity of the payoffs (and enables to use the optimality principle).
- It is also important because it leads to stationarity: the game starting tomorrow is the same as today's game (which is not true in finite games or in games with time dependent discount factors).

We now want to extend this study to the case of non standard signalling, more precisely, to the case of public equilibria of games with signals.

2.2.2 Application to Public Signals

Consider the component of the signals that is public and denote by Y the corresponding set of *public signals.* It can be defined through the finest σ-algebra that is included the information of any of the players. This means that a signal is in Y if it is the maximum signal one can give to all the players, in addition to their own signal, without changing their information. Denote by ϕ the function from S to Y that defines for each profile of moves the public component of the signal.
Consider the following example: player 1 chooses a row and player 2 chooses a column. In the matrix the letters a, b, c, d represent outcomes.

$$\begin{pmatrix} a & b \\ c & d^\star \end{pmatrix}$$

Each player is informed only of his own choice if the outcome is a, b, or c, and he knows his action and $\star$ if the outcome is d. Hence this game is a game with private signals, and the public component of the signal is $\star$ or not $\star$.
The previous result extends to public equilibria of the game with non standard signalling, in the following sense.

Definition: A *public strategy* for player i is a strategy σ^i that depends only on the public part of the signals.
A *public equilibrium* is a public strategy profile such that no public strategy can be a unilateral profitable deviation.

Lemma 2.3: *A public equilibrium is an equilibrium, i.e. if everyone uses public strategies, the best reponse is a public strategy.*
Proof: To find a best response to some strategies s^{-i}, one can look only for strategies that are measurable with respect to the finest σ-algebra generated by s^{-i}. (The argument is the same as in Blackwell's Theorem in dynamic programming.) ∎

Notations: $F^\star(W)$ is now the set of functions $f : Y \to W$, where W is a fixed bounded subset of $I\!R^m$. Let $G^\star(f, \lambda)$ be the one shot game with the same sets of players and moves, and payoffs $\lambda g + (1 - \lambda) f \circ \phi$; $E_f^\star(\lambda)$ is its set of Nash equilibrium payoffs. $E_\lambda'^\star$ is the set of *public* perfect equilibrium payoffs of G_λ. Let $T_\lambda^\star$ be the following operator: $T_\lambda^\star(W) = \cup_{f \in F^\star(W)} E_f^\star(\lambda)$.

Theorem 2.4: *$E_\lambda'^\star$ is the largest fixed point of $T_\lambda^\star$*
Hence this result gives a characterization of all subgame perfect public equilibria in the discounted game. It is not true for all equilibria, the word public is crucial here.
Idea of the proof:
If one restricts oneself to public histories, i.e. to the sequence of the public component of the signal at each stage, the other parts are not used, so one can forget them. One can easily check that the same proof as in theorem 2.2 holds. ∎

The previous result is true for all λ. We would like to get more insight of what happens when λ is close to 0, as we do in the folk theorem.

2.2.3 Asymptotic Results

The following results have been obtained in a series of papers by Fudenberg, Levine and Maskin (see [11] and [12]).
The idea is to study the geometry of $E_\lambda'^\star$ as λ goes to 0. Recall that if f is in $E_\lambda'^\star$, there is a one stage payoff g, and a constellation of payoffs according to the first period public signal, $f^\star$ ($f^\star \in F^\star(E_\lambda'^\star)$), such that f is an equilibrium payoff in the game with payoff $\lambda g + (1 - \lambda) f^\star \circ \phi$.

Definitions: A point f in a half space W of $I\!R^m$ is *W-λ-stable* if there exists $f^\star \in F^\star(W)$, and σ equilibrium in $G^\star(f^\star, \lambda)$ such that $f = E_\sigma[\lambda g(s) + (1 -$

$\lambda)f^\star \circ \phi(s)]$. In words, f is the equilibrium payoff of a game where the payoff tomorrow is in W.

A half space W is λ-*reproducing* if:

$$\forall f \in W, \ f \text{ is } W\text{-}\lambda\text{-stable.}$$

Notation: A half space W is defined by a vector $\alpha \in I\!R^m$, such that $\|\alpha\| = 1$ and a number $k \in I\!R$, such that $W = \{u| \ < u, \alpha > \leq k\}$.
$W_\alpha(\lambda)$ is the maximal (for the inclusion) λ-reproducing half space defined by α and some real $k = k(\alpha, \lambda)$.

Lemma 2.5: *W is λ-reproducing iff there is an f in its boundary that is W-λ-stable.*

Proof:
We only have to prove the sufficiency. So let f be such that:
$\exists f' \in F^\star(W), \ \sigma \in E^\star_{f'}(\lambda), \ f = E_\sigma[\lambda g(s) + (1-\lambda)f' \circ \phi(s)]$.
Take any ℓ in W. Then it is easy to see that associated to the same strategy σ, ℓ is an equilibrium payoff in the game $G^\star(f' + (\ell - f)/(1-\lambda), \lambda)$ ($\ell - f$ is a constant).
As f is on the boundary of W, $f' + (\ell - f)/(1-\lambda)$ is in $F^\star(W)$. Hence ℓ is W-λ-stable, and W is λ-reproducing. ∎

Lemma 2.6: *$W_\alpha(\lambda)$ is independent of λ.*
Proof:
Let W be a λ-reproducing half space, of direction α, and let f be a point in W. Write $f = E_\sigma[\lambda g + (1-\lambda)f^\star \circ \phi]$ with $f^\star \in F^\star(W)$ and $\sigma \in E^\star_{f^\star}(\lambda)$.
For any μ, define

$$\mu f^\star(y) + (1-\mu)f = \ell(y)$$

This equation defines a function ℓ in $F^\star(W)$ if μ is in [0,1].
For $\mu = [\lambda'(1-\lambda)]/[\lambda(1-\lambda')]$ (μ is between 0 and 1 iff $\lambda > \lambda'$), one can write:

$$f = E_\sigma[\lambda' g(s) + (1-\lambda')\ell \circ \phi(s)].$$

One has now to prove that $\sigma \in E^\star_\ell(\lambda')$. Since $\sigma \in E^\star_{f^\star}(\lambda)$,

$$E_{\sigma^{-i},t}[\lambda g + (1-\lambda)f^\star \circ \phi] \leq E_\sigma[\lambda g + (1-\lambda)f^\star \circ \phi]. \quad (1)$$

Let us prove the same kind of inequality, for ℓ and λ'.

$$\begin{aligned} E_{\sigma^{-i},t}[\lambda' g + (1-\lambda')\ell \circ \phi] &= E_{\sigma^{-i},t}[\lambda' g + (1-\lambda')(\mu f^\star + (1-\mu)f)] \circ \phi] \\ &\leq f \qquad \text{because of (1)} \\ &\leq E_\sigma[\lambda' g + (1-\lambda')\ell \circ \phi] \end{aligned}$$

Hence σ is in $E_h^\star(\lambda')$, so that W is reproducing for $\lambda' < \lambda$, and $W_\alpha(\lambda) \subset W_\alpha(\lambda')$, for $\lambda' < \lambda$.
Let us now prove the converse.
It is enough to take f on the boundary of W (see lemma 2.5).
$f^\star \circ \phi(s)$ is in W, and as μ is bigger than 1, $f^\star \circ \phi(s)$ is between f and $\ell \circ \phi(s)$, for all s. Hence, ℓ is in $F^\star(W)$.
One can prove as before that σ is in $E_\ell^\star(\lambda')$, and deduce that for $\lambda' > \lambda$, $W_\alpha(\lambda) \subset W_\alpha(\lambda')$.
Hence, $\forall \lambda,\ \lambda',\ W_\alpha(\lambda) = W_\alpha(\lambda')$. ∎

Notations: Let us simply denote $W_\alpha(\lambda)$ by W_α and let Q be $\bigcap_\alpha W_\alpha$.
We have the following result :

Theorem 2.7: *If the interior of Q is nonempty, then*

$$\lim_{\lambda \to 0} E_\lambda'^\star = Q$$

Sketch of the proof :

a) Proof of $E_\lambda'^\star \subset Q$

Let E be the convex hull of $E_\lambda'^\star$, f an extreme point of E, H a supporting hyperplane containing f and W the half space of boundary H containing E.
As f is an extreme point of E, it is in $E_\lambda'^\star$. So there is a $f^\star$ in $F^\star(E_\lambda'^\star)$ and $\sigma \in E_{f^\star}^\star$ such that $f = E_\sigma[\lambda g(s) + (1-\lambda) f^\star \circ \phi(s)]$.
Hence f is W-λ-stable, so that W is λ-reproducing and if α is the direction of W, $W \subset W_\alpha$. Finally $E \subset W_\alpha$.
This is true for all f, extreme point of E. Hence E is included in all the sets W_α, such that α is the direction of a supporting hyperplane of E, through an extreme point of E. This implies $E \subset Q$.

b) Proof of $Q \subset \lim_{\lambda \to 0} E_\lambda'$

The interior of Q is nonempty, Q can thus be approximated by convex compact sets with a C^2 boundary. Let Q_ϵ be such an approximation.
Let f be on the boundary of Q_ϵ. We can consider the tangent hyperplane to Q_ϵ at f, H, and denote by W the half space of boundary H containing Q_ϵ. W is included in some W_α, by definition of Q. Let f' be the continuation payoff expressing that f is W_α-λ-stable. For all s in S, $f - f'(s)$ is of the order of λ and the distance between H and Q_ϵ is of the order of λ^2, so for λ small enough, $f'(s)$ is in Q for all s.
Hence $Q \subset T_\lambda(Q)$ for λ small enough, and we have already seen that this implies that $Q \subset E_\lambda'^\star$. ∎

Comments

a) In Fudenberg and Levine [11], the result is shown in the framework of a game with short run and long run players; if L is the number of long run players, the result is: if $\dim(Q) = L$, then $\lim_{\lambda \to 0} E_\lambda' = Q$.

This is due to the fact that the short run players (that live only for one day) play one shot best responses to the strategies of the long run players. So you can compute, for each strategy profile of the long run players, the strategies of the short run players, and hence the payoffs. You are then reduced to the case with only long run players and upper semi continuous payoff correspondence.

b) The result relates $\lim_{\lambda\to 0} E'^{*}_{\lambda}$ to a quantity that depends only on the one shot game as in the folk theorem. But in general, Q is different from F. In fact there are two issues:
- Is the set of public equilibria equal to the set of equilibria? The answer is no. It is clearly shown by a game with 3 players: player 3 is a dummy and has no signal on the moves, while the other two players observe the moves. There is no public signal, and there are obviously some equilibria where player 1 and player 2 correlate their moves, using their information.
- Is the set of equilibrium payoffs equal to F?
In a game where there is no signal (except, for each player, his own move), the set of equilibrium payoffs is the convex hull of E_1 (all you can do is play a one shot equilibrium at each stage). Hence the answer is no.
The example of the partnership game in Fudenberg and Levine [11] shows that the answer is no even if the players have some information. Signals may introduce a lack of Pareto optimality.

c) The same results hold if the public component of the signal is random.

2.3 Private Signals

This section follows the work of Lehrer (see [20]). There are many other results by Lehrer concerning different kinds of equilibria. For more details, we refer to [18], [19], [21], [22].
We want to study the case of private signals and look for results concerning equilibria in games without discounting. We consider the two player case.
Recall that after each move s, player i is told $h^i(s)$. We assume that i knows his own move, and that signalling is nontrivial in the following sense:

$$\forall i,\ \exists s^i\ \exists s^{-i},\ \exists t^{-i},\ \text{such that } h^i(s^i,s^{-i}) \neq h^i(s^i,t^{-i}).$$

Hence the players are able to communicate through actions. Otherwise, at least one player has no information at all on his opponent's behavior and the analysis is straightforward.
Since the signalling is not standard, there are usually no subgames, because past histories are not common knowledge. Hence we will not deal with subgame perfection. In a Nash equilibrium with standard signalling, given any history,

the future behavior of the players is a Nash equilibrium, on the equilibrium path. But in the present situation, the players don't have the same information about the past, so it is not the case.
After any history, each player has a private signal on the random variable describing the future history for a given strategy profile. Then, on a Nash equilibrium path of the original game, the strategies induce a correlated equilibrium of the game starting after some history. Correlated equilibria appear naturally in this framework (for more precise definition of this notion we refer to the chapter on "Communication, correlation and cooperation" by this author in the present book), and they are easier to characterize. This is why we will focus on correlated equilibria rather than on Nash equilibria.
We first define a few relations between moves.

Definition: Two moves of player i, s^i and t^i are *equivalent* ($s^i \sim t^i$) if:

$$j \neq i,\ \forall s^j,\ h^j(t^i, s^j) = h^j(s^i, s^j)$$

i.e. player j cannot distinguish between s^i and t^i.

Definition: The move s^i is *at least as informative* as t^i ($s^i \succ t^i$) if:

$$s^i \sim t^i \text{ and } j \neq i,\ \forall s^j,\ \forall t^j, [h^i(t^i, s^j) \neq h^i(t^i, t^j)] \Rightarrow [h^i(s^i, s^j) \neq h(s^i, t^j)]$$

This means that player j cannot distinguish s^i from t^i but also that s^i is at least as informative as t^i. If i is asked what he knows about his opponent's move, he can give as good an answer if he is playing s^i as if he was playing t^i. So, if i deviates from t^i to s^i, player j will never detect it even by asking questions to i about the signals he received.

The following example shows the difference between the two definitions. Take a game with two players, where each of them has two possible strategies. The following matrices are the signalling matrices of the two players:

$$H_1 = \begin{pmatrix} a & a \\ a & b \end{pmatrix} \qquad H_2 = \begin{pmatrix} b & a \\ b & a \end{pmatrix}$$

If player 1 is supposed to play Bottom, and deviates and plays Top, the information of player 2 is the same, whatever he does. Hence Top$\sim$Bottom. But if 1 plays Top he cannot distinguish the situation where player 2 plays Right from the one where he plays Left, whereas if he had played Bottom he could have done it.
This notion is difficult to extend to more than two players: if there are three players, and if player 1 knows that 2 has deviated, can he tell it to 3, in order for them to correlate their moves to punish 2?
It doesn't extend immediately either to the case of random signals.

Definition:
$C^i = \{Q;\ Q \in \Delta(S^1 \times S^2),\ \forall s^i \in S^i,\ \forall t^i \in S^i$ such that $t^i \succ s^i$,
$\sum_{s^j} Q(s^i, s^j) g(s^i, s^j) \geq \sum_{s^j} Q(t^i, s^j) g(t^i, s^j)\}$
The interpretation is that Q is a correlation device (see Chapter "Communication, correlation and cooperation"in this book) according to which a signal is chosen. The players are asked to follow the recommendation given by the signal they receive (player i receives the signal s^i with probability $Q^i(s)$). Q is in C^i if a deviation from the strategies prescribed by the signal is detectable (eventually through questions about the information of the deviating player) or non profitable.

Theorem 2.8: *The set of correlated equilibrium payoffs in the infinitely repeated game is* $g(\bigcap_{i \in N} C^i) \cap IR$.

Remark: Recall that with private signals, there is a problem even to define the individually rational level; but in the two player case, there is no problem of correlation for punishing, so the definition is the same as in the case of standard signalling.

Sketch of the proof :

a) Proof that any correlated equilibrium payoff is in $g(\bigcap_{i \in N} C^i) \cap IR$.
As in the usual proof of the folk theorem, if a plan leads to a non-IR payoff, player i can deviate to get at least v^i at each stage.
If a plan leads to a payoff that is not in $g(\bigcap_i C^i)$, with positive probability, by definition of C^i, a player i has a non detectable, profitable deviation. In fact, in a set of stages that has a positive density in the set of stages, there exists a move that induces a better payoff and the same signal. Moreover, this move is at least as informative as the recommended one. Suppose a player deviates and plays this move. He can construct a fictitious history as follows: he can compute for each stage, the signals he would have received if he had played as recommended, and hence the moves he would have played. If he is asked a question about his information at any stage, he can answer according to this fictitious history. Hence, with positive probability a player has a non detectable profitable deviation.

b) Proof of the converse.
The proof is involved and introduces a lot of new ideas, so we can only indicate some of the main points.
Let x be a payoff in $g(\bigcap_i C^i) \cap IR$ induced by some Q.
The idea is to choose the moves at random at each stage according to Q and the players are supposed to follow this recommendation . Then if one player deviates, the deviation will be detected and he will be punished. To detect a deviation that may be equivalent to the proposed strategy (but less informative), the players have to communicate, and to ask each other questions about what

they know. They communicate through actions to ask these questions, hence non trivial signalling is used here.
More precisely, a profile of pure moves s is chosen according to Q. The players are supposed to play according to this choice s. In order to detect deviations, a player must be able to check that the move his opponent played is equivalent to the recommended one. This may not be possible if Q is not of full support. Even if Q is of full support, a player may not be able to check that the move his opponent played is as informative as the recommended one if he never knows the recommendation the other player received. The idea is to define a sequence Q_n converging to Q, such that each Q_n is of full support and with a small probability (small enough not to affect the payoff because of the new possible deviations it introduces) a player is informed of s and not only of s^i.
The strategies are as follows:
A profile of pure moves s is chosen according to Q_n for the block n. The players are supposed to play according to their recommendation for a large number of stages, say 2^n. Then they begin a communication phase of length $2n$ where each player chooses at random a number p smaller than 2^n and tells it to the other player (which can be done in n periods) who must tell in response what he played at stage p, and what he knows of his opponent's move at stage p. If player i cannot give all the information he should have if he had played according to the recommendation, (that is known to player $-i$ with some small probability) i is punished for ever. As the game is infinitely repeated, a profitable deviation implies deviating at infinitely many stages. Hence, with probability one, if a player deviates and uses a strategy that is not at least as informative as the recommended one, the deviation will be detected. Hence the only deviations that may be profitable are the ones that cannot be punished in this way, i.e. that cannot be detected by the above procedure. The only possibly profitable deviations for player i, from a strategy s^i, are t^i that are less informative than s^i. By definition of C^i such deviations are unprofitable.
The communication phase is short as compared to the other, hence it does not influence the payoff.
Hence altogether, x is a correlated equilibrium payoff. ∎

Conclusion

In repeated games, folk-theorem-like results are numerous and sometimes involved, in particular when non standard signalling is concerned. These results connect the one shot game with the equilibria of the repeated game.
More precisely, under standard signalling, they express the set of equilibrium payoffs of the repeated game as the set of feasible and individually rational payoffs, hence relate a strategic notion (equilibrium) to a cooperative notion (feasible payoffs). The results show that repetition gives the possibility, under

some conditions to achieve any individually rational Pareto efficient payoff, i.e. to cooperate; but there are also many other equilibrium payoffs that are not optimal and that can be achieved as well. Repetition *may* lead to cooperation. A further question is whether these "most cooperative" payoffs (Pareto optimal ones) are in some sense more likely to arise.

Achieving cooperation may fail if full monitoring is not assumed. In fact, if there is not enough public information, two phenomena arise: first it is more difficult to define a public plan (which is one of the cooperative aspects: social norm), moreover it is harder to make it self enforcing (since deviations are less likely to be observed, suspicion will occur). Hence in a non standard signalling framework, cooperation may be more complicated to obtain and a lack of Pareto optimality is more likely to occur.

References

[1] Abreu, D. (1988): "On the theory of infinitely repeated games with discounting", *Econometrica*, 56, 383-396.

[2] Abreu, D., D. Pearce and E. Stacchetti (1990): "Toward a theory of discounted repeated games with imperfect monitoring", *Econometrica*, 58, 1041-1063.

[3] Abreu, D., P.K. Dutta and L. Smith (1994): "The folk theorem for repeated games: a NEU condition", *Econometrica*, 62, 939-948.

[4] Aumann, R.J. (1986): "Repeated Games", in G.R. Feiwel (ed.) "Issues in contemporary microeconomics and welfare", 209-242, Macmillan.

[5] Aumann, R.J. and L.S. Shapley (1994): "Long term competition-a game theoretic analysis", in Meggido (ed.), "Essays in Game Theory in Honor of Michael Maschler", 1-15, Springer Verlag.

[6] Benoît, J.P. and V. Krishna (1985): "Finitely repeated games", *Econometrica*, 53, 905-922.

[7] Benoît, J.P., and V. Krishna (1987): "Nash equilibria of finitely repeated games", *International Journal of Game Theory*, 16, 197-204.

[8] Forges, F., J.F. Mertens and A. Neyman (1986): "A counterexample to the folk theorem with discounting", *Economic Letters* , 20, 7.

[9] Fudenberg, D. (1992): "Explaining cooperation and commitment in repeated games", in J.J. Laffont (ed.), "Advances in Economic Theory, sixth world congress", vol. 1, 89-131, Cambridge University Press.

[10] Fudenberg, D. and D. Levine (1991): "An approximate folk theorem with imperfect private information", *Journal of Economic Theory*, 9, 1-12.

[11] Fudenberg, D. and D. Levine (1994): "Efficiency and observability with long run and short run players", *Journal of Economic Theory*, 62, 103-135.

[12] Fudenberg, D., D. Levine and E. Maskin (1994): "The folk theorem with imperfect public information", *Econometrica*, 62, 997-1039.

[13] Fudenberg, D. and E. Maskin (1986): "The folk theorem in repeated games with discounting and with incomplete information", *Econometrica*, 54, 533-554.

[14] Gossner, O. (1992): "Folk theorem with mixed strategies", to appear in *International Journal of Game Theory*.

[15] Gossner, O. (1993): "Overlapping generations games with mixed strategies", preprint.

[16] Kandori, M. (1992): "Repeated games played by overlapping generations of players", *Review of Economic Studies*, 59, 81-92.

[17] Krishna, V. (1988): "The folk theorems for repeated games", in "Proceedings of the NATO-ASI conference: Models of incomplete information and bounded rationality, Anacapri, 1987", to appear.

[18] Lehrer, E. (1990): "Nash equilibria of n player repeated games with semi standard information", *International Journal of Game Theory*, 13, 191-217.

[19] Lehrer, E. (1991): "Internal correlation in repeated games", *International Journal of Game Theory*, 19, 431-456.

[20] Lehrer, E. (1992): "Correlated equilibria of two player repeated games with non observable actions", *Mathematics of Operations Research*, 17, 175-199.

[21] Lehrer, E. (1992): "Two players repeated games with nonobservable actions and observable payoffs", *Mathematics of Operations Research*, 17, 200-224.

[22] Lehrer, E. (1992): "On the equilibrium payoffs set of two player repeated games with imperfect monitoring", *International Journal of Game Theory* , 20, 211-226.

[23] Meggido, N. (1994): (ed.) "Essays in game theory in honor of Michael Maschler", Springer Verlag .

[24] Mertens, J.F. (1987): "Repeated games", in "Proceedings of the international congress of mathematicians, Berkeley, 1986", 1528-1577, American Mathematical Society.

[25] Mertens, J.F., S. Sorin and S. Zamir (1994): "Repeated games", Part A, Core Discussion Paper 9420.

[26] Pearce, D.G. (1992): "Repeated games: cooperation and rationality", in J.J. Laffont (ed.) "Advances in Economic Theory, sixth world congress", vol. 1, 132-174, Cambridge University Press.

[27] Rubinstein, A. (1994): "Equilibrium in Supergames", in Meggido (ed.), "Essays in Game Theory in Honor of Michael Maschler", 17-27, Springer Verlag.

[28] Rubinstein, A. (1992): "Comments on the interpretation of repeated games theory", in J.J. Laffont (ed.) "Advances in economic theory, sixth world congress", vol. 1, 175-181, Cambridge University Press.

[29] Rubinstein, A. and A. Wolinski (1992): "Remarks on infinitely repeated extensive games", to appear in *Games and Economic Behavior.*

[30] Salant, D.J (1991): "A repeated game with finitely lived overlapping generations of players", *Games and Economic Behavior*, 3, 244-259.

[31] Smith, L. (1992): "Folk theorem in overlapping generations games", *Games and Economic Behavior*, 4, 426-449.

[32] Sorin, S. (1986): "On repeated games with complete information", *Mathematics of Operations Research*, 11, 147-160.

[33] Sorin, S. (1990): "Supergames", in T. Ichiishi, A. Neyman and Y. Tauman (eds.), "Game theory and applications", 46-63, Academic Press.

[34] Sorin, S. (1992): "Repeated Games with complete information", in R.J. Aumann and S. Hart (eds.) "Handbook of game theory with economic applications", vol. 1, 71-107, North Holland.

[35] Sorin, S (1994): "A note on repeated extensive form games", to appear in *Games and Economic Behavior.*

[36] Wen, Q. (1994): "The folk theorem for repeated games with complete information", *Econometrica*, 62, 949-954.

Communication, Correlation and Cooperation

Sylvain Sorin

Université Paris X-Nanterre and Ecole Normale Supérieure, Paris, France[1]

Keywords: correlation, correlated equilibrium, extensive form correlated equilibrium, communication, communication equilibrium.

The purpose of this presentation is to introduce models of extension of games with preplay or intraplay information and communication. These extensions will allow us to define new notions of equilibria. The relevant question is to see how the outcomes change when communication between players is allowed, or when they are given some kind of preplay information.
This section will be divided in two parts: the first one about concepts and the second one about mechanisms. We first define the basic tools (correlation device, correlated equilibrium, communication equilibrium...). Then, we provide different mechanisms of communication that lead to these equilibria.
The concepts studied here have been introduced by Aumann [1] and Forges [4]. The mechanisms presented in this section are essentially the works of Bárány [3], Forges [6] and Lehrer [10]. For related surveys and further results see [12], [16] and [13].

1 Concepts

We are going to introduce three concepts: correlated equilibrium, extensive form correlated equilibrium and communication equilibrium. All these concepts are extensions of the notion of equilibrium in a game. These new equilibria are Nash equilibria of some new games, that are extensions of the original one. They consist in two parts, one related to the problems of correlation and communication among the players, independently of the initial game, and the other one to the strategic play in the game itself.

1.1 Correlated Equilibrium

The concept is due to Aumann [1]. We will first illustrate the idea with two examples.

---

[1]Notes written by Dinah Rosenberg and revised by the author (December 1994).

1.1.1 Examples and Intuition

Example 1

Consider the battle of sexes. There are two players, He and She. He plays the rows and She plays the columns. She wants to go to the theater and He wants to go to the movies. T and L stand for the movies and B and R for the theater. The payoffs are as follows :

$$\begin{array}{c} \\ T \\ B \end{array} \begin{array}{c} \begin{array}{cc} L & R \end{array} \\ \begin{pmatrix} (2,1) & (0,0) \\ (0,0) & (1,2) \end{pmatrix} \end{array}$$

The problem of this game is how the players can correlate their moves. Their common interest is obviously to be in the same place at the same time, so they would like to correlate their choices in order to achieve it.
Suppose the two players are told an integer between 0 and 9 chosen at random. They decide to go to the movie if the integer is less than 4, and to go to the theater if it is greater than 5. One can check that whatever number is announced, no one has an incentive to deviate from this plan. Moreover, the expected payoff is (3/2, 3/2), which is not feasible in the one shot game. In this case, the players coordinate through public messages.
This coordination leads to a distribution on the cells of the payoff matrix : the cell (2,1) is reached with probability 1/2, the cell (1,2), with probability 1/2, and the cells (0,0) with probability 0.

The following example is an example of correlation where, unlike example 1, the players play according to private signals.

Example 2

Consider the following game G :

$$\begin{array}{c} \\ T \\ B \end{array} \begin{array}{c} \begin{array}{cc} L & R \end{array} \\ \begin{pmatrix} (7,2) & (0,0) \\ (6,6) & (2,7) \end{pmatrix} \end{array}$$

Suppose there are three messages: white, grey, black, that appear with probability (1/3, 1/3, 1/3). Suppose player I can distinguish white from grey or black, and player II can distinguish white or grey from black. They get respectively the signals W, GB, or WG, B. In that case the signals are not public. The probability of joint messages can be represented by the following matrix:

$$\begin{array}{c} \\ W \\ GB \end{array} \begin{array}{c} \begin{array}{cc} WG & B \end{array} \\ \begin{pmatrix} 1/3 & 0 \\ 1/3 & 1/3 \end{pmatrix} \end{array}$$

The game is played as follows:
- a colour is chosen at random;
- each player receives the private signal corresponding to this choice;
- the players choose actions in G.

Consider the following strategies :
Player I : if the signal is W, play Top, and if the signal is GB, play Bottom.
Player II : if the signal is WG, play Left, and if the signal is B, play Right.
Then no player has an incentive to deviate from these strategies.

The induced distribution on the matrix leads to the payoff (5,5) which is outside the convex hull of the set of Nash equilibrium payoffs, which are:
$\{(7,2),(2,7),(14/3,14/3)\}$.

These are examples of correlated equilibria. Let us come to the formal definition expressing that given the signal and the ex-post probability it induces on the other players' signals, hence on their moves, the strategies are best responses to one another.

1.1.2 Definition and Properties

Let G be a game in normal form, defined by a finite set of players N of cardinality m, a finite set of pure moves for player i, S^i, and a payoff function, $g : S = \prod_{i\in N} S^i \to \mathbb{R}^m$. As usual $-i$ denotes the set of players in N except i and $S^{-i} = \prod_{j\neq i} S^j$. Let $\mathcal{C}$ be a probability space $(\Omega, \mathcal{A}, p)$, (Ω is the space of states of nature). There are no restrictions about $\mathcal{C}$.
Let θ^i be a measurable function from $\mathcal{C}$ to the finite set of signals of player i, A^i. θ^i is the signalling function of player i.

Remark: the game G is fixed, it does not depend on the state of nature. The probability space defines only a signal in order for the players to make non independent choices; but the actual game they play is known to everybody, this being the difference with a Bayesian game.

Formally we have:
Definition: A *correlation device* is a $n+1$-uplet of the form $\gamma = (\mathcal{C}, \theta^1, \ldots, \theta^m)$

Then we introduce:
Definition: Let us define a new game G_γ, *G extended by γ*, as follows (it is a two stage game):
Stage 1 : $\omega \in \Omega$ is chosen according to p. $\theta^i(\omega)$ is told to player i.
Stage 2 : each player chooses an action in G.

A strategy for player i is a function $\sigma^i : \Omega \to \Delta(S^i)$, such that if $\theta^i(\omega) = \theta^i(\xi)$, then $\sigma^i(\omega) = \sigma^i(\xi)$, i.e. σ^i is measurable with respect to i's information partition induced by θ^i on Ω.

If each player i plays the mixed move $\sigma^i(\omega)$, and if ω is chosen, the resulting profile of mixed moves is $\sigma(\omega) = (\sigma^1(\omega), \ldots, \sigma^m(\omega))$, and the payoff of a player j is $g^j(\sigma(\omega))= E_{\sigma(\omega)}(g^j(s))$. The payoff of player j in G_γ, if the players play the strategies σ is thus $\phi^j(\sigma) = \int_\Omega g^i(\sigma(\omega))p(d\omega)$.
The players' actions depend on what they know about ω. As their information about it is correlated, the signal about ω enables them to correlate their moves, though strategic independence is kept. Correlation is achieved through an exterior signal about which two players are at least partially informed.

Definition: A Nash equilibrium of G_γ is a *correlated equilibrium of* G; let C be the set of all correlated equilibrium payoffs of G obtained by letting the correlation device γ vary. $E(G_\gamma)$ denotes the set of Nash equilibrium payoffs of G_γ, then $C = \cup_\gamma E(G_\gamma)$.

For the connection between this notion and the notion of sunspot equilibria, we refer to Peck [15].
Comparison between two correlated equilibria is difficult because since the spaces $\mathcal{C}$ and the signals may be different, the set of strategies may also differ. So, to compare two correlated equilibria, one looks at the distribution they induce on S.

Definition: If σ is a profile of pure strategies in G_γ (i.e. for all ω, $\sigma(\omega)$ belongs to S) and s is a profile of pure strategies, $Q_\sigma(s)$ is the probability of the event $\{w \; ; \; \sigma(\omega) = s\}$.
This definition can be extended to mixed strategies : if σ is a profile of mixed strategies, and $\rho_{\sigma(\omega)}(s)$ is the probability of the pure move s, if $\sigma(\omega)$ is played, namely, $\rho_{\sigma(\omega)}(s) = \prod_{i \in N} \sigma^i(\omega)(s^i)$, one has

$$Q_\sigma(s) = \int_\Omega \rho_{\sigma(\omega)}(s) \; p(d\omega).$$

Q represents the way the players correlate their moves through the signals. In the examples we saw that they achieve in this way some probability distributions on outcomes that are not achievable if the players choose independently mixed strategies.
Q can be represented by a matrix on S:
In example 1,

$$Q \text{ is } \begin{pmatrix} 1/2 & 0 \\ 0 & 1/2 \end{pmatrix}$$

In example 2,

$$Q \text{ is } \begin{pmatrix} 1/3 & 0 \\ 1/3 & 1/3 \end{pmatrix}$$

The problem is to characterize the set of distributions that can be obtained from a correlated equilibrium. We have a well defined map that associates to any game G and any device γ, a game G_γ; then one can consider its Nash equilibria, and associate to each of them the distribution it induces over S. Let us denote by CED the set of *correlated equilibrium distributions*, i.e. the set of distributions Q over S such that there exists a correlated device γ and an equilibrium of G_γ inducing Q. We define now a specific class of correlated equilibria :

Definitions:
i) A *canonical correlation device*, is a correlation device such that $\Omega = S$, $\mathcal{A} = \mathcal{P}(S)$ (subsets of S) and $\theta^i(s) = s^i$ (ω is a profile of moves and the signal of each player is his own move).
ii) A *canonical correlated equilibrium* is a correlated equilibrium such that :
- the correlation device is canonical;
- player i's equilibrium strategy is σ^i, with $\sigma^i(s) = s^i$, (if i received the signal s^i, then he plays s^i).

Recall that σ^i is actually a function of i's signal, hence σ is the identity.

Theorem 1.1 *(canonical representation)*: *The set of correlated equilibrium distributions is equal to the set of canonical correlated equilibrium distributions.*

Remark: This theorem is a version of the revelation principle (see [14]).

Proof : Let us denote by CCED the set of canonical equilibrium distributions.
-We have to show that CED is included in CCED.
Take p in CED. Let us denote by H the original game, and by G an extension that induces p. Define G' as a canonical extension of H such that the probability distribution on S is p. Let us prove that for each player, following the recommendation is an equilibrium strategy.
In G' each player has less information than in G. Hence if there is a profitable deviation in G', it was also available in the original extension of the game since in G', the player knows only the move and not the signal he would have received in G.
As the payoff depends only on p and not on the correlation device γ, the deviation was also profitable in the original game.
So, if p is in CED, there can be no profitable deviation from the recommendation given by the signal given in G'. Hence p is in CCED. ∎

Comments:
- To determine the set of correlated equilibria one can restrict oneself to the following framework: each player first receives a recommendation about what

action to choose (the recommendations for the players are chosen jointly and at random according to a fixed probability), then each player chooses an action, and it is an equilibrium to have each of them follow the recommendation.
- This theorem allows us to restrict the analysis to canonical equilibria, and hence makes computations more tractable.
- Nevertheless, in constructive proofs, it may be useful not to restrict ourselves to canonical equilibria, and to use the great diversity of possible correlation devices.

We now come to some properties of correlated equilibria.

Property 1.2: *Q is in CED iff for all i in N and for all pair of moves, (s^i, t^i),*

$$\sum_{s^{-i} \in S^{-i}} Q(s^i, s^{-i}) g^i(s^i, s^{-i}) \geq \sum_{s^{-i} \in S^{-i}} Q(s^i, s^{-i}) g^i(t^i, s^{-i})$$

Proof:
let Q be a correlated equilibrium distribution. It can be induced by a canonical correlated equilibrium.
The expected payoff of i, receiving signal s^i and playing t^i is

$$\sum_{s^{-i}} Q(s^i, s^{-i}) g^i(t^i, s^{-i})$$

(if the other players follow the recommendations i.e. play s^{-i}).
The equilibrium condition says that for every i, for every s^i, and every t^i,

$$\sum_{s^{-i}} Q(s^i, s^{-i}) g^i(s^i, s^{-i}) \geq \sum_{s^{-i}} Q(s^i, s^{-i}) g^i(t^i, s^{-i}).$$

■

The set of correlated equilibrium payoffs will be denoted by CP.

Property 1.3: *CED and CP are a convex polyhedra.*
Proof :
The equations expressing that Q is in CED are linear, and one can therefore concentrate on pure strategies. We then have a finite collection of linear inequalities defining CED, which is hence a convex polyhedron.
Since the map that associates to a distribution on S the corresponding payoff is linear, CP is a polyhedron as well. ■

Property 1.4: *The set of Nash equilibrium distributions is contained in the set of correlated equilibrium distributions, which is hence nonempty.*

Remark: As CP is a convex polyhedron, there is an elementary proof based on the separation theorem of convex polyhedra showing that the set of correlated equilibria is nonempty (see [9]).
In particular, it implies that if the problem one starts with has rational parameters, there are correlated equilibria with rational parameters. Hence, if all Nash equilibrium distributions have irrational parameters, there is a correlated equilibrium distribution that is not a Nash equilibrium distribution.

In the following example, due to Moulin and Vial, there exists a correlated equilibrium payoff that strictly dominates all Nash equilibrium payoffs.
Example 3: The game is represented by the following matrix:

$$\begin{pmatrix} (5,4) & (4,5) & (0,0) \\ (0,0) & (5,4) & (4,5) \\ (4,5) & (0,0) & (5,4) \end{pmatrix}$$

The unique Nash equilibrium has both players playing (1/3,1/3,1/3), and induces a payoff of (3,3).
We now exhibit a canonical correlated equilibrium that induces the (stricly better) payoff (9/2,9/2). It is defined by the canonical device represented by the following matrix :

$$\begin{pmatrix} 1/6 & 1/6 & 0 \\ 0 & 1/6 & 1/6 \\ 1/6 & 0 & 1/6 \end{pmatrix}$$

We now come to Aumann's approch of correlated equilibria in terms of bayesian maximisation.

1.1.3 Aumann's Theorem

The issue of this section has been studied by Aumann in [2].
The framework is the following. Assume you have a collection of players and a probability distribution P on Ω that represents the beliefs of a player about all players' actions. We assume that P is the same for all the players, and that each player is bayesian rational, and that the players play according to P. Then the initial probability is a correlated equilibrium distribution.
Rationality and bayesian players imply that the players play according to P and that it is a best response to the situation where the others play according to P.
The proof is very simple but the result is striking: it expresses the emergence of correlated equilibria as the expression of the rationality of bayesian players with consistent beliefs.

1.2 Extensive Form Correlated Equilibrium

Take an extensive form game. You can reduce it to a normal form game, and find the correlated equilibria. They define the correlated equilibria of the extensive form game.
But in the original formulation you may do more: before each stage of the game, a state of nature is chosen and a signal is given. You extend the game by a family of such correlation devices, and consider the Nash equilibria of the extended game. They define the extensive form correlated equilibria.

Definition: Let G be a finite game played in T stages (1,...,T). An *extensive form correlation device* for G is a collection of sets M_t^j for each player j and each period t, and a probability function p_t for each period t on $M_t^1 \times \ldots \times M_t^m$.

The extensive form game is here a game that can be played in stages: there is a public calendar for all the players. It is an extensive form game in the sense of Von Neuman.
The game G_γ, G extended by the extensive form correlation device γ is the game played in $2T$ stages as follows ($k \in \{1, \ldots, T\}$):
- stage $2k-1$: a message $(m_k^1, \ldots, m_k^m)$ in $M_k^1 \times \ldots \times M_k^m$ is chosen according to the probability p_k, and player i is informed of m_k^i.
- stage $2k$: knowing their messages, the players who have to play at stage k of the game G, choose an action available at that stage k of G.

Definition: An *extensive form correlated equilibrium* of game G is a Nash equilibrium of an extension of G by an extensive form correlation device.

You can define as before a canonical extensive form correlated equilibrium, an get a canonical representation theorem.
A *canonical extensive form correlation device* is a correlation device such that at each time t, the signal is a profile of moves s, feasible at stage t and each player i's signal is the component s^i of this signal.
A *canonical extensive form correlated equilibrium* is an extensive form correlated equilibrium such that the correlation device is a canonical extensive form correlation device, and such that at any time, if the last signal of player i has been s^i, his best response to it is s^i (it is an equilibrium strategy to follow the recommendation given by the signal)

Theorem 1.5: *The set of extensive form correlated equilibrium distributions is equal to the set of canonical extensive form correlated equilibrium distributions.*

The following example due to Myerson [14], shows that the set of extensive form correlated equilibrium distributions may strictly exceed the set of correlated equilibrium distributions.

Example 4: We focus on the canonical extensive form correlated equilibria of this game, and on the canonical correlated equilibria.
The game is as follows:

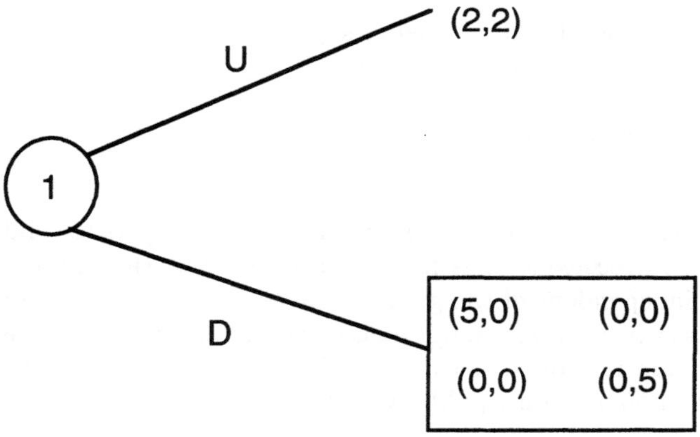

The game is played in two stages:

Stage 1: we call the node at this stage a; player I must choose between U and D; if he chooses U the payoff is (2,2); if he chooses D, we go to stage 2.

Stage 2 : we call the node at this stage b; the players then play the following normal form game :

$$\begin{array}{cc} & \begin{array}{cc} L & R \end{array} \\ \begin{array}{c} T \\ B \end{array} & \begin{pmatrix} (5,0) & (0,0) \\ (0,0) & (0,5) \end{pmatrix} \end{array}$$

Suppose a message is sent at point b. The signal is (T, L) with probability $1/2$, and (B, R) with probability $1/2$. The first component is the message received by player I, and the second component is the message received by player II.
The strategies of player I are, at b :

- if the message is T, play Top;
- if the message is B, play Bottom;

The strategies of player II are :

- if the message is L, play Left;
- if the message is R, play Right.

These strategies define a correlated equilibrium of the subgame starting at b, leading to the payoff (5/2, 5/2), and hence, player I will play D at point a. We have thus an extensive form correlated equilibrium with payoff (5/2, 5/2).

If messages can only be sent at the begining of the game, player I will never follow a recommendation telling him to play D followed by B since this strategy is strictly dominated. So player II will never get more than 2 in a normal form correlated equilibrium. Hence in this example, there is an extensive form correlated equilibrium payoff that is not achievable as a normal form correlated equilibrium payoff.

We will consider now a more general notion, where the players can communicate, i.e. where they can influence the signals they receive.

1.3 Communication Equilibrium

The concepts of this section are due to Forges [4]. We now allow players to send signals to a machine and to receive signals from this machine at each period of time. The machine is independent of the game, because if the players could choose the machine, an informed player could manipulate the information. We extend the game by this information structure and the Nash equilibria of the extended game are the communication equilibria of the original game.

Remark: If players are allowed only to receive messages from the machine, but not to send messages, the definition reduces to the previous one of extensive form correlated equilibrium, and if they can only receive a message at the first period, the definition reduces to the one of normal form correlated equilibrium.

We come now to the formal definition.

Definitions: Let G be a game played in T stages.

i) A *communication device* d for the game G is a collection of sets of inputs for any player j at any time t, I_t^j, and of set of outputs for any player j at any time t, O_t^j, and of eventually random functions p_t from $\prod_j I_t^j$ to $\prod_j O_t^j$. $d = (I_t^j, O_t^j, p_t)_{j \in N, t \in \{1,\dots,T\}}$

ii) The *extension* G_d *of the game* G *by the communication device* d is the following game: each stage t is composed of three parts; during the first one each player j sends to the machine an input i_t^j in I_t^j; then he receives the output $p_t^j(i_t^1, \dots, i_t^m) = o_t^j$ from the machine; knowing o_t^j, he chooses an action as he would have done at stage t of game G.

iii) A *communication equilibrium* of the game G is a Nash equilibrium of a game G_d extended by a communication device d.

Notice that the communication device does not depend on the game. If it were the case, the strategic issues would be totally different because the players could try to influence the machine. This would change the strategic structure of the game and the result would not be an extension of the original game but rather another strategically different game.

We have a theorem of canonical representation :

Definition: A *canonical communication device* is a communication device such that I_t^j is the set of information of player j at time t, and O_t^j is the set of moves of player j at time t.

Theorem 1.6: *The set of distributions over strategies that can be induced by a communication equilibrium is equal to the set of distributions of strategies that can be induced by a canonical communication equilibrium.*

We have the same convexity property:

Property 1.7: *The set of correlated equilibrium payoffs, the set of extensive form correlated equilibrium payoffs, and the set of communication equilibrium payoffs are convex polyhedra.*

Proposition 1.8: *The set of communication equilibrium distributions may be stricly bigger than the set of extensive form correlated equilibrium distributions.*

This is due to the fact that in the extension of a game by a communication device, the players can both correlate their moves through the signals they receive, and transmit information to one another through the signals they send to the machine.

The proof is given by the following example :

The game is a game of incomplete information. One of the following games G_1 and G_2 is chosen with probability 1/2, and player I is informed of the true game while player II is not.

$$G_1 = \begin{array}{c} \\ T \\ B \end{array} \begin{array}{c} \begin{array}{cc} L & R \end{array} \\ \begin{pmatrix} (1,1) & (0,0) \\ (1,1) & (0,0) \end{pmatrix} \end{array}$$

$$G_2 = \begin{array}{c} \\ T \\ B \end{array} \begin{array}{c} \begin{array}{cc} L & R \end{array} \\ \begin{pmatrix} (0,0) & (1,1) \\ (0,0) & (1,1) \end{pmatrix} \end{array}$$

The difference between the set of communication equilibrium distributions and the set of extensive form correlated equilibrium distributions is due to the possibility to send messages that allow for communication. Hence the informed player, who is a dummy, can transmit his information to the uninformed player. Let us describe the machine and the strategies:

- the machine sends back all the messages it receives;
- player I sends the message T if $k = 1$, and B if $k = 2$; when player II receives the message he is hence informed of the state of the world.
- player II plays L if the message he receives is T and R if the message is B.

These strategies are obviously a communication equilibrium and they lead to the payoff (1,1).

In any extensive form correlated equilibrium, player II has no information about the true game , and hence he plays in the following average game (as player 1 is a dummy, his information has no impact):

$$\begin{pmatrix} (1/2,1/2) & (1/2,1/2) \\ (1/2,1/2) & (1/2,1/2) \end{pmatrix}$$

The payoff is (1/2,1/2), hence the result. ∎

We have defined three different notions of equilibria, that take into account the fact that players may communicate, or use some tools to correlate their actions. We have also seen some of their properties. The three notions of equilibria we have defined here introduce more and more opportunities of communication through machines. But one must be careful not to introduce too many opportunities in order not to affect too much the structure of the game. For example if there were a different correlation device for each node of an extensive game, it would affect the information sets of the players.

We now want to see if one can achieve these equilibria without relying on some outside machine, but only on what the players can do by themselves.

2 Mechanisms

We are looking for mechanisms through which the players can generate by themselves the different class of equilibria discussed above.

2.1 Correlated Equilibrium Through the Phone

This part follows the work of Bárány [3]. The idea is to allow the players to have private conversations, i.e. to communicate through the phone. In that case we will see that they can achieve any correlated equilibrium.
More precisely, any correlation device can be achieved through private conversation if there are at least four players.

The framework is as follows: There are m players. Each player i has a finite set of pure strategies S^i. S is the product of $S^1, \ldots, S^m$. The players want to correlate their choices, i.e. to choose $s \in S$ with a predetermined probability p, each player k knowing only s^k.

This can be done with a machine that chooses s in S with probability p, and sends the message s^k to k, i.e. through a correlation device. Can the players find a protocol that enables them to do this without the help of a machine?

Definition: A *protocol* is a finite set of rules describing, for each step r, which player is active and what action this active player should take at this stage.

Definition: The *information of player k at stage r* is composed of all the messages he sent and received, and all the choices he made up to that stage. This set of information is denoted by I_r^k.

Definition: An *action of player k at stage r* is one of the following things :
- Make a random choice z_r^k knowing I_r^k, compute a message m_r^k (the rules give functions g_r^k, such that $g_r^k(I_r^k) = m_r^k$), and send it to a player specified by the rules (call him and tell him m_r^k).
- Compute s^k knowing I_r^k (the rules give functions f_r^k such that $f_r^k(I_r^k) = s^k$).
- If for some I_r^k there is no sequence of random choices of the other players that leads to this set of information, send to everybody the message "Deviation".

Definition: The player k *deviates from the rules* if for some sequence z_r^k, (for r such that k is active), there is an r^* such that the message k sends is not in $g_{r^*}^k(I_{r^*}^k)$.

Definition: A *sure protocol* is a protocol such that :
(1)
$$p(s^1, \ldots, s^m) = Prob(f^1(I^1), \ldots, f^m(I^m))$$

(2)
$$p(s^1, \ldots, s^m | s^k) = Prob(f^1(I^1), \ldots, f^m(I^m) | I^k)$$

(3) A deviation from the rules is detected with probability 1.
(4) A deviation consisting of choosing z_r^k with a probability different from the one prescribed by the rules does not influence (1) and (2).

This means that the random device mimics p and that the information the players get is given by p. It means also that a deviation is detectable or innocuous. We can now state the result:

Theorem 2.1: *For any probability distribution p over S, that is rational valued, if $m \geq 4$, there is a sure protocol that induces the correlation device p.*

An immediate corollary of this theorem says that any rational correlated equilibrium of a game G can be achieved as a Nash equilibrium of an extension of the following kind :

Stage 1: the players make phone calls;
Stage 2 : the players play the game G.

The proof is quite long and composed of many steps, and we refer to the paper of Bárány [3]. Here are some brief ideas.
As p is rational valued, you can restrict yourself to the case where you have to implement a uniform probability over a given set. For instance to choose a number in $\{1, 2, 3\}$, with 1 having probability 1/2 and 2 and 3 probability 1/4, you can choose a number in $\{1, 1, 2, 3\}$ uniformly.
In order to achieve Q, the players are supposed to make random choices, and to communicate them to other players. It may happen that one of the players prefers some issues to some others. In that case, he may want to deviate: he may want to lie when he is asked to send some information to another player.
The idea is that each random choice is known by at least two players; they both tell it to a third one who can check whether he received the same information from both; in that way he can check that no one has lied. Every message is always received at the same time from two sources, to permit checking.
In order to get the same message from two sources, a player always sends the same message to two different interlocutors. They can then send it to a fourth player, that will receive two similar messages.
Moreover, no one should know everything. That is the reason why you need at least four players.
A second aspect is the following: to prevent the players from cheating you must not give them too much information. That is why the signals are encoded through permutations of the sets of actions, chosen with uniform probability by the players.
For instance, if the players want to choose a number at random in a finite set X, they can ask player 1 to choose a number x in X at random and to tell it to players 3 and 4. Then player 2 chooses a permutation of X at random, q, and tells it to 3 and 4. Players 3 and 4 then compute $q(x)$, which is the requested number. Players 1 and 2 cannot influence the final distribution. This is a "jointly controlled lottery" and it is much easier to obtain if simultaneous moves are available.

Recall that this result relates correlated equilibria of a game to Nash equilibria of an extension of it.

2.2 Communication Equilibrium with Incomplete Information

We want to generate any communication equilibrium distribution as correlated equilibria of the game extended by plain conversation as in Forges [6].

Let G be a game with incomplete information (see Harsanyi, [8]). Given m players, we denote by K_i the finite set of possible types of player i, and by S^i his finite set of actions. K denotes the product of $K_1, \ldots, K_m$, and S denotes the product of $S^1, \ldots, S^m$.
Let p be a probability on K, and g the payoff function from $K \times S$ to $\mathbb{R}^m$. g depends on the real vector of types, i.e. the game depends on the state of nature.

The game is played in the following way :
- Step 1 : k is chosen in K according to p; player i is informed of k_i in K_i.
- Step 2 : each player i chooses a move s_i in S^i.

The extension of G through plain conversation is the following game :
- Step 1 : k is chosen in K according to p; player i is informed of his signal k_i.
- Step 2 : each player may send public messages.
- Step 3 : each player i chooses a move s_i in S^i.

Remark: This extension is a particular case of an extension by a communication device.

We have the following result (Forges [6]):

Theorem 2.2: *If $m \geq 4$, the set of communication equilibrium distributions of G is the set of correlated equilibrium distributions of G extended by plain conversation.*
More precisely, any communication equilibrium distribution can be achieved in the following way:
- *step 1: k is chosen in K according to p, and player i is informed of k_i;*
- *step 2: the players send public messages, twice, from a finite set of messages;*
- *step 3: the players choose actions.*

Comments:
- The mechanism is independent of the game, the priors, etc...
- For three players, the result is true if infinite sets of messages are allowed at the second time where the players send messages.

Sketch of the proof:
Only one inclusion has to be proved (the second one is obvious).
Let Q be a canonical communication equilibrium distribution. It induces a probability Q' on mappings from K to S: let f be such a mapping, then $Q'(f) = \prod_k Q(f(k)|k)$.
The first idea is the following:
Let f be selected according to Q', and f_i be told to player i. Then each player

learns his type k_i, and player i sends the public message k_i. Then $f(k)$ is played. This mechanism would lead to the right probability Q. But players get extra information that they may use to find new profitable deviations.
The new idea is thus to use this kind of mechanisms but to encode both f_i and k_i. For the detailed proof, we refer to the paper of Forges [6]. Hence permutations of K_i and $K \times S^i$ are chosen with uniform probabilibty by the players (so as to prevent deviations as in the previous mechanism) to encode the elements of K_i and $K \times S^i$. ∎

Starting with a game of incomplete information, one can apply Theorem 2.2, and then Theorem 2.1, to get the following result :

Theorem 2.3: *In a rational game with incomplete information, and at least 4 players, any communication equilibrium distribution can be realized as a Nash equilibrium of the game extended by two phases of conversation : one (private) before the types are announced, and one (public) afterwards.*

2.3 Mediated Talk

We now want to generate a correlation device with no restriction to 4 players and no private conversation or signal. This section follows Lehrer [10] and Lehrer and Sorin [11].

Let G be a m player game. We extend G by a pre-play communication phase. In this phase player i selects possibly randomly a signal a_i from a finite set A_i, and sends it to a mediator. The mediator announces a public and deterministic message $\mu(a_1, \ldots, a_m)$.
Then each player chooses an action that may obviously depend on $\mu(a)$, and on a_i.
We have the following result.

Theorem 2.4: *Let Q be a correlated equilibrium distribution of G. Assume that all the probabilities of Q are rational numbers. Then there exists a "public mediated talk" extension of G, as defined above, that has a Nash equilibrium that induces Q.*

The fact that all the coefficients are rational permits us to restrict the study to uniform distributions. We will show how the proof goes on an example.
Consider the following 2×2 game with 2 players :

$$\begin{pmatrix} (7,7) & (3,8) \\ (8,3) & (0,0) \end{pmatrix}$$

We want to generate the following correlated equilibrium :

$$\begin{pmatrix} 1/2 & 1/4 \\ 1/4 & 0 \end{pmatrix}$$

There are four different signals (a, b, c, d), each of them corresponding to a cell of the matrix (a corresponds to Top-Left, b to Top-Right, c to Botttom-Left, d to Bottom- Right). Suppose that the mediator sends the signals to the players according to the following matrix denoted by $M(a,b,c,d)$:

$$\begin{array}{c} \\ 1 \\ 2 \\ 3 \\ 4 \end{array} \begin{array}{c} \begin{array}{cccc} 1 & 2 & 3 & 4 \end{array} \\ \begin{pmatrix} a & a & b & c \\ c & a & a & b \\ b & c & a & a \\ a & b & c & a \end{pmatrix} \end{array}$$

If the signal is a it means that player I is told to play Top and player II to play Left.
Then, if the players choose the numbers at random, they get the signals with the right probability.
But, we also want the information of the players to correspond to the correlation device: for instance, if player I has chosen 1, and the signal is a, we want him to think that player II will play Left with probability $2/3$ and Right with probability $1/3$, and we want him to play Top.
If we consider the signals as a recommendation, and if we suppose that the players will follow the recommendations, using the above matrix player I knows what player II will play if the signal is a.
That is why we have to duplicate the matrix. We encode the message that will be received later. So, we construct a big 2×2 matrix according to which the mediator will compute the message he sends to the players.
Player I chooses at random a couple in $\{T, B\} \times \{1, ..., 4\}$. Player II chooses at random a couple in $\{L, R\} \times \{1, ..., 4\}$.
Each cell of the big matrix is a signalling matrix as defined above : in the cell (T,L) put $M(a,b,c,d)$, in (T,R) put $M(b,a,d,c)$, in (B,L) put $M(c,d,a,b)$, and in (B,R) put $M(d,c,b,a)$. The matrix is thus the following :

$$\begin{pmatrix} \begin{pmatrix} a & a & b & c \\ c & a & a & b \\ b & c & a & a \\ a & b & c & a \end{pmatrix} & \begin{pmatrix} b & b & a & d \\ d & b & b & a \\ a & d & b & b \\ b & a & d & b \end{pmatrix} \\ \begin{pmatrix} c & c & d & a \\ a & c & c & d \\ d & a & c & c \\ c & d & a & c \end{pmatrix} & \begin{pmatrix} d & d & c & b \\ b & d & d & c \\ c & b & d & d \\ d & c & b & d \end{pmatrix} \end{pmatrix}$$

The first component of the messages (the pair of letters) determines the matrix chosen and the second one (the pair of numbers) the cell in the submatrix. In the big matrix player I chooses a row and player II a column.

The mediator sends as a public message the letter corresponding to the choice of the players : a if the choice was $(T, 1, L, 1)$, b if the choice was $(T, 1, R, 1)$ and so on... $T, B, L, R,$ are codes. Each block of the matrix gives the right probability to each signal, and the signification of each signal in terms of action is encoded through the block to which it belongs.
Note that each player has only a partial information about this block.

The strategies are the following :
Player 1 :
If he sent T and the signal is a or b then he plays T.
If he sent T and the signal is c or d then he plays B.
If he sent B and the signal is a or b then he plays B.
If he sent B and the signal is a or b then he plays T.
Player 2 :
If he sent R and the signal is a or c then he plays R.
If he sent R and the signal is b or d then he plays L.
If he sent L and the signal is a or c then he plays L.
If he sent L and the signal is a or d then he plays R.

Now the signalling matrices give the right ex-post probabilities on the signal the other player sent, and hence on his action : if player I sent $(T, 1)$ and the signal is a, he knows we are on the first row. Hence with probability $1/3$, player II sent $(R, 1)$, with probability $1/3$, $(R, 2)$, and with probability $1/3$, (L,3). So the outcome (Top, Left) will appear with probability $2/3$, and (Top, Right) with probability $1/3$, if player II follows the recommendation. The matrix leads to the desired information structure.
If one of the players does not choose his message uniformly, the probability to get a given signal will be the same, and also the information he gets. So there can be no profitable deviation at that stage. Given the messages and the information they give, the fact that Q was initially a correlated equilibrium distribution implies that there is no profitable deviation at the action stage.

In fact the result is more precise: given any correlation device there is a "public mediated talk" device that mimics it, i.e. that induces the same outcome and the same information (compare with a sure protocol, but here there is no "deviation" message).

Conclusion

In this lecture, we saw new notions of equilibrium related to the possibility for the players to correlate their moves and to communicate.
Pre-play information and communication give the players larger strategy sets, and hence permits to achieve outcomes that are Pareto prefered to all Nash equilibrium outcomes.
One has to define precisely the kind of communication that is allowed: it has a great impact on the nature of the equilibria one obtains.
Cooperation is a notion that is concerned with the behavior of the players, namely their moves. But we have seen that it may come from another level: through signals and information, the players can enrich their strategy spaces and generate correlation that, in fine, induce the requested consequence on their moves.

References

[1] Aumann, R.J. (1974): "Subjectivity and correlation in randomized strategies", *Journal of Mathematical Economics*, 1, 67-96.

[2] Aumann, R.J. (1987): "Correlated equilibrium as an expression of bayesian rationality", *Econometrica*, 55, 1-18.

[3] Bárány, I. (1992): "Fair distribution protocols or how the players replace fortune", *Mathematics of Operations Research*, 17, 327-340.

[4] Forges, F. (1986): "An approach to communication equilibria", *Econometrica*, 54, 1375-1385.

[5] Forges, F. (1988): "Communication equilibria in repeated games with incomplete information", *Mathematics of Operations Research*, 13, 191-231.

[6] Forges, F. (1990): "Universal mechanisms", *Econometrica*, 58, 1341-1364.

[7] Forges, F. (1992): "Repeated games of incomplete information - non zero sum", in R.J. Aumann and S. Hart (eds), "Handbook of Game Theory with Economic Applications", vol 1, 155-177, North Holland.

[8] Harsanyi, J.C. (1967):" Games with incomplete information played by "bayesian" players, Part I-III", *Management Science*, 14, 150-182, 320-334, 486-502.

[9] Hart, S. and D. Schmeidler (1989): "Existence of correlated equilibria", *Mathematics of Operations Research*, 14, 18-25.

[10] Lehrer, E. (1994): " Mediated talk", Northwestern University Discussion Paper.

[11] Lehrer, E. and S. Sorin (1994): "One shot public mediated talk", mimeo.

[12] Mertens, J.F. (1994): "Correlated and communication equilibria", in J.F. Mertens and S. Sorin (eds), "Game Theoretic Methods in General Equilibrium Analysis", 243-248, Kluwer Academic Publishers.

[13] Mertens, J.F., S. Sorin and S. Zamir (1994): "Repeated games", Part A, Core Discussion Paper 9420.

[14] Myerson, R. (1986): "Multistage games with communication", *Econometrica*, 54, 323-358.

[15] Peck, J. (1994): " Notes on correlated equilibrium and sunspot equilibrium", in J.F. Mertens and S. Sorin (eds), "Game Theoretic Methods in General Equilibrium Analysis", 249-260, Kluwer Academic Publishers.

[16] Sorin, S. (1994): "Implementation with plain conversation", in J.F. Mertens and S. Sorin (eds), "Game Theoretic Methods in General Equilibrium Analysis", 261-268, Kluwer Academic Publishers.

Rationality and Bounded Rationality [1]

Robert J. Aumann

The Hebrew University of Jerusalem

CONTENTS

1. Introduction

Economists have for long expressed dissatisfaction with the complex models of strict rationality that are so pervasive in economic theory. There are several objections to such models. First, casual empiricism or even just simple introspection lead to the conclusion that even in quite simple decision problems, most economic agents are not in fact maximizers, in the sense that they do not scan the choice set and consciously pick a maximal element from it. Second, such maximizations are often quite difficult, and even if they wanted to, most people (including economists and even computer scientists) would be unable to carry

---

[1]Research partially supported by the National Science Foundation under grant IRI-8814953. Previous to its presentation at the NATO ASI at Stony Brook, versions of this lecture were presented as the 1986 Nancy L. Schwartz Memorial Lecture, at a 1989 workshop on bounded rationality at the Institute for Mathematical Studies in the Social Sciences (Economics) at Stanford University, and at the Fourth Conference on Theoretical Aspects of Reasoning About Knowledge, Monterey, March, 1992. We are grateful to the editors of the Schwartz Lecture Series publications for permission to reprint.

them out in practice. Third, polls and laboratory experiments indicate that people often fail to conform to some of the basic assumptions of rational decision theory. Fourth, laboratory experiments indicate that the conclusions of rational analysis (as distinguished from the assumptions) sometimes fail to conform to "reality." And finally, the conclusions of rational analysis sometimes seem unreasonable even on the basis of simple introspection.

From my point of view, the last two of the above objections are more compelling than the first three. In science, it is more important that the conclusions be right than that the assumptions sound reasonable. The assumption of a gravitational force seems totally unreasonable on the face of it, yet leads to correct conclusions. "By their fruits shall ye know them" (Matthew).

In the sequel, though, we shall not hew strictly to this line; we shall examine various models that, between them, address all the above issues.

To my knowledge, this area was first extensively investigated by Herbert Simon (1955, 1972). Much of Simon's work was conceptual rather than formal. For many years after this initial work, it was recognized that the area was of great importance, but the lack of a formal approach impeded its progress. Particular components of Simon's ideas, such as satisficing, were formalized by several workers, but never led to an extensive theory, and indeed did not appear to have significant implications that went beyond the formulations themselves.

There is no unified theory of bounded rationality, and probably never will be. Here we examine several different but related approaches to the problem, which have evolved over the last ten or fifteen years. We will not survey the area, but discuss some of the underlying ideas. For clarity, we may sometimes stake out a position in a fashion that is more one-sided and extreme than we really feel; we have the highest respect and admiration for all the scientists whose work we cite, and beg them not to take offense.

From the point of view of the volume of research, the field has "taken off" in the last half dozen years. An important factor in making this possible was the development of computer science, complexity theory, and so on, areas of inquiry that created an intellectual climate conducive to the development of the theory of bounded rationality. A significant catalyst was the experimental work of Robert Axelrod (1984) in the late seventies and early eighties, in which experts were asked to prepare computer programs for playing the repeated prisoners' dilemma. The idea of a computer program for playing repeated games presaged some of the central ideas of the later work; and the winner of Axelrod's tournament — tit-for-tat — was, because of its simplicity, nicely illustrative of the bounded rationality idea. Also, repeated games became the context of much of the subsequent work.

The remainder of this lecture is divided into five parts. First we discuss the evolutionary approach to optimization — and specifically to game theory — and some of its implications for the idea of bounded rationality, such as the development of truly dynamic theories of games, and the idea of "rule rationality" (as opposed to "act rationality"). Next comes the area of "trembles," including equilibrium refinements, "crazy" perturbations, failure of common knowledge of rationality, the limiting average payoff in infinitely repeated games as an expression of bounded rationality, ε-equilibria, and related topics. Section 3 deals with players who are modeled as computers (finite state automata, Turing machines), which has now become perhaps the most active area in the field. In Section 4 we discuss the work on the foundations of decision theory that deals with various paradoxes (such as Allais (1953) and Ellsberg (1961)), and with results of laboratory experiments, by relaxing various of the postulates and so coming up with a weaker theory. Section 5 is devoted to one or two open problems.

Most of this lecture is set in the framework of non-cooperative game theory, because most of the work has been in this framework. Game theory is indeed particularly appropriate for discussing fundamental ideas in this area, because it is relatively free from special institutional features. The basic ideas are probably applicable to economic contexts that are not game-theoretic (if there are any).

2. Evolution

2.1 Nash Equilibria as Population Equilibria

One of the simplest, yet most fundamental ideas in bounded rationality — indeed in game theory as a whole — is that no rationality at all is required to arrive at a Nash equilibrium; insects and even flowers can and do arrive at Nash equilibria, perhaps more reliably than human beings. The Nash equilibria of a strategic (normal) form game correspond precisely to population equilibria of populations that interact in accordance with the rules — and payoffs — of the game.

A version of this idea — the evolutionarily stable equilibrium — was first developed by John Maynard Smith (1982) in the early seventies and applied by him to many biological contexts (most of them animal conflicts within a species). But the idea applies also to Nash equilibria — not only to interaction within a species, but also to interactions between different species. It is worthwhile to give a more precise statement of this correspondence.

Consider, then, two populations — let us first think of them as different species — whose members interact in some way. It might be predator and prey, or cleaner and host fish, or bees and flowers, or whatever. Each interaction between an individual of population A and one of population B results in an increment (or decrement) in the fitness of each; recall that the *fitness* of an individual is defined as the expected number of its offspring (I use "its" on purpose, since strictly speaking, reproduction must be asexual for this to work). This increment is the payoff to each of the individuals for the encounter in question. The payoff is determined by the genetic endowment of each of the interacting individuals (more or less aggressive or watchful or keen-sighted or cooperative, etc.). Thus one may write a bimatrix in which the rows and columns represent the various possible genetic endowments of the two respective species (or rather those different genetic endowments that are relevant to the kind of interaction being examined), and the entries represent the single encounter payoffs that we just described. If one views this bimatrix as a game, then the Nash equilibria of this game correspond precisely to population equilibria; that is, under asexual reproduction, the proportions of the various genetic endowments within each population remain constant from generation to generation if and only if these proportions constitute a Nash equilibrium.

This is subject to the following qualification: in each generation, there must be at least a very small proportion of *each* kind of genetic endowment; that is, each row and column must be represented by at least *some* individuals. This minimal presence, whose biological interpretation is that it represents possible mutations, is to be thought of as infinitesimal; specifically, an encounter between *two* such mutants (in the two populations) is considered impossible.

A similar story can be told for games with more than two players, and for evolutionary processes other than biological ones; e.g., economic evolution, like the development of the QWERTY typewriter keyboard, studied by the economic historian Paul David (1986). It also applies to learning processes that are perhaps not strictly analogous to asexual reproduction. And though it does not apply to sexual reproduction, still one may hope that roughly speaking, similar ideas may apply.

One may ask who are the "players" in this "game"? The answer is that the two "players" are the two populations (i.e., the two species). The individuals are definitely *not* the "players"; if anything, each individual corresponds to the pure strategy representing its genetic endowment (note that there is no sense in which an individual can "choose" its own genetic endowment). More accurately, though, the pure strategies represent kinds of genetic endowment, and not individuals. Individuals indeed play no explicit role in the mathematical model; they are swallowed up in the proportions of the various pure strategies.

Some biologists object to this interpretation, because they see it as implying group or species selection rather than individual selection. The player is not the species, they argue; the individual "acts for its own good," not the good of the group, or of the population, or of the species. Some even argue that it is the gene (or rather the allele) that "acts for its own good," not the individual. The point, though, is that *nothing* in this model really "acts for its own good"; nobody "chooses" anything. It is the process as a whole that selects the traits. The most we can do is ask what it is that corresponds to the player in the mathematical model, and this is undoubtedly the population.

A question that at first seems puzzling is what happens in the case of interactions within a species, like animal conflicts for females, etc. Who are the players in this game? If the players are the populations, then this must be a one-person game, since there is only one population. But that doesn't look right, either, and it certainly doesn't correspond to the biological models of animal conflicts.

The answer is that it is a two-person symmetric game, in which both players correspond to the same population. In this case we look not for just any Nash equilibria, but for symmetric ones only.

2.2 Evolutionary Dynamics

The question of developing a "truly" dynamic theory of games has long plagued game theorists and economic theorists. (If I am not mistaken, it is one of the conceptual problems listed by Kuhn and Tucker (1953) in the introduction to Volume II of "Contributions to the Theory of Games" — perhaps the last one in that remarkably prophetic list to be successfully solved.) The difficulty is that ordinary rational players have foresight, so they can contemplate all of time from the beginning of play. Thus the situation can be seen as a one-shot game each play of which is actually a long sequence of "stage games," and then one has lost the dynamic character of the situation.

The evolutionary approach outlined above "solves" this conceptual difficulty by eliminating the foresight. Since the process is mechanical, there is indeed no foresight; no strategies for playing the repeated game are available to the "players."

And indeed, a fascinating dynamic theory does emerge. Contributions to this theory have been made by Young (1993), Foster and Young (1990), and Kandori, Mailath, and Rob (1993). A book on the subject has been written by Hofbauer and Sigmund (1988) and there is an excellent chapter on evolutionary dynamics in the book by van Damme (1987) on refinements of Nash equilibrium. Many others have also contributed to the subject.

It turns out that Nash equilibria are often unstable, and one gets various kinds of cycling effects. Sometimes the cycles are "around" the equilibrium, like in "matching pennies," but at other times one gets more complicated behavior. For example, the game in Figure 1 has ((1/3,1/3,1/3),(1/3,1/3,1/3)) as its only Nash equilibrium; the evolutionary dynamics does not cycle "around" this point, but rather confines itself (more or less) to the strategy pairs in which the payoff is 4 or 5. This suggests a possible connection with correlated equilibria; this possibility has recently been investigated by Foster and Vohra (1994).

Thus evolutionary dynamics emerges as a form of rationality that is bounded in that foresight is eliminated.

0,0	4,5	5,4
5,4	0,0	4,5
4,5	5,4	0,0

Figure 1

2.3 "Rule Rationality" vs. "Act Rationality"

In a famous experiment conducted by Güth et al. (1982) and later repeated, with important variations, by Binmore et al. (1985), two players were asked to divide a considerable sum of money (ranging as high as DM 100). The procedure was that P1 made an offer, which could be either accepted or rejected by P2; if it was rejected, nobody got anything. The players did not know each other and never saw each other; communication was a one-time affair via computer.

"Rational" play would predict a 99-1 split, or 95-5 at the outside. Yet in by far the most trials, the offered split was between 50-50 and 65-35. This is surprising enough in itself. But even more surprising is that in most (all?) cases in which P2 was offered less than 30 percent, he actually *refused*. Thus, he *preferred* to walk away from as much as DM 25 or 30. How can this be reconciled with ordinary notions of utility maximization, not to speak of game theory?

It is tempting to answer that a player who is offered five or ten percent is "insulted." Therefore, his utilities change; he gets positive probability from "punishing" the other player.

That's alright as far as it goes, but it doesn't go very far; it doesn't explain very much. The "insult" is treated as exogenous. But obviously the "insult" arose from the situation. Shouldn't we treat the "insult" itself endogenously, somehow explain *it* game-theoretically?

I think that a better way of explaining the phenomenon is as follows: ordinary people do not behave in a consciously rational way in their day-to-day activities. Rather, they evolve "rules of thumb" that work in general, by an evolutionary process like that discussed at 2.1 above, or a learning process with similar properties. Such "rules of thumb" are like genes (or, rather, alleles). If they work well, they are fruitful and multiply; if they work poorly, they become rare and eventually extinct.

One such rule of thumb is "Don't be a sucker; don't let people walk all over you." In general, the rule works well, so it becomes widely adopted. As it happens, the rule doesn't apply to Güth's game, because in that particular situation, a player who refuses DM 30 does not build up his reputation by the refusal (because of the built-in anonymity). But the rule has not been consciously chosen, and will not be consciously abandoned.

So we see that the evolutionary paradigm yields a third form of bounded rationality: rather than consciously maximizing in each decision situation, players use rules of thumb that work well "on the whole."

3. Perturbations of Rationality

3.1 Equilibrium Refinements

Equilibrium refinements — Selten (1975), Myerson (1978), Kreps and Wilson (1982), Kalai and Samet (1984), Kohlberg and Mertens (1986), Basu and Weibull (1991), Van Damme (1984), Reny (1992), Cho and Kreps (1989), and many others — don't really sound like bounded rationality. They sound more like super-rationality, since they go beyond the basic utility maximization that is inherent in Nash equilibrium. In addition to Nash equilibrium, which demands rationality on the equilibrium path, they demand rationality also off the equilibrium path. Yet all are based in one way or another on "trembles" — small departures from reality.

The paradox is resolved by noting that in game situations, one man's irrationality requires another one's superrationality. *You* must be superrational in order to deal with *my* irrationalities. Since this applies to all players, taking account of possible irrationalities leads to a kind of superrationality for all. To be superrational, one must leave the equilibrium path. Thus, a more refined concept of rationality cannot feed on itself only; it can only be defined in the context of irrationality.

3.2 Crazy Perturbations

An idea related to the trembling hand is the theory of irrational or "crazy" types, as propounded first by the "gang of four" (Kreps, Milgrom, Roberts, and Wilson (1982)), and then taken up by Fudenberg and Maskin (1986), Aumann and Sorin (1989), Fudenberg and Levine (1989), and no doubt others. In this work there is some kind of repeated or other dynamic game set-up; it is assumed that with high probability the players are "rational" in the sense of being utility maximizers, but that with a small probability, one or both play some one strategy, or one of a specified set of strategies, that are "crazy" — have no a priori relationship to rationality. An interesting aspect of this work, which differentiates it from the "refinement" literature, and makes it particularly relevant to the theory of bounded rationality, is that it is usually the crazy type, or a crazy type, that wins out — takes over the game, so to speak. Thus, in the original work of the gang of four on the prisoner's dilemma, there is only one crazy type, who always plays tit-for-tat, no matter what the other player does; and it turns out that the rational type must imitate the crazy type, he must also play tit-for-tat, or something quite close to it. Also, the "crazy" types, while irrational in the sense that they do not maximize utility, are usually by no means random or arbitrary (as they are in refinement theory). For example, we have already noted that tit-for-tat is computationally a very simple object, far from random. In the work of Aumann and Sorin, the crazy types are identified with bounded recall strategies; and in the work of Fudenberg and Levine (1989), the crazy types form a denumerable set, suggesting that they might be generated in some systematic manner, e.g., by Turing machines. There must be method to the madness; this is associated with computational simplicity, which is another one of the underlying ideas of bounded rationality.

3.3 Epsilon-Equilibria

Rather than playing irrationally with a small probability, as in 3.1 and 3.2 above, one may deviate slightly from rationality by playing so as almost, but not quite, to maximize utility; i.e., by playing to obtain a payoff that is within ε of the optimum payoff. This idea was introduced by Radner (1980) in the context of repeated games, in particular of the repeated prisoners' dilemma; he showed that in a long but finitely repeated prisoners' dilemma, there are ε-equilibria with small ε in which the players "cooperate" until close to the end (though, as is well-known, all exact equilibria lead to a constant stream of "defections").

3.4 Infinitely Repeated Games with Limit-of-the-Average Payoff

There is an interesting connection between ε-equilibria in finitely repeated games and infinitely repeated games with limit of the average payoff ("undiscounted"). The limit of the average payoff has been criticized as not representing any economic reality; many workers prefer to use either the finitely repeated game or limits of payoffs in discounted games with small discounts. Radner, Myerson and Maskin (1986), Forges, Mertens and Neyman (1986), and perhaps others, have demonstrated that the results of these two kinds of analysis can indeed be quite different.

Actually, though, the infinitely repeated undiscounted game is in some ways a simpler and more natural object than the discounted or finite games. In calculating equilibria of a finite or discounted game, one must usually specify the number n of repetitions or the discount rate d; the equilibria themselves depend crucially on these parameters. But one may want to think of such a game simply as "long," without specifying how long. Equilibria in the undiscounted game may be thought of as "rules of thumb," which tell a player how to play in a "long repetition," independently of *how* long the repetition is. Whereas limits of finite or discounted equilibrium payoffs tell the players approximately how much payoff to expect in a long repetition, analysis of the undiscounted game tells him approximately how to play.

Thus, the undiscounted game is a framework for formulating the idea of a duration-independent strategy in a repeated game. Indeed, it may be shown that an equilibrium in the undiscounted game is an approximate equilibrium simultaneously in all the n-stage truncations, the approximation getting better and better as n grows. Formally, a strategy profile ("tuple") is an equilibrium in the undiscounted game if and only if for some sequence of ε_n tending to zero, each of its n-stage truncations is an ε_n-equilibrium (in the sense of Radner described above) in the n-stage truncation of the game.

3.5 Failure of Common Knowledge of Rationality

In their paper on the repeated prisoners' dilemma, the Gang of Four pointed out that the effect they were demonstrating holds not only when one of the players believes that with some small probability, the other is a tit-for-tat automaton, but also if one of them only believes (with small probability) that the other believes this about him (with small probability). More generally, it can be shown that many of the perturbation effects we have been discussing do not require an actual departure from rationality on the part of the players, but only a lack of common knowledge of rationality (Aumann 1992).

4. Automata, Computers and Turing Machines

We come now to what is probably the "mainstream" of the newer work in bounded rationality, namely, the theoretical work that has been done in the last four or five years on automata and Turing machines playing repeated games. The work was pioneered by A. Neyman (1985) and A. Rubinstein (1986), working independently and in very different directions. Subsequently, the theme was taken up by Ben-Porath (1993), Kalai and Stanford (1988), Zemel (1989), Abreu and Rubinstein (1988), Ben-Porath and Peleg (1987), Lehrer (1988), Papadimitriou (1992), Stearns (1989), and many others, each of whom made significant new contributions to the subject in various different directions. Different branches of this work have been started by Lewis (1985) and Binmore (1987 and 1988), who have also had their following.

It is impossible to do justice to all this work in a reasonable amount of space, and we content ourselves with brief descriptions of some of the major strands. In one strand, pioneered by Neyman, the players of a repeated game are limited to using mixtures of pure strategies, each of which can be programmed on a finite automaton with an exogenously fixed number of states. This is reminiscent of the work of Axelrod, who required the entrants in his experiment to write the strategies in a fortran program not exceeding a stated limit in length. In another strand, pioneered by Rubinstein, the size of the automaton is endogenous; computer capacity, so to speak, is considered costly, and any capacity that is not actually used in equilibrium play is discarded. The two approaches lead to very different results. The reason is that Rubinstein's approach precludes the use of "punishment" or "trigger" strategies, which swing into action only when a player departs from equilibrium, and whose sole function is precisely to prevent such departures. In the evolutionary interpretation of repeated games, Rubinstein's approach may be more appropriate when the stages of the repeated game represent successive generations, whereas Neyman's may be more appropriate when each generation plays the entire repeated game (which would lead to the evolution of traits having to do with reputation, like "Don't be a sucker").

The complexity of computing an optimal strategy in a repeated game, or even just a best response to a given strategy, has been the subject of works by several authors, including Gilboa (1988), Ben-Porath (1989), and Papadimitriou (1989). Related work has been done by Lewis (1992), though in the framework of recursive function theory (which is related to infinite Turing machines) rather than complexity theory (which has to do with finite computing devices). Roughly speaking, the results are qualitatively similar: finding maxima is hard. Needless to say, in the evolutionary approach to games, nobody has to find the maxima; they are picked out by evolution. Thus, the results of complexity theory again underscore the importance of the evolutionary approach.

Binmore (1987 and 1988) and his followers have modeled games as pairs (or n-tuples) of Turing machines in which each machine carries in it some kind of idea of what the other "player" (machine) might look like.

Other important strands include work by computer scientists who have made the connection between distributed computing and games ("computers as players," rather than "players as computers"). For a survey, see Linial (1995).

5. Relaxation of Rationality Postulates

A not uncommon activity of decision, game, and economic theorists since the fifties has been to call attention to the strength of various postulates of rationality, and to investigate the consequences of relaxing them. Many workers in the field — including the writer of these lines — have at one time or another done this kind of thing. People have constructed theories of choice without transitivity, without completeness, violating the sure-thing principle, and so on. Even general equilibrium theorists have engaged in this activity, which may be considered a form of limited rationality (on the part of the agents in the model). This kind of work is most interesting when it leads to outcomes that are qualitatively different — not just weaker — from those obtained with the stronger assumptions; but I don't recall many such cases. It can also be very interesting and worthwhile when one gets roughly similar results with significantly weaker assumptions.

6. An Open Problem

We content ourselves with one open problem, which is perhaps the most challenging conceptual problem in the area today: to develop a meaningful formal definition of rationality in a situation in which calculation and analysis themselves are costly and/or limited. In the models we have discussed up to now, the problem has always been well defined, in the sense that an absolute maximum is chosen from among the set of feasible alternatives, no matter how complex a process that maximization may be. The alternatives themselves involve bounded rationality, but the process of choosing them does not.

Here, too, an evolutionary approach may eventually turn out to be the key to a general solution.

References

Abreu, D. and A. Rubinstein (1988), "The Structure of Nash Equilibrium in Repeated Games with Finite Automata," *Econometrica*, 56, 1259-1281.

Allais, M. (1953), "Le Comportement de l'Homme Rationnel devant le Risque: Critiques des Postulats et Axioms de l'Ecole Americaine," *Econometrica*, 21, 503-546.

Aumann, R. J. (1992), "Irrationality in Game Theory," in *Economic Analysis of Markets and Games (Essays in Honor of Frank Hahn)*, edited by P. Dasgupta, D. Gale, O. Hart, and E. Maskin, Cambridge and London: MIT Press, 214-227.

Aumann, R. J. and S. Sorin (1989), "Cooperation and Bounded Recall," *Games and Economic Behavior*, 1, 5-39.

Axelrod, R. (1984), *The Evolution of Cooperation*, New York: Basic Books.

Basu, K. and J. W. Weibull (1991), "Strategy Subsets Closed Under Rational Behavior," *Economics Letters*, 36, 141-146.

Ben-Porath, E. and B. Peleg (1987), "On the Folk Theorem and Finite Automata," Hebrew University discussion paper.

Ben-Porath, E. (1993), "Repeated Games with Finite Automata," *Journal of Economic Theory*, 59, 17-32.

Ben-Porath, E. (1990), "The Complexity of Computing Best Response Automata in Repeated Games with Mixed Strategies," *Games and Economic Behavior*, 2, 1-12.

Binmore, K. G. (1987), "Modelling Rational Players, I," *Economics and Philosophy*, 3, 179-214.

Binmore, K. G. (1988), "Modelling Rational Players II," *Economics and Philosophy*, 4, 9-55.

Binmore, K., A. Shaked and J. Sutton (1985), "Testing Noncooperative Bargaining Theory: A Preliminary Study," *American Economic Review*, 75, 1178-1180.

Cho, I.-K. and D. Kreps (1987), "Signaling Games and Stable Equilibria," *Quarterly Journal of Economics*, 102, 179-221.

David, P. A. (1986), "Understanding the Economics of QWERTY: The Necessity of History," Chapter 4, *Economic History and the Modern Economist*, edited by W. N. Parker, New York, Basil Blackwell.

Ellsberg, D. (1961), "Risk, Ambiguity and the Savage Axioms," *Quarterly Journal of Economics*, 75, 643-669.

Forges, F., J.-F. Mertens and A. Neyman (1986), "A Counter Example to the Folk Theorem with Discounting," *Economics Letters*, 20, 7.

Foster, D. and H. P. Young (1990), "Stochastic Evolutionary Game Dynamics," *Theoretical Population Biology*, 38, 219-232.

Foster, D. and R. Vohra (1994), "Calibrated Learning and Correlated Equilibrium," University of Pennsylvania.

Fudenberg, D. and D. K. Levine (1989), "Reputation and Equilibrium Selection in Games with a Patient Player," *Econometrica*, 57, 759-779.

Fudenberg, D. and E. Maskin (1986), "The Folk Theorem in Repeated Games with Discounting and Incomplete Information," *Econometrica*, 54, 533-554.

Gilboa, I. (1988), "The Complexity of Computing Best Response Automaton in Repeated Games," *Journal of Economic Theory*, 45, 342-352.

Güth, W., R. Schmittberger and B. Schwarze (1982), "An Experimental Analysis of Ultimatum Bargaining," *Journal of Economic Behavior and Organization*, 3, 367-388.

Hofbauer, J. and K. Sigmund (1988), *Theory of Evolution and Dynamical Systems*, Cambridge University Press.

Kalai, E. and W. Stanford (1988), "Finite Rationality and Interpersonal Complexity in Repeated Games," *Econometrica*, 56, 397-410.

Kalai, E. and D. Samet (1984), "Persistent Equilibria," *International Journal of Game Theory*, 13, 129-144.

Kandori, M., G. Mailath and R. Rob (1993), "Learning, Mutation, and Long Run Equilibria in Games," *Econometrica*, 61, 29-56.

Kohlberg, E. and J.-F. Mertens, (1986), "On the Strategic Stability of Equilibria," *Econometrica*, 54, 1003-37.

Kreps, D. and R. Wilson (1982), "Sequential Equilibria," *Econometrica*, 50, 863-894.

Kreps, D., P. Milgrom, J. Roberts, and R. Wilson (1982), "Rational Cooperation in the Finitely Repeated Prisoners' Dilemma," *Journal of Economic Theory*, 27, 245-252.

Kuhn, H. W. and A. W. Tucker (eds.) (1953), *Contributions to the Theory of Games*, Vol. II, Annals of Mathematics Studies, 28, Princeton: Princeton University Press.

Lehrer, E. (1988), "Repeated Games with Stationary Bounded Recall Strategies," *Journal of Economic Theory*, 46, 130-144.

Lewis, A. (1985), "On Effectively Computable Realizations of Choice Functions," *Mathematical Social Sciences*, 10, 43-80.

Lewis, A. (1992), "Some Aspects of Effectively Constructive Mathematics that are Relevant to the Foundations of Neoclassical Mathematical Economics and the Theory of Games," *Mathematical Social Sciences*, 24, 209-236.

Linial, N. (1995), "Game Theoretic Aspects of Computing," Chapter 38, *Handbook of Game Theory with Applications to Economics*, Vol. 2, edited by R. J. Aumann and S. Hart, Amsterdam: North Holland.

Maynard Smith, J. (1982), *Evolution and the Theory of Games*, Cambridge: Cambridge University Press.

Myerson, R. B. (1978), "Refinements of the Nash Equilibrium Concept," *International Journal of Game Theory*, 7, 73-80.

Neyman, A. (1985), "Bounded Complexity Justifies Cooperation in the Finitely Repeated Prisoners' Dilemma," *Economics Letters*, 19, 227-229.

Papadimitriou, C. H. (1992), "On Players with a Bounded Number of States," *Games and Economic Behavior*, 4, 122-131.

Radner, R. (1980), "Collusive Behavior in Noncooperative Epsilon-equilibria of Oligopolies with Long but Finite Lives," *Journal of Economic Theory*, 22, 136-154.

Radner, R., R. Myerson and E. Maskin (1986), "An Example of a Repeated Partnership Game with Discounting and with Uniformly Inefficient Equilibria," *Review of Economic Studies*, 53, 59-69.

Reny, P. J. (1992), "Backwards Induction, Normal Form Perfection and Explicable Equilibria," *Econometrica*, 60, 627-649.

Rubinstein, A. (1986), "Finite Automata Play the Repeated Prisoners' Dilemma," *Journal of Economic Theory*, 39, 83-96.

Selten, R. (1975), "Reexamination of the Perfectness Concept for Equilibrium Points in Extensive Games," *International Journal of Game Theory*, 4, 25-55.

Simon, H. (1955), "A Behavioral Model of Rational Choice," *Quarterly Journal of Economics*, 64, 99-118.

Simon, H. (1972), "Theories of Bounded Rationality," in *Decision and Organization*, McGuire and Radner (eds.), Amsterdam: North Holland.

Stearns, R. E. (1989), "Memory-Bounded Game Playing Computing Devices," Technical Report No. 547, IMSSS, Stanford University.

van Damme, E. (1984), "A Relation Between Perfect Equilibria in Extensive Form Games and Proper Equilibria in Normal Form Games," *International Journal of Game Theory*, 13, 1-13.

van Damme, E. (1987), *Stability and Perfection of Nash Equilibria*, Berlin: Springer-Verlag.

Young, H. P. (1993), "The Evolution of Conventions," *Econometrica*, 61, 57-84.

Zemel, E. (1989), "Small Talk and Cooperation: A Note on Bounded Rationality," *Journal of Economic Theory*, 49, 1-9.

Cooperation, Repetition, and Automata

Abraham Neyman

The Hebrew University, Jerusalem, and S.U.N.Y. at Stony Brook

Abstract This chapter studies the implications of bounding the complexity of players' strategies in long term interactions. The complexity of a strategy is measured by the size of the minimal automaton that can implement it.

A finite automaton has a finite number of states and an initial state. It prescribes the action to be taken as a function of the current state and its next state is a function of its current state and the actions of the other players. The size of an automaton is its number of states.

The results study the equilibrium payoffs per stage of the repeated games when players' strategies are restricted to those implementable by automata of bounded size.

1 Introduction

The simplest strategic game quickly gives rise to a game of formidable complexity when one considers a finitely-repeated version of it.This is because the number of pure strategies in the repeated game grows as a double exponential of the number of repetitions. To just write down in decimal form the number of pure strategies available to a player in the hundred-times repeated prisoner's dilemma would require more digits than the number of all the letters in all of the books in the world. This chapter examines the implications of restricting the set of strategies to those that are implementable by finite automata of bounded size. Such restrictions place a bound on the complexity of strategies and they can (dramatically) alter the equilibrium play of a repeated game.

When we try to argue that an outcome is or is not an equilibrium in a game there are direct references to all possible strategies in that game. In the case of the hundred-times repeated Prisoner's Dilemma it would obviously be an impossible task to merely write out all of the strategies, let alone construct the huge matrix which would constitute the explicit representation of this game in normal (strategic) form. Moreover, many of the strategies in the finitely or infinitely repeated game are extremely complicated. They may involve actions contingent on so many possible past events that it would be nearly impossible to describe them; even writing them down or writing a program to execute some of these strategies is practically impossible. We consider here a theory that limits the strategies available to a player in a repeated game. The restriction is to those strategies that are implementable by bounded size automata- the simplest theoretical model of a computer. It turns out that the equilibria of the resulting strategic game can be dramatically different from those of the original game.

In the finitely repeated Prisoner's Dilemma, it is well known that all equilibria, and all correlated equilibria or communication equilibria, result in the repeated play of (defect, defect). This is in striking contrast to the experimental observation that real players do not choose always the dominant action of defecting, but in fact achieve some mode of cooperation.

The present approach justifies cooperation in the finitely repeated prisoner's dilemma, as well as in other finitely repeated games, without departing from the hypothesis of strict utility maximization, but under the added assumption that there are bounds (possibly very large) on the complexity of the strategies that players may use.

There are other methods of restricting strategies. I am not going to advocate here that the avenue we are taking is superior. Each one of the possible approaches has its pros and cons.

The paper surveys results about the equilibrium payoffs of repeated games when players' strategies in the repeated game are resricted. It contains also several new results, e.g., Propositions 2, 3, 4, 5, 6, and 7. There is no attempt here to survey all results related to the title, and therefore several important and related papers are not covered in this survey.

2 The Model

2.1 Strategic Games

Let G be an n-person game, $G = (N, A, r)$, where $N = \{1, 2, \ldots, n\}$ is the set of players, $A = \times_{i \in N} A_i$,A_i is a finite set of actions for player i, $i = 1, \ldots, n$, and $r = (r^i)_{i \in N}$ where $r^i : A \to I\!R$ is the payoff function of player i. The set A_i is called also the set of pure strategies of player i. We denote by $r : A \to I\!R^N$ the vector valued function whose ith component is r^i, i.e., $r(a) = (r^1(a), \ldots, r^n(a))$. We use also the more detailed description of G, $G = (N; (A_i)_{i \in N}; (r^i)_{i \in N})$, or $G = ((A_i)_{i \in N}; (r^i)_{i \in N})$ for short. For any finite set (or measurable space) B we denote by $\Delta(B)$ the set of all probability distributions on B. For any player i and any n-person game G, we denote by $v^i(G)$ his individual rational payoff in the mixed extension of the game G, i.e., $v^i(G) = \min \max r^i(a^i, \sigma^{-i})$ where the max ranges over all pure strategies of player i, and the min ranges over all $N \setminus \{i\}$-tuples of mixed strategies of the other players, and r^i denotes also the payoff to player i in the mixed extension of the game. We denote by $u^i(G)$ the individual rational payoff of player i in pure strategies, i.e., $u^i(G) = \min \max r^i(a^i, a^{-i})$ where the max ranges over all pure strategies of player i, and the min ranges over all $N \setminus \{i\}$ -tuples of pure strategies of the other players. Obviously $u^i(G) \geq v^i(G)$. We denote by $w^i(G)$ the max min of player i where he maximizes over his mixed strategies and the min is over the pure strategies of the other players, i.e., $w^i(G) = \max_{x \in \Delta(A_i)} \min_{a^{-i} \in A_{-i}} r^i(x, a^{-i})$ where $A_{-i} = \times_{j \neq i} A_j$. Recall that the minimax theorem asserts that for a two person game G, $v^i(G) = w^i(G)$. For any game G in strategic form we denote by $E(G)$ the set of all equilibrium payoffs in the game G, and by $F(G)$ the convex hull of all payoff vectors in the one shot game, i.e., $F(G) = co(r(A))$. Given a 2-person 0-sum game G we denote by $\mathrm{Val}(G)$ the minimax value of G, i.e., $\mathrm{Val}(G) = v^1(G)$.

2.2 The repeated games G^T and G^*

Given an n-person game, $G = ((A_i)_{i \in N}; (r^i)_{i \in N})$, we define a new game in strategic form $G^T = ((\Sigma^i(T))_{i \in N}; (r^i_T)_{i \in N})$ which models a sequence of T plays of G, called *stages*. After each stage, each player is informed of what the others did at the previous stage, and he remembers what he himself did and what he knew at previous stages. Thus, the information available to each player before choosing his action at stage t is all past actions of the players in previous stages of the game. Formally, let H_t, $t = 1, \ldots, T$, be the cartesian product of A by itself $t-1$ times, i.e., $H_t = A^{t-1}$, with the common set theoretic identification $A^0 = \{\emptyset\}$, and let $H = \cup_{t=1}^T H_t$. A pure strategy σ^i of player i in G^T is a function $\sigma^i : H \to A_i$. Obviously, H is a disjoint union of H_t, $t = 1, \ldots, T$ and therefore one often defines $\sigma^i_t : H_t \to A_i$ as the restriction of σ to H_t. We denote the set of all pure strategies of player i in G^T by $\Sigma^i(T)$. The set of pure strategies of player i in the infinitely repeated game G^* is denoted by Σ^i, i.e., $\Sigma^i = \{\sigma^i : \cup_{t=1}^\infty H_t \to A_i\}$.

Any N-tuple $\sigma = (\sigma^1, \ldots, \sigma^n) \in \times_{i \in N} \Sigma^i(T)$ (Σ^i) of pure strategies in G^T (in G^*) induces a play $\omega(\sigma) = (\omega_1(\sigma), \ldots, \omega_T(\sigma))$ $(\omega(\sigma) = (\omega_1(\sigma), \omega_2(\sigma), \ldots))$ defined by induction: $\omega_1(\sigma) = (\sigma^1(\emptyset), \ldots, \sigma^n(\emptyset)) = \sigma(\emptyset)$ and $\omega_t(\sigma) = \sigma(\omega_1(\sigma), \ldots, \omega_{t-1}(\sigma))$ or in other words $\omega^i_1(\sigma) = \sigma^i(\emptyset)$ and $\omega^i_t(\sigma) = \sigma^i(\omega_1(\sigma), \ldots, \omega_{t-1}(\sigma)) = \sigma^i_t(\omega_1(\sigma), \ldots, \omega_{t-1}(\sigma))$.

Set

$$r_T(\sigma) = \frac{r(\omega_1(\sigma)) + \ldots + r(\omega_T(\sigma))}{T}.$$

We define R_T or R for short to be the function from plays of G^T to the associated payoffs, i.e., $R_T : A^T \to \mathbf{R}^N$ is given by

$$R_T(a_1, \ldots, a_T) = \frac{r(a_1) + r(a_2) + \ldots + r(a_T)}{T}.$$

Two pure strategies σ^i and τ^i of player i in G^T (in G^*) are *equivalent* if for every $N \setminus \{i\}$ tuple of pure strategies $\sigma^{-i} = (\sigma^j)_{j \in N \setminus \{i\}}$, $\omega_t(\sigma^i, \sigma^{-i}) = \omega_t(\tau^i, \sigma^{-i})$ for every $1 \leq t \leq T$ ($1 \leq t$). The equivalence classes of pure strategies are called *reduced strategies*. For $i \in N$ let $A_{-i} = \times_{j \neq i} A_j$. Then an equivalence class of pure strategies is naturally identified with a function $\bar{\sigma}^i : \cup_{t=0}^\infty (A_{-i})^t \to A_i$.

2.3 Finite Automata

We will consider strategies of the repeated games which are described by means of automata, (which are also sometimes referred to as Moore machines or exact automata). An *automaton* for player i consists of a finite state space M; an initial state $q_1 \in M$; a function f that describes the action to be taken as a function of the different states of the machine, $f : M \to A_i$, where A_i denotes the set of actions of player i; and a transition function g that determines the next state of the machine as a function of its present state and the action of the other players, i.e., $g : M \times A_{-i} \to M$. Thus, an automaton of player i is represented by a 4-tuple $\langle M, q_1, f, g\rangle$. The *size* of an automaton is the number of states.

This machine, the automaton, will change its state in the course of playing a repeated game. At every state $q \in M$, f determines what action it will take. The next state of the automaton is determined by the current state and the action taken by the other players. We can think of such an automaton as playing a repeated game. It starts in its initial state q_1, and plays at the first stage of the game the action assigned by the action function f, $f(q_1)$. Thus, $f(q_1) = a_1^i$ is the action of the player at stage 1. The other players' action at this stage is $b_1 = a_1^{-i} \in A_{-i}$. Thus the history of play before the start of stage 2 is the n-tuple of actions, (a_1^i, a_1^{-i}), played at the first stage of the game . As a function of the present state, and the other players' actions, the machine is transformed into a new state which is given by the transition function g. The new state of the machine is $q_2 = g(q_1, b_1)$. The action that player i takes at stage 2, a_2^i, is described by the function $f : f(q_2) = f(g(q_1, b_1))$, and denoting by a_2^{-i} the action of the other players in stage 2, $(f(q_2), a_2^{-i})$ is the pair of actions played in the second stage of the repeated game, and so on.

What is the state of the machine at stage t of the game? The machine moves to a new state which is a function of the state of the machine in the previous stage and the action played by the other players. Thus $q_t = g(q_{t-1}, a_{t-1}^{-i})$, is the new state of the automaton at stage t, and player i takes at stage t the action $f(q_t) = f(g(q_{t-1}, a_{t-1}^{-i}))$ and so on.

Define inductively,

$$g(q, b_1, \ldots, b_t) = g(g(q, b_1, \ldots, b_{t-1}), b_t),$$

where $b_j \in A_{-i}$. The action prescribed by the automaton for player i at stage t is $f(g(q_1, a_1^{-1}, \ldots, a_{t-1}^{-i}))$ where a_j^{-i}, $1 \le j < t$, is the $N\backslash\{i\}$ tuple of actions at stage j. Therefore, any automaton $\alpha = \langle M, q_1, f, g\rangle$ of player i induces a strategy σ_α^i in G^T that is given by $\sigma_\alpha^i(\emptyset) = f(q_1)$ and

$$\sigma_\alpha^i(a_1, \ldots, a_{t-1}) = f(g(q_1, a_1^{-i}, \ldots, a_{t-1}^{-i})).$$

Note also that an automaton α of player i induces also a strategy σ_α^i of player i in the infinitely repeated game G^*. A strategy σ^i of player i in G^* (in G^T) is *implemented* by the automaton α of player i if σ^i is equivalent to σ_α^i, i.e., if for every $\sigma^{-i} \in \times_{j\neq i}\Sigma^j$ $(\Sigma^j(T))$, $\omega(\sigma^i, \sigma^{-i}) = \omega(\sigma_\alpha^i, \sigma^{-i})$.

A finite sequence of actions $a_1, \ldots, a_t$ is *compatible* with the pure strategy σ^i of player i in G^*, if for every $1 \leq s \leq t$, $\sigma^i(a_1, \ldots, a_{s-1}) = a^i_s$. Given a strategy σ^i of player i in G^*, any sequence of actions $a_1, \ldots, a_t$, induces a strategy $(\sigma^i | a_1, \ldots, a_t)$ in G^*, by

$$(\sigma^i | a_1, \ldots, a_t)(b_1, \ldots, b_s) = \sigma^i(a_1, \ldots, a_t, b_1, \ldots, b_s).$$

Proposition 1 *The number of different reduced strategies that are induced by a given pure strategy σ^i of player i in G^* and all σ^i-compatible sequences of actions equals the size of the smallest automaton that implements σ.*

2.4 Repeated Games with Finite Automata

Given a strategic game G and positive integers $m_1, \ldots, m_n$, we define $\Sigma^i(T, m_i)$ ($\Sigma^i(m_i)$) to be all pure strategies in $\Sigma^i(T)$ (in Σ^i) that are induced by an automaton of size m_i. Note that if a strategy is induced by an automaton of size m_i and $m'_i \geq m_i$ then it is also induced by an automaton of size m'_i. The game $G^T(m_1, \ldots, m_n)$ is the strategic game $(N; (\Sigma^i(T, m_i))_{i \in N}; r_T)$ where r_T here is the restriction of our earlier payoff function r_T to $\times_{i \in N} \Sigma^i(T, m_i)$.

The play in the supergame G^* which is induced by an n-tuple of strategies $\sigma = (\sigma^i)_{i \in N}$ with $\sigma^i \in \Sigma^i(m_i)$ enters a cycle of length $d \leq \prod_{i \in N} m_i$ after a finite number of stages. Indeed, if at stages t and s the n-tuple of states of the automata coincide, then for every nonnegative integer r, $\omega_{t+r}(\sigma) = \omega_{s+r}(\sigma)$. As the number of different n-tuples of automata states is bounded by $\prod_{i \in N} m_i$ the periodicity follows. Therefore, the limiting average payoff per stage is well defined whenever all players are restricted to strategies which are implemented by finite automata. The game $G^*_\infty(m_1, \ldots, m_n)$ or $G(m_1, \ldots, m_n)$ for short, is the strategic game $(N; (\Sigma^i(m_i))_{i \in N}; r_\infty)$ where r_∞ is defined as the limit of our earlier payoff function r_T as $T \to \infty$.

3 Zero-Sum Games with Finite Automata

In this section we present results of the value of 2-person 0-sum repeated games with finite automata. Results concerning zero-sum games are important for the study of the non-zero sum case by specifying the individual rational payoffs and thus the effective "punishments."

Consider the two-person zero-sum game of matching pennies:

1	−1
−1	1

Assume that player 1, the row player, and player 2, the column player, are restricted to play strategies that are implemented by automata of size m_1 and m_2 respectively. Recall that we are considering the mixed extension of the game in which the pure strategies of player i are those implemented by an automaton of size m_i. An easy observation is that for every m_1 there exists a sufficiently large m_2, a pure strategy $\tau \in \Sigma^2(m_2)$ and a positive integer T such that for any $t \geq T$ and $\sigma \in \Sigma^1(m_1)$, $r^1(\omega_t(\sigma, \tau)) = -1$. Therefore, we conclude in particular that for the above matching pennies game $G = (A, B, h)$, for every m_1 there exists m_2 such that $\mathrm{Val}(G(m_1, m_2)) = \max_A \min_B h(a, b)$. Moreover this statement is valid for any two-person zero-sum game $H = (A, B, h)$. Theorem 1 of Ben-Porath (1993) asserts that if $m_2 \geq m_1 |\Sigma^1(m_1)|$, where for a set X, $|X|$ denotes the number of elements in X, then

$$\mathrm{Val}\,(H(m_1, m_2)) = \max_{a \in A} \min_{b \in B} h(a, b).$$

Note that $|\Sigma^1(m)|$ is of the order of an exponential function of $m \log m$. However, it turns out that if the larger bound m_2 is subexponential in m_1, player 2 is unable to use effectively in the long run his larger bound. Indeed,

Theorem 1 *(Ben-Porath, 1986, 1993). Let $H = (A, B, h)$ be a two person 0-sum game in strategic form, and let $(m(n))_{n=1}^{\infty}$ be a sequence of positive integers with*

$$\lim_{n \to \infty} \frac{\log m(n)}{n} = 0.$$

Then,

$$\liminf_{n \to \infty} \mathrm{Val}\,(H(n, m(n))) \geq \mathrm{Val}\,(H).$$

Proof. W.l.o.g. we assume that $n \leq m(n)$. For every sequence $a = (a_1^1, a_2^1, \ldots)$ of actions of player 1 we denote by σ^a the pure strategy of player 1 with $\sigma_t^a(*) = a_t^1$. Note that if a is k-periodic then $\sigma^a \in \Sigma^1(k)$. For every k, $\sigma^1(k)$ denotes the mixed strategy σ^X of player 1 where $X = (X_1, X_2, \ldots)$ is a random k-periodic sequence of actions of player 1, with $X_1, X_2, \ldots, X_k$ i.i.d and the distribution of X_t is an optimal strategy of player 1 in the one shot game. It follows that for every pure strategy τ of player 2 and every $t \leq k$, $E_{\sigma^1(k),\tau}(h(a_t, b_t)|\mathcal{H}_t) \geq \mathrm{Val}(H)$, where $\mathcal{H}_t$ denotes the algebra generated by the actions $a_1, b_1, \ldots, a_{t-1}, b_{t-1}$ in stages $1, \ldots, t-1$. Therefore $Prob_{\sigma^1(k),\tau}(\sum_{t=1}^{k} h(a_t, b_t)/k < \mathrm{Val}(H) - \varepsilon) \leq e^{-C(\varepsilon)k}$ with $C(\varepsilon) > 0$. Therefore for every finite set $\mathcal{T} \subset \Sigma^2$,

$$Prob_{\sigma^1(n)}(\min_{\tau \in \mathcal{T}} h_n(\sigma^a, \tau) \leq \mathrm{Val}(H) - \varepsilon) \leq |\mathcal{T}| \exp(-C(\varepsilon)n). \tag{1}$$

Let τ be a pure strategy of player 2 which is implemented by an automaton of size $m(n)$, and set $\mathcal{T} = \{(\tau \mid b_1, \ldots, b_t)\}$. Then for every n periodic sequence a and every positive integer s,

$$\sum_{t=s+1}^{s+n} h(\omega_t(\sigma^a, \tau)) \geq \min_{\tau \in \mathcal{T}} \sum_{t=s+1}^{s+n} h(\omega_t(\sigma^a, \tau)).$$

As $|\mathcal{T}| \leq m(n)$, and $\sigma^1(n)$ is a mixture of (at most $|A|^n$) pure strategies of the form $\sigma^a \in \Sigma^1(n)$, the result follows from (1). ■

It is worth mentioning that the proof implies a stronger result. Setting $\Sigma^i_g(m)$ to be all strategies σ^i such that for each t $|\{(\sigma^i \mid b_1, \ldots, b_t) : b_j \in B\}| \leq m$, and $\sigma^1(n)$ as constructed in the proof, we conclude that under the same condition as in the theorem,

$$\liminf_{n\to\infty} Val\ H(\{\sigma^1(n)\}, \Sigma^2_g(m(n))) \geq \text{ Val } (H).$$

This stronger result implies that whenever $\lim_{n\to\infty} \log m(n)/n = 0$, for every n there exists a *random* n-periodic sequence of actions of player 1, (σ^X), which guarantees approximately the value $\text{Val}(H)$ against any strategy in $\Sigma^2_g(m(n))$. Note that for every pure strategy σ of player 1, there exists a strategy $\tau \in \Sigma^2_g(1)$ with $h_t(\sigma,\tau) \leq \max_{a\in A} \min_{b\in B} h(a,b)$. The next result asserts that when $m(n)\log m(n) = o(n)$ as $n \to \infty$, then there is a *deterministic* n-periodic sequence of actions of player 1, a, such that σ^a guarantees approximately $\text{Val}(H)$ when player 2 is restricted to strategies in $\Sigma^2(m(n))$.

Proposition 2 *Let $m : I\!N \to I\!N$ with $\lim_{n\to\infty} \frac{m(n)\log m(n)}{n} = 0$. Then for every n there exists an n-periodic sequence of actions of player 1, a, such that*

$$\lim_{n\to\infty} (\inf\{h_t(\sigma^a,\tau) \mid \tau \in \Sigma^2(m(n)), t \geq n\}) = \text{Val}(H).$$

Proof. Note that there is a positive constant K such that $|\Sigma^2(m(n))| \leq m(n)^{Km(n)}$. Let $k : I\!N \to I\!N$ be such that $\lim_{n\to\infty} \frac{m(n)\log m(n)}{k(n)} = 0$, and $\lim_{n\to\infty} k(n)/n = 0$. $X = (X_1, \ldots, X_{k(n)}, \ldots)$ be a random n-periodic sequence of actions of player 1, where $X_1, \ldots, X_{k(n)}$ are i.i.d each distributed according to the distribution of an optimal mixed strategy of player 1 in the one shot game, and $(X_1, \ldots, X_n)$ is $k(n)$-periodic. As $\lim_{n\to\infty} \frac{m(n)\log m(n)}{k(n)} = 0$, it follows that for every positive constant $C > 0$,

$$\lim_{n\to\infty} |\Sigma^2(m(n))| \exp\left(-Ck(n)\right) = 0,$$

and therefore it follows from (1) that

$$\lim_{n\to\infty} \Pr(\min_{\tau\in\Sigma^2(m(n))} h_{k(n)}(\sigma^X,\tau) \leq \text{ Val}(H) - \varepsilon) = 0$$

and therefore there is an n-periodic sequence of actions a such that

$$\lim_{n\to\infty} (\inf\{h_t(\sigma^a,\tau) \mid \tau \in \Sigma^2(m(n)), t \geq n\}) \geq Val(H).$$

■

The next result follows from the proof of the result of Ben-Porath (1993), and is used in the proof of theorems 5 and 6.

Theorem 2 *For every $\varepsilon > 0$ sufficiently small, if*

$$\exp(\varepsilon^2 m_1) \geq m_2 > 1,$$

then for every positive integer T,

$$\text{Val}\left(H^T(m_1, m_2)\right) \geq \text{Val}\left(H\right) - \varepsilon.$$

The next corollary is a restatement of Theorems 1 and 2 which provides a lower bound for equilibrium payoffs in nonzero sum repeated games with finite automata.

Corollary 1 *For every strategic game $G = (N, A, r)$, $i \in N$, and $\varepsilon > 0$ sufficiently small, if*

$$exp(\varepsilon^2 m_i) \geq m_j > 1 \qquad \text{for every } j \neq i,$$

then for every $x \in E(G^T(m_1, \ldots, m_n))$, or $x \in E(G(m_1, \ldots, m_n))$,

$$x^i \geq w^i(G) - \varepsilon.$$

The next result asserts that if the bound on the sizes of the automata of player 2 is larger than an exponential of the sizes of the automata of player 1, then player 2 could hold player 1 down to his maxmin in pure strategies.

Theorem 3 *For every 2-person 0-sum game $H = (\{1,2\}; (A, B); h)$, and every positive constant K with $K > \ln |A|$, if $m(n) \geq exp(Kn)$, then*

$$\text{Val}\,(H(n, m(n)) \to \max_{a \in A} \min_{b \in B} h(a, b) \text{ as } n \to \infty.$$

Proof. Let $K > \ln |A|$. It is sufficient to prove that for every $\varepsilon > 0$ there exists n_0 such that for every $n \geq n_0$ and every $m \geq \exp(Kn)$,

$$\text{Val}\ (H(n, m)) \leq \max_{a \in A} \min_{b \in B} h(a, b) + \varepsilon.$$

Note that for every positive constant C there exists n_0 such that for every $n \geq n_0$, $\exp(Kn) \geq Cn^2|A|^n$. Therefore the theorem follows from the next Lemma. ∎

Lemma 1 *For every $\varepsilon > 0$, there is a sufficiently large integer $K = K(\varepsilon)$, such that for every $m \geq Kn^2|A|^n$ there exists a strategy $\tau^* \in \Delta(\Sigma^2(m))$ such that for every $T \geq K^2n^3|A|^n$, and any strategy $\sigma \in \Sigma^1(n)$,*

$$h_T(\sigma, \tau^*) \leq \max_{a \in A} \min_{b \in B} h(a, b) + \varepsilon.$$

and therefore,

$$\text{Val}\,(G^T(n, m)) \leq \max_{a \in A} \min_{b \in B} h(a, b) + \varepsilon.$$

and

$$\text{Val}\,(G(n, m)) \leq \max_{a \in A} \min_{b \in B} h(a, b) + \varepsilon.$$

Proof. The idea of the proof is as follows: for every pure strategy $\sigma \in \Sigma^1(n)$ of player 1, there is $1 \leq k \leq n$ and a sequence of actions $b = b_1, \ldots, b_n, b_{n-k+1}, \ldots$ with $b_i = b_j$ whenever $i > j > n - k$ and $i - j = 0(mod\ k)$, such that the strategy τ^b of player 2 which plays the sequence b results in payoff $\leq \max_{a \in A} \min_{b \in B} h(a, b)$ in seach stage $t > m - k$. Such a strategy is implemented by an automaton of size n, and the number of such strategies σ^b is bounden by $n|B|^n$. The strategy of player 2 immitates choosing at random a pair k, σ^b, and if the resulting payoffs are not sufficiently small, it attempts another randomly chosen pair k, σ^b. The sufficiently large number of states of the automaton of player 2, gaurantees that with high probability the induced play will eventually enter a cycle with payoffs $\leq \max_{a \in A} \min_{b \in B} h(a, b)$ in each stage .

Formally, let $b : A \to B$ be a selection from the best reply correspondence of player 2. Construct the following mixed strategy of player 2, τ^*, which is implemented by an automaton with state space

$$M^2 = \{1, \ldots, n\} \times \{1, \ldots, \ell\}.$$

where $\ell = Kn|A|^n$. The initial state of the automaton of player 2 is $(1, 1)$. Let $a : M^2 \to A$ be a random function, each such function equally likely, i.e., for every $1 \leq i \leq n$, and every $1 \leq j \leq \ell$, $a(i, j)$ is a random element of A each one equally likely, and the various random elements $a(i, j)$ are independent. We define now the random action function of the automaton.

$$f^2(i, j) = b(a(i, j)).$$

The transition function of the automaton depends on a random sequence $k = k_1, \ldots, k_\ell$, $1 \leq k_j \leq n$, each such sequence equally likely and the sequence k is independent of the function a. We are ready now to define the transition function which depends on the functions b and a and on the random sequence k.

$$g^2((i, j), c) = \begin{cases} (i + 1, j) & \text{if } \; i < n \text{ and } c = a(i, j) \\ (k_j, j) & \text{if } \; i = n \text{ and } c = a(i, j) \\ (1, j + 1) & \text{if } \; j < \ell \text{ and } c \neq a(i, j) \\ (1, 1) & \qquad \text{otherwise.} \end{cases}$$

Let σ be a pure strategy of player 1 that is implemented by an automaton of size n. Let $x_1, x_2, \ldots$ where $x_t = (a_t, b_t)$ be the random play induced by the strategy pair σ and τ^*, and let $q_1^i, q_2^i, \ldots$ be the random sequence of states of the automaton of player 2. Fix $1 \leq j \leq \ell$ and let $t = t_j$ be the random time of the first stage t with $q_t^2 = (1, j)$. Note that

$$Prob(a_{t+s} = a(s + 1, j) \;\; \forall\, 0 \leq s < n) = \frac{1}{|A|^n}.$$

and if $a_{t+s} = a(s+1, j)$ $\forall\, 0 \leq s < n$ then there exists $0 \leq s < n$ such that the state of the automaton of player 1 at stage $t + n$, q_{t+n}^1 coincides with its state at stage $t + s$. Therefore if $k_j = s + 1$, the play will enter a cycle in which the payoff to player 1 is at most $\max_{a \in A} \min_{b \in B} h(a, b)$. Therefore the conditional probability, given the history of play up to stage t_j, that the payoff to player 1 in any future stage is at most $\max_{a \in A} \min_{b \in B} h(a, b)$, and that $t_{j+1} = \infty$ is at least $1/(|A|^n n)$. Otherwise, if $t_{j+1} < \infty$, $t_{j+1} \leq t_j + n^2$. Therefore, either $t_\ell = \infty$ and then for every stage $t > \ell n^2$, the payoff to player 1 is at most

$\max_{a \in A} \min_{b \in B} h(a,b)$, or $t_\ell < \infty$. However, the previous inequalities imply that,

$$Prob(t_\ell = \infty) \geq 1 - (1 - 1/(n|A|^n))^{\ell-1} \to 1 \text{ as } K \to \infty,$$

which completes the proof of the lemma. ∎

It is of interest to bridge between the results of this section concerning the infinitely repeated 2-person 0-sum games, by providing asymptotic results of the value $\text{Val}(H(m_1, m_2))$ when $m_1 \to \infty$ and m_2 is approximately a fixed exponential function of m_1. Given a 2-person 0-sum game $H = (A, B, h)$, it will be interesting to find the largest (smallest) monotonic nondecreasing functions $\bar{v} : (0, \infty) \to \mathbb{R}$ ($v : (0, \infty) \to \mathbb{R}$) such that if $\frac{\ln m_2(m)}{m} \to \alpha > 0$ as $m \to \infty$ then

$$v(\alpha) \leq \liminf_{m \to \infty} \text{Val}(H(m, m_2)) \leq \limsup_{m \to \infty} \text{Val}(H(m, m_2)) \leq \bar{v}(\alpha).$$

Theorems 1 asserts that $\lim_{\alpha \to 0} \bar{v}(\alpha) = \lim_{\alpha \to 0} v(\alpha) = \text{Val}(H)$, and Theorem 3 asserts that for $\alpha > \ln|A|$, $\bar{v}(\alpha) = \max_{a \in A} \min_{b \in B} h(a,b)$. We conjecture that the two functions $\bar{v}$ and v are continuous with $\bar{v} = v$ for all values of $\alpha > 0$ with the possible exception of one critical value.

The next two conjectures address the number of repetitions needed for a unrestricted player to use his advantage over bounded automata. The positive resolutions of each of the conjectures have implications on the equilibrium payoffs of finitely repeated games with automata. A positive resolution of the next conjecture, will provide a positive answer to conjecture 3.

Conjecture 1 *For every $\varepsilon > 0$, if $m : \mathbb{N} \to \mathbb{N}$ satisfies $m(T) \geq \varepsilon T$, then*

$$\lim_{T \to \infty} \text{Val}(H^T(m(T), \infty)) = \text{Val}(H).$$

The truth of the above conjecture implies that there is a function $m : \mathbb{N} \to \mathbb{N}$ with $\lim_{T \to \infty} m(T)/T = 0$, and such that

$$\lim_{T \to \infty} \text{Val}(H^T(m(T), \infty) = \text{Val}(H).$$

An interesting open problem is to find the "smallest" such function. The next conjecture specifies a domain for such a function.

Conjecture 2 *If $m : \mathbb{N} \to \mathbb{N}$ obeys $\lim_{T \to \infty}(T/\log T)/m(T) = 0$, then*

$$\lim_{T \to \infty} \text{Val}(H^T(m(T), \infty)) = \text{Val}(H).$$

If $m : \mathbb{N} \to \mathbb{N}$ obeys $\lim_{T \to \infty} m(T)/(T/\log T) = 0$, then

$$\lim_{T \to \infty} \text{Val}(H^T(m(T), \infty)) = \max_{a^1 \in A_1} \min_{a^2 \in A_2} h^1(a^1, a^2).$$

4 Equilibrium Payoffs of the Supergame G^*_∞

We state here a result, due to Ben-Porath, which is a straightforward corollary of his result in the 2-person 0-sum case. All convergence of sets is with respect to the Hausdorff topology. Recall that for a sequence of subsets, E_n, of a Euclidean space $\mathbb{R}^k$,

$$\liminf_{n\to\infty} E_n = \{x \in \mathbb{R}^k \mid \forall\varepsilon > 0, \exists N \text{ s.t. } \forall n \geq N,\ d(x, E_n) < \varepsilon\}$$

where $d(x, E)$ denotes the distance of the point x from the set E,

$$\limsup_{n\to\infty} E_n = \{x \in \mathbb{R}^k \mid \forall\varepsilon > 0,\ \forall N,\ \exists n \geq N \text{ with } d(x, E_n) < \varepsilon\},$$

and $\lim_{n\infty} E_n = E$ if $E = \liminf E_n = \limsup E_n$.

Theorem 4 *(Ben-Porath 1986, 1993). Let $G = (N; (A_i)_{i\in N}; (r^i)_{i\in N})$ be a strategic game, and $m_i(k)$, $i \in N$, sequences with $\lim_{k\to\infty} m_i(k) = \infty$ and*

$$\lim_{k\to\infty} \frac{\log(\max_{i\in N} m_i(k))}{\min_{i\in N} m_i(k)} = 0.$$

Then,

$$\{x \in F \mid x^i \geq v^i(G)\} \subseteq \liminf_{k\to\infty} E(G(m_1(k), \ldots, m_n(k))),$$

and

$$\limsup_{k\to\infty} E(G(m_1(k), \ldots, m_n(k))) \subseteq \{x \in F \mid x^i \geq w^i(G)\}.$$

Note that in two-person games $v^i(G) = w^i(G)$ and therefore the above theorem provides exact asymptotics for two-person games. An interesting open problem is to find the asymptotic behavior of $E(G(m_1(k), \ldots, m_n(k)))$ as $k \to \infty$ and $\lim_{k\to\infty}\{\log(\max_{i\in N} m_i(k))/ \min_{i\in N} m_i(k)\} = 0$. Such questions lead to the study of the asymptotics of

$$v^i(G(m_1(k), \ldots, m_n(k))) = \min_{\tau^{-i}} \max_{\sigma^i} r^i_\infty(\sigma^i, \tau^{-i}),$$

where the min ranges over all $\tau^{-i} \in \times_{j\neq i}\Delta(\Sigma^j(m_j(k)))$ and the max is over $\sigma^i \in \Sigma^i(m_i(k))$ and where $m_i(k)$, $i \in N$, is a sequence with $\lim_{k\to\infty} m_i(k) = \infty$ and $\lim_{k\to\infty}(\log(\max_{i\in N} m_i(k)))/(\min_{i\in N} m_i(k)) = 0$. W.l.o.g. assume that $m_1(k) \leq m_2(k) \leq \ldots \leq m_n(k)$ and $\lim_{n\to\infty} \log m_n(k)/m_1(k) = 0$, and let $i < n$. Set $v^i(k) = v^i(G(m_1(k), \ldots, m_n(k)))$. We denote by $Q(i)$, or Q for short, the set of all probability measures on A_{-i} whose marginal distribution on $\times_{j<i}A_j$ is a product measure. The following is a partial answer to the study of the asymptotics of $v^i(k)$.

Proposition 3 *(a) If $\lim_{k\to\infty} \frac{m_1(k)\log m_1(k)}{m_2(k)} = 0$, then*

$$\limsup_{k\to\infty} v^1(k) \leq \min_{q\in Q} \max_{a^1\in A_1} \sum_{a^{-1}\in A_{-1}} q(a^{-1})r^1(a^1, a^{-1}).$$

(b) If for a fixed player $1 < i < n$, $\lim_{k\to\infty} \frac{\log m_{i+1}(k)}{\log m_i(k)} = \infty$, *then*

$$\limsup_{k\to\infty} v^i(k) \le \min_{q\in Q} \max_{a^i\in A_i} \sum_{a^{-i}\in A_{-i}} q(a^{-i}) r^i(a^i, a^{-i}).$$

Proof. Part (a) follows from Proposition 2. We turn to the proof of part (b). Let $(N(k))_{k=1}^{\infty}$ be a sequence of positive integers with $\lim_{k\to\infty} N(k)/\log m_i(k) = \infty$ and $\lim_{k\to\infty} N(k)/\log m_{i+1}(k) = 0$. The constructed $N \setminus \{i\}$ tuple of minimax strategies, $(\sigma^j)_{j\neq i}$, will enter a cycle of length $N(k)$, following the first $N(k)(n-1)$ stages. The cycle play, $X_1, \ldots, X_{N(k)}$, is a sequence of i.i.d. actions in A_{-i} with each X_t distributed according to a minimizing probability $q \in Q$. For every k let $q^*(k) \in Q$, or q for short, attain the minimum and let q_j be the marginal distribution of q on A_j, $j < i$. Let σ^j, $j < i$, be the strategy $\sigma^{j,X}$ which plays a random $N(k)$-periodic sequence of actions $X_1^j, X_2^j, \ldots$ where $X_1^j, \ldots, X_{N(k)}^j$ are i.i.d. and the distribution of each X_t^j is q_j. We define next the strategy σ^j for $j > i$, which is a mixture of pure strategies, each implemented by an automaton of size $iN(k)|A^{N(k)}|$ which for sufficiently large k is $\le m_{i+1}(k)$. For every $b = (b_1, \ldots, b_{i-1}, b_{i+1}, \ldots, b_n) \in \times_{j<i} A_j$, denote by b^j the projection of b on $\times_{i\neq j'<j} A_{j'}$, and let $q^{i+1}(b)$ denote the marginal of the conditional probability $(q|b^j)$ on A_j. The automaton includes $|A|^{N(k)} + (j-i-1)N(k)$ states which are used to record the realization of the choices of all players in stages $N(k)(j-i-1)+1, \ldots, N(k)(j-i)$ stages. Thereafter, player j plays an $N(k)$-periodic sequence $a_1^j, \ldots, a_{N(k)}^j$ which is a realization of a sequence of independent actions $X_1^j, \ldots, X_{N(k)}^j$ with the distribution of X_t^j being $(q^{i+1}|b^j_{t+(j-i-1)N(k)})$. One verifies that following the first $(n-1)N(k)$ stages the play of players $j \neq i$ enters an $N(k)$ cycle of i.i.d actions each distributed according to q. ■

In the above proof we can also construct the minimaxing strategies of players $j > i$ to be pure strategies, as in Proposition 2. Consider the following 3 player game G.

0,0,0	8,0,4
0,0,8	0,0,8

0,0,8	0,0,8
0,8,4	0,0,0

Player 1 chooses the row, player 2 the column, and player 3 chooses the matrix. Note that $v^1(G) = 0 = v^2(G)$ and $v^3(G) = 5$. However, $w^3(G) = 4$ (and $w^1(G) = w^2(G) = 0$). Therefore we can not deduce from Theorem 4 whether or not the vector payoff $(4,4,4)$ is approximated by equilibrium payoffs of the restricted games $G(m_1(k), m_2(k), m_3(k))$ for sufficiently large k and where $k < m_i(k)$ are sequences with $\log \max m_i(k)/\min m_i(k) \to 0$ as $k \to \infty$. However, Proposition 3 characterizes for this game the limit of the equilibrium payoffs provided that we assume in addition that $\lim_{k\to\infty} \log \max(m_1(k), m_2(k))/\log m_3(k) = \infty$. In particular, it follows in this case that $(4,4,4)$ is in the limit of the equilibrium payoffs.

We state now a result which provides a partial answer to the asymptotic behavior of the set of equilibrium payoffs of repeated games with bounded automata. Denote by

$$\mathbf{d}^i = \min_{q\in Q(i)} \max_{a^i\in A_i} \sum_{a^{-i}\in A_{-i}} q(a^{-i})r^i(a^i, a^{-i})$$

and

$$\mathbf{F} = \{x \in F(G) | x^i > \mathbf{d}^i\}$$

Proposition 4 *Assume that* $m_1(k) \leq \ldots \leq m_n(k)$, $\lim_{k\to\infty} \frac{\log m_n(k)}{m_1(k)} = 0$, $\lim_{k\to\infty} \frac{m_1(k)\log m_1(k)}{m_2(k)} = 0$ *and that for* $i > 1$ $\lim_{k\to\infty} \frac{\log m_i(k)}{\log m_{i+1}(k)} = 0$. *Then,*

$$\liminf_{k\to\infty} E(G(m_1(k),\ldots,m_n(k))) \supset \mathbf{F}$$

5 Cooperation in Finitely Repeated Games

The results in this section address the asymptotic behavior of the sets of equilibrium payoffs, $E(G^T(m_1,m_2))$, of the games $G^T(m_1,m_2)$, as T, m_1 and m_2 go to ∞. All convergence of sets is with respect to the Hausdorff topology. In each one of the theorems in the present section we assume that $G = (\{1,2\}, A, r)$ is a fixed 2-person strategic game, $F = F(G)$ stands for the feasible payoffs in the infinitely repeated game, i.e., $F = co(r(A))$ and that $(T(n), m_1(n), m_2(n))_{n=1}^{\infty}$ is a sequence of triples. For simplicity, the statements of the theorems are nonsymmetric with respect to the two players, and therefore we assume in addition that $m_2(n) \geq m_1(n)$. We also suppress often the dependence on n; no confusion should result.

Theorem 5 *Let* $G = (\{1,2\}, A, r)$ *be a two person game in strategic form, and assume that there is* $x \in F(G))$ *with* $x^1 > v^1(G)$, *and* $x^2 > u^2(G)$. *Then, if* $m_1(n) \to \infty$ *and* $\frac{\log m_1(n)}{T(n)} \to 0$ *as* $n \to \infty$,

$$\liminf_{n\to\infty} E(G^T(m_1,m_2)) \supseteq \{x \in F \mid x^1 \geq v^1(G) \text{ and } x^2 \geq u^2(G)\}.$$

Special cases of the above theorem have been stated in previous publications. Neyman (1985) states that in the case of the finitely repeated prisoner's dilemma G, for any positive integer k, there is T_0 such that if $T \geq T_0$ and $T^{1/k} \leq \min(m_1,m_2) \leq \max(m_1,m_2) \leq T^k$, then there is a mixed strategy equilibrium of $G^T(m_1,m_2)$ in which the payoff is $1/k$-close to the "cooperative" payoff of G. Papadimitriou and Yannakakis (1994) state the special case of theorem 5 obtained by assuming that the payoffs of the underlying game are rational numbers and replacing $F(G)$ in the statement of the theorem with $\{x \in r(A)$ with $x^i > v^i(G)\}$. They also state a result for a subset of F with the additional assumption that the bounds on both automata are subexponential in the number of repetitions.

The conclusion of the theorem fails if we replace in the assumptions of the theorem the strict inequality $x^1 > v^1(G)$ by the weak inequality $x^1 \geq v^1(G)$. For example in the game

0,4	1,3
1,1	1,0

the only equilibrium payoff in $G^T(m_1, m_2)$ with $m_2 \geq 2^T$ is (1,1).

The next theorem relates the equilibrium payoffs of $G^T(m_1, m_2)$ to the equilibrium payoffs of the undiscounted infinitely repeated game G^*_∞. Recall that the Folk Theorem asserts that

$$E(G^*_\infty) = \{x \in F | \, x^1 \geq v^1(G) \text{ and } x^2 \geq v^2(G)\}.$$

Theorem 6 *Let $G = (\{1,2\}, A, r)$ be a two person game in strategic form , and let $(T, m_1(T), m_2(T))_{T=1}^{\infty}$ be a sequence of triples of positive integers with $m_1(T) \leq m_2(T)$ and $m_1(T) \to \infty$ as $T \to \infty$, and*

$$\lim_{T\to\infty} \frac{\log m_2(T)}{\min(m_1(T), T)} = 0.$$

Then,

$$\lim_{T\to\infty} E(G^T(m_1(T), m_2(T))) = E(G^*_\infty).$$

The limiting assumption $\lim_{T\to\infty} \frac{\log m_2(T)}{m_1(T)} = 0$ in theorem 6, could probably be replaced by an alternative lower bound, as a function of T, on $m_1(T)$, provided that we also assume that there is $x \in F$ with $x^1 > v^1(G)$. One example of such a result is presented in the following conjecture.

Conjecture 3 *Let $G = (\{1,2\}, A, r)$ be a two person game in strategic form, and assume that there is $x \in F$ with $x^1 > v^1(G)$. Then, if*

$$\liminf_{T\to\infty} m_1(T)/T > 0,$$

and

$$\lim_{T\to\infty} \frac{\log m_1(T)}{T} = 0,$$

Then

$$\lim_{T\to\infty} E(G^T(m_1, m_2)) = \{x \in F \mid x^i \geq v^i(G)\}.$$

The next theorem is straightforward and very easy. We state it as a contrast to the previous results. It shows that the subexponential bounds on the sizes of the automata as a function of the number of repetitions is essential to obtain equilibrium payoffs that differ from those of the finitely repeated game G^T.

Theorem 7 *For every game G in strategic form there exists a constant c such that if $m_i \geq \exp(cT)$ then*

$$E(G^T(m_1, \ldots, m_n)) = E(G^T).$$

6 Repeated Games with Bounded Recall

Aumann (1981) mentioned two ways of modeling a player with bounded rationality: with finite automata and with bounded recall strategies. There are two alternatives to define strategies with bounded recall. The first one (see e.g. Aumann and Sorin, 1989) considers strategies with bounded recall which choose an action as a function of the recalled opponents' actions, and the second alternative (see e.g. Kalai and Stanford, 1988, or Lehrer 1988) allows a player to rely on his opponents' actions as well as on his own. The following are results on repeated games with bounded recall of the second type which are closely related to those presented for finite automata. Let $BR^i(m)$ denote all strategies of player i in a repeated game that choose an action as a function of all players action in the last m stages. Each pure strategy $\sigma^i \in BR^i(m)$ is thus represented by a function $f^i : A^m \to A_i$ and a fixed element, initial memory, $e = (e_1, \ldots, e_m) \in A^m$; for $t > m$, $\sigma^i(a_1, \ldots, a_{t-1}) = f^i(a_{t-m}, \ldots, a_{t-1})$ and for $t \leq m$, $\sigma^i(a_1, \ldots, a_{t-1}) = f^i(e_t, \ldots, e_m, a_1, \ldots, a_{t-1})$. Given a strategy $\sigma^i = (e, f^i) \in BR^i(m)$, the automaton $\langle A^m, e, f^i, g^i\rangle$ where $g^i(x, y)$, $x = (x_1, \ldots, x_m) \in A^m$ and $y \in A_{-i}$, equals $(x_2, \ldots, x_m, (f^i(x), y))$, implements the strategy σ^i. Thus each strategy in $BR^i(m)$ is implemented by an automaton of size $|A|^m$, or in symbols and identifying a strategy with its equivalence class, $BR^i(m) \subset \Sigma^i(|A|^m)$. Given a fixed two-person zero-sum game $G = (A, B, h)$, we denote by V_{m_1,m_2} the value of the undiscounted infinitely repeated game G where player i is restricted to mixed strategies with support in $BR^i(m_i)$. Lehrer (1988) proves the following result which is related and has a spirit similar to the result of Ben-Porath (1986,1993).

Theorem 8 *(Lehrer, 1988). For every function* $m : I\!N \to I\!N$ *with* $\log m(n)/n \to 0$ *as* $n \to \infty$,

$$\liminf V_{n,m(n)} \geq \ Val\,(G).$$

Proof. Note that, by identifying a strategy with its equivalence class, $BR^2(m(n)) \subset \Sigma^2(|A \times B|^{m(n)})$. Let $k : I\!N \to I\!N$ with $\lim_{n\to\infty} m(n)/k(n) = 0$ and $\lim_{n\to\infty} \log k(n)/n = 0$. E.g., $k = m^2$. Let $X = X_1, X_2, \ldots, X_{k(n)}, \ldots$ be a random $k(n)$-periodic sequence of actions, with $X_1, \ldots, X_{k(n)}$ i.i.d and X_t an optimal strategy of player 1 in the one shot game. W.l.o.g. we assume that the support of X_t has at least two elements. Consider the mixed strategy $\sigma^1(k(n))$ of player 1 which plays the realization of X. The proof of Theorem 1 shows that for any sequence of strategies $\tau(n) \in \Sigma^2(|A \times B|^{m(n)})$, $\liminf_{n\to\infty} r_\infty(\sigma^1(k(n)), \tau(n)) \geq \mathrm{Val}(G)$ as $n \to \infty$. (Note that $\sigma^1(k(n)) \notin \Delta(BR^1(n))$). It is thus sufficient to prove that the norm distance of $\sigma^1(k(n))$ from $\Delta(BR^1(n))$ tends to zero as $n \to \infty$, i.e., that for most realizations of X, the implied pure strategy is in $BR^1(n)$. Note that for any $0 \leq s < t < k(n)$ there are positive integers s' and t' with $t \leq t' \leq t+n-[n/3]$, and $s \leq s' \leq s+n-[n/3]$ such that $X_{s'+1}, \ldots, X_{s'+[n/3]}, X_{t'+1}, \ldots, X_{t'+[n/3]}$ are independent and $(X_{s+1}, \ldots, X_{s+n}) = (X_{t+1}, \ldots, X_{t+n})$ only if $(X_{s'+1}, \ldots, X_{s'+[n/3]}) = (X_{t'+1}, \ldots, X_{t'+[n/3]})$. Indeed, if $\min\{t-s, s+k(n)-t\} \geq [n/3]$ set $s' = s$ and $t' = t$; if $t < s + [n/3]$ set $s = s'$ and $t' - s$ is the smallest multiple of

$t-s$ which is $\geq [n/3]$; and if $s+k(n) < t+[n/3]$ set $t=t'$ and $s'-t$ is the smallest multiple of $s+k(n)-t$ which is $\geq [n/3]$. There is a constant $0<\alpha<1$ that depends on the optimal strategy of player 1 in the one shot game, such that $\Pr(X_{s'+i}=X_{t'+i}) \leq \alpha$. (e.g., if p is the probability vector associated with the optimal strategy in the one shot game $\alpha=\sum p_i^2$).Therefore, $\Pr(\forall i,\ 1\leq i\leq [n/3],\ X_{s'+i}=X_{t'+i}) \leq \alpha^{[n/3]}$. Therefore

$$\Pr(\exists\ s,t,\ 0\leq s<t<k(n) \text{ s.t } \forall\, i, 0<i\leq n,\ X_{s+i}=X_{t+i}))$$

$$< k^2(n)\alpha^{[n/3]} \to_{n\to\infty} 0.$$

Note that the strategy $\sigma^*(n)$ which is defined as the strategy $\sigma^1(k(n))$ conditional on $\{\forall\ s,t,\ 0\leq s<t<k(n), \exists 1\leq i\leq n \text{ s.t.} X_{s+i}\neq X_{t+i}\}$ is in $\Delta(BR^1(n))$ with $d(\sigma^*(n),\sigma(k(n)))\to 0$ as $n\to\infty$, where $d(\sigma^*(n),\sigma(k(n)))$ denotes the norm distance between the mixed strategies (viewed as distributions) $\sigma^*(n)$ and $\sigma(k(n))$. Therefore

$$\liminf_{n\to\infty} \operatorname{Val}(BR^1(n),\Sigma^2(|A\times B|^{m(n)}),h_\infty) \geq \operatorname{Val}(G).$$

which completes the proof. ∎

The next result is an analog of Proposition 2.

Proposition 5 *For every 2-player 0-sum game $H=(A,B,h)$ there is a positive constant K such that if $m:\mathbb{N}\to\mathbb{N}$ with $n>Km(n)$, then for every n there exists a strategy $\sigma(n)\in BR^1(n)$ such that*

$$\lim_{n\to\infty}(\inf\{h_t(\sigma(n),\tau)\mid \tau\in BR^i(m(n)),\ t\geq \exp(n)\}) = \operatorname{Val}(H).$$

An interesting straightforward corollary of this proposition is the following. Let $G=(A,r)$ be an n-person game, $A=\times_{1\leq i\leq n}A_i$ and $r=(r^i)_{1\leq i\leq n}$, and assume that $m_1(k)\leq\ldots\leq m_n(k)$.

Corollary 2 *There is a constant K such that if $Km_1(k)\leq m_2(k)$, then for every k there exist an $(n-1)$-tuple of strategies $\tau(k)=(\tau_2,\ldots,\tau_n)\in \times_{1<i\leq n}BR^i(m_i(k))$ such that for any strategy $\sigma(k)\in BR^1(m_1(k))$,*

$$\limsup_{k\to\infty} r^1_\infty(\sigma(k),\tau(k)) \leq \max_{q\in\Delta(A^1)}\ \min_{a^{-1}\in A_{-1}} \sum_{a^i\in A^i} q(a^i)r^1(a^1,a^{-1}).$$

The next result is an analog of Proposition 3. Assume that $m_1(k)\leq\ldots\leq m_n(k)$ with $\lim_{k\to\infty}\log m_n(k)/m_1(k)=0$. For every $1\leq i\leq n$ denote by $v^i(m_1(k),\ldots,m_n(k))$, or $v^i(k)$ for short, the minimax payoff to player i, i.e., $\min_{\tau^{-i}}\max_{\sigma^i} r^i_\infty(\sigma^i,\tau^{-i})$, where the min ranges over all $\tau^{-i}\in\times_{j\neq i}\Delta(BR^j(m_j(k)))$ and the max ranges over all $\sigma^i\in BR^i(m_i(k))$. As in section 4, we denote by $Q(i)$, or Q for short, the set of all probability measures on A_{-i} whose marginal distribution on $\times_{j<i}A_j$ is a product measure.

Proposition 6 *If for a fixed player* $1 \leq i < n$, $\lim_{k\to\infty} m_{i+1}/m_i(k) = \infty$, *then*

$$\limsup_{k\to\infty} v^i(k) \leq \min_{q\in Q} \max_{a^i\in A_i} \sum_{a^{-i}\in A_{-i}} q(a^{-i}) r^i(a^i, a^{-i}).$$

We state now a result which provides a partial answer to the asymptotic behavior of the set of equilibrium payoffs of repeated games with bounded recall. It is an analog of Proposition 4.

Proposition 7 *There is a constant* $K > 0$ *such that if* $m_1(k) \leq \ldots \leq m_n(k)$, $\lim_{k\to\infty} \frac{\log m_n(k)}{m_1(k)} = 0$, $m_2(k) > K m_1(k)$, *and* $\lim_{k\to\infty} \frac{m_i(k)}{m_{i+1}(k)} = 0$ *for* $i > 1$, *then*

$$\liminf_{k\to\infty} E(G(BR^1(m_1(k)), \ldots, BR^n(m_n(k)))) \supset \mathbf{F}$$

The next proposition and conjecture address the advantage of an unrestricted player over a player restricted to bounded recall strategies in finitely repeated 2-player 0-sum games. For a fixed two-person zero-sum game $H = (A, B, h)$, we denote by $V_{n,\infty}^T$, or V_n^T for short, the value of the finitely repeated game $H^T(BR^1(n), \Sigma^2)$, i.e. the value of the T-repeated game in which player 1 is restricted to use strategies in $BR^1(n)$ while player 2 can use any strategy in Σ^2. The following proposition asserts that if the duration T is shorter then some exponential function of n then the unrestricted player has no advantage.

Proposition 8 *There exists a constant* $K > 0$ *such that if* $T : I\!N \to I\!N$ *satisfies* $T(n) \leq \exp(Kn)$, *then*

$$\lim_{n\to\infty} V_n^{T(n)} = Val\ H.$$

Proof. It is sufficient to prove the result in the case that any optimal strategy of player 1 in the one shot game is not pure. Let $X_1, X_2, \ldots$ be a sequence of i.i.d optimal strategies in the one shot game. The stochastic process $X_1, \ldots$ induces a strategy $\sigma \in \Delta(BR^1(n))$ as follows. For each realization of the random sequence define the initial memory $e = (X_1, \ldots, X_n)$ and the action function $f : (A \times B)^n \to A$ is defined as follows: for every $(a_1, b_1), \ldots, (a_n, b_n)$ define the stopping time S as the smallest value of t such that $(a_1, \ldots, a_n) = (X_{t-n}, \ldots, X_{t-1})$ and define

$$f((a_1, b_1), \ldots, (a_n, b_n)) = X_S.$$

Note that the strategy induced by each realization of the random sequence $X_1, \ldots$ consists of a deterministic sequence (which enters eventually a cycle) of actions of player 1 and that the sequence is independent of the strategy of player 2. It is easy to verify that there is a positive constant $K > 0$ such that

$$\lim_{n\to\infty} \text{Prob}_\sigma(a_t(\sigma) = X_{t+n}\ \forall t \leq \exp Kn) = 0$$

and therefore the norm distance between the strategies σ and σ^X goes to zero as n goes the infinity. As σ^X is an optimal strategy in H^T the result follows.
■

The conjecture below claims that there is an exponential function of n, $\exp(Kn)$ such that if the number of repeatition T is larger then $\exp(Kn)$, the values V_n^T converge to the maxmin in pure strategies.

Conjecture 4 *There is a constant K such that if $T : \mathbb{N} \to \mathbb{N}$ satisfies $T(n) \geq \exp(Kn)$, then*

$$\lim_{n\to\infty} V_n^{T(n)} = \max_{a\in A} \min_{b\in B} h(a,b).$$

As $BR^1(n) \subset \Sigma^1(|A \times B|^n)$ a positive answer to the second part of Conjecture 2 provides also a positive answer to Conjecture 4. It is of interest to study the asymptotics of $V_n^{T(n)}$ where $T(n)$ is approximately a fixed exponential function of n. This would close the gap between Proposition 8 and Conjecture 4. Given a 2-person 0-sum game $H = (A, B, h)$, it will be interesting to find the largest (smallest) function $\bar{u} : (0, \infty) \to \mathbb{R}$ ($u : (0, \infty) \to \mathbb{R}$) such that if $\frac{\ln T(n)}{n} \to \alpha$ as $n \to \infty$ then

$$u(\alpha) \leq \liminf_{n\to\infty} V_n^{T(n)} \leq \limsup_{n\to\infty} V_n^{T(n)} \leq \bar{u}.$$

Proposition 8 asserts that there is a constant $K_1 > 0$ such that $u(K_1) = \text{Val } H$ and the conjecture claims that there is a constant K_2 with $\bar{u}(K_2) = \max_{a\in A} \min_{b\in B} h(a,b)$. It is interesting to find the sup and inf of K_1 and K_2 respectively. We conjecture that the two functions $\bar{u}$ and u are continuous with $\bar{u} = u$ for all values of α with the possible exception of one critical value. We do not exclude the possibility of the existence of a positive constant K such that $u(K) = \text{Val } (H)$ and $\bar{u}(K) = \max_{a\in A} \min_{b\in B} h(a,b)$.

7 Variations and Extensions

We have studied here some topics in the theory of repeated games with deterministic automata. There are several variants of the concept of automata which merits study in the context of repeated games broadly conceived, i.e., including repeated games with incomplete information and stochastic games. The variations of the concept of an automaton are in several independent dimensions. E.g., we can allow transitions that depend on the actions of all players and also allow for probabilistic actions and/or transitions, and moreover we can consider transition and/or action functions which are time dependent. A *full automaton* for player i is a 4-tuple $\langle M, q_1, f, g\rangle$ where the set of states M , the initial state $q_1 \in M$ and the action function $f : M \to A_i$ are as in a (standard) automaton , and the transition function $g : M \times A \to M$ specifies the next state as a function of the current state and the n-tuple of actions of all players. The strategy σ_α^i induced by a full automaton α= $\langle M, q_1, f, g\rangle$ is defined naturally by $\sigma_1^i = f(q_1)$ and for every $a_1, \ldots, a_{t-1}$

in A, $\sigma^i(a_1,\ldots,a_{t-1}) = f(q_t)$ where q_t is defined inductively for $t > 1$ by $q_t(a_1,\ldots,a_{t-1}) = g(q_{t-1},a_{t-1})$. Obviously, every strategy which is induced by a full automaton of size m is equivalent to a strategy induced by an automaton of size m. Therefore when the actions and transitions are deterministic, allowing transitions to depend on your own action is not affecting the equilibrium theory. However it does have implications in the study of subgame perfect equilibrium of repeated games (see e.g. Kalai and Stanford (1988) and Ben-Porath and Peleg (1987)). Kalai and Stanford (1988) show that given a pure strategy σ^i of a player in the repeated game, the number of different strategies induced by it and any finite history $(a_1,\ldots,a_t)$ equals the size of the full automaton that induces σ^i. A *mixed action automaton* for player i is a 4-tuple $\langle M, q_1, f, g\rangle$ where M is a finite set, $q_1 \in M$ is the initial state, $f : M \to \Delta(A_i)$ is a function specifying a mixed action as a function of the state, and $g : M \times A \to M$ is the transition function. Each mixed action automaton induces a behavioral strategy σ^i as follows. $\sigma_1^i = f(q_1)$. Define inductively $q_t(a_1,\ldots,a_{t-1}) = g(q_{t-1},a_{t-1})$, and

$$\sigma_t^i(a_1,\ldots,a_{t-1}) = f(q_t(a_1,\ldots,a_{t-1})).$$

Denote by $\Sigma_p^i(m_i)$ all equivalence classes of behavioral strategies which are induced by a mixed action automata of size m_i. Two mixed (or behavioral) strategies, σ^i and τ^i, of player i are *equivalent* if for any $N \setminus \{i\}$-tuple of pure strategies σ^{-i}, (σ^i,σ^{-i}) and (τ^i,σ^{-i}) induce the same distribution on the play of the repeated game. Note that $\Sigma^i(m_i) \subset \Sigma_p^i(m_i)$ and that $\Sigma_p^i(1) \setminus \Delta(\cup_{m=1}^\infty \Sigma^i(m)) \neq \emptyset$. A stationary behavioral strategy in a repeated game with complete information is induced by a mixed action automaton with one state. Given a behavioral strategy $\sigma^i = (\sigma_t^i)_{t=1}^\infty$, the number of equivalence classes of the behavioral strategies of the form $(\sigma^i \mid a_1,\ldots,a_t)$ where $a_1,\ldots,a_t$ ranges over all histories which are consistent with σ^i, (i.e., $(\sigma^i \mid a_1,\ldots,a_s)(a_{s+1}^i) > 0$ for every $s < t$), equals the size of the smallest mixed action automaton that implements σ^i. A *probabilistic transition automaton* is a 4-tuple $\langle M, q_1, f, g\rangle$ where M is a finite set, $q \in M$ is the initial state, $f : M \to A_i$ is the action, and $g : M \times A_{-i} \to \Delta(M)$. Each probabilistic transition automaton induces a mixed strategy σ^i as follows. $\sigma_1^i = f(q)$. Then the automaton changes its states stochastically in the course of playing the repeated game. If its state in stage t is q_t and the other players action in stage t is a_t^{-i} the conditional probability of q_{t+1}, given the past is $g(q_t,a_t^{-i})$, and its action in stage t is $f(q_t)$. Denote by $\Sigma_t^i(m_i)$ all equivalence classes of strategies which are induced by probabilistic transition automata of size m_i. Note that $\Sigma_p^i(m) \subset \Sigma_t^i(m|A_i|)$.

Repeated games with complete information. The theory of finitely or infinitely repeated 2-person 0-sum games with complete information and either mixed action or probabilistic transition automata is trivial and not of much interest. However, the asymptotic behavior of the set of equilibrium payoffs of n-person ($n \geq 3$) infinitely repeated or 2-person finitely repeated games with either mixed action or probabilistic transition automata is unknown and of interest. The difficulties in the study of equilibrium payoffs of n-person infinitely repeated games is the asymptotics of the minmax payoffs which is unknown. As for 2-player finitely repeated games with either mixed action or probabilistic transition automata, it seems that our constructed equilibrium (Neyman 1995) in the finitely repeated games, remains an equilibrium in the game in which players are restricted to play with either mixed action or probabilistic automata with the same bounds. Note that as $\Sigma^i(m)$ is a proper subset of $\Sigma^i_p(m)$ (and of $\Sigma^i_t(m)$), the assertion that $\sigma^* \in \Delta(\Sigma^1(m_1)) \times \Delta(\Sigma^2(m_2))$ is an equilibrium in $(\{1,2\}; \Sigma^1_p(m_1), \Sigma^2_p(m_2); r_T)$ (in $(\{1,2\}; \Sigma^1_t(m_1), \Sigma^2_t(m_2); r_T)$) is stronger than the assertion that it is an equilibrium in $G^T(m_1, m_2)$. Moreover, in this setup, holding a player down to his individual rational payoff requires just one state (a fixed finite number of states) and therefore in the theorems there is only a need to bound the size of one of the automata.

Repeated games with incomplete information. The theory of repeated games with incomplete information and either probabilistic or deterministic action function is of interest. Here the initial state is allowed to be a function of the initial information, or equivalently, the initial move of nature is part of the input at stage 0. Alternatively, one allows the action function to depend on the state of the machine and the information about the state of nature. It is relatively easy to verify that in the case of 2-person 0-sum repeated games with incomplete information on one side the value of the "restricted game" $\Gamma(p, m_1, m_2)$ converges to $\lim v_n(p)$ as $m_i \to \infty$ and $(\log \max\{m_1, m_2\})/\min\{m_1, m_2\} \to 0$. It is of interest to find whether in repeated games with incomplete information on both sides and under the above asymptotic conditions on m_1 and m_2 the values of $\Gamma(p, m_1, m_2)$ converge to a limit and whether this limit equals $\lim v_n(p)$.

Stochastic Games. My initial interest in the theory of repeated games with finite automata stemmed from my work with J.-F. Mertens on the existence of a value in stochastic games. The ε-optimal strategies exhibited there are behavioral strategies which are not implemented by any finite state mixed action automaton. Blackwell and Ferguson (1968) show that in the "Big Match" there are no stationary (i.e. implemented by a mixed action automaton of size 1) ε-optimal strategies, and it can be shown further that there are no ε-optimal strategies which are implemented by a mixture of mixed action automata of finite size. However, when both players are restricted to strategies that are implemented by either deterministic or mixed action automata of sizes m_1 and m_2 we are faced with a 2-person 0-sum game $G(m_1, m_2)$ in normal form which has a value $V(m_1, m_2)$. It is of interest to study the asymptotic behavior of $V(m_1, m_2)$ as $m_i \to \infty$. Consider the "Big Match" (Blackwell and Ferguson 1968) which is an example of a 2-person 0-sum stochasic game.

1	0
0*	1*

The value of the (unrestricted) undiscounted game is 1/2. Mor Amitai (1989) showed that for this game there is a polynomial function $m : I\!N \to I\!N$ such that the value of the restricted "Big Match" where player 1 and player 2 are resricted to strategies which are implemented by automata of sizes $m(n)$ and n respectively converges as $n \to \infty$ to 1.

Another generalization of automata suggested by the theory of stochastic games is a time dependent probabilistic action automaton. In a time dependent automaton the action of the automaton depends on its internal state and the stage. This generalizes also the concept of a Markov strategy. Blackwell and Ferguson (1968) have shown that in the "Big Match" there are no ε-optimal Markov strategies. This leads to the natural question as to whether or not there are ε-optimal strategies which are implemented by a finite state time dependent probabilistic automaton. When raising this question in a seminar in Stanford University, Jerry Green has pointed out to me the work of T. Cover (1969) which illustrates a statistical decision problem in which a difference between the stationary finite automata and the time dependent automata emerges. My attention to the topic of repeated games with bounded automata was recalled in discussions I had with Alan Hoffman during his visit to Jerusalem in the spring of 1983. Hoffman informed me that in the early fifties, when engineers at SEAC were actually playing tick-tack-toe on the SEAC, he was concerned on how game theorists will view/study the fact that a 2-person 0-sum fair (value 0) game, becomes an unfair game when players are restricted by their "programs". This triggered my attention to pose the problem settled by Ben-Porath and later to the study of the possible cooperation in finitely repeated games with bounded automata. I am indepted to each one of the above mentioned individuals for their influence, either directly or indirectly, on my working on repeated games with finite automata.

Mor Amitai proved the following interesting result concerning the maxmin of stochastic games with probabilistic transition automata: for any stochastic game there exists a constant m such that for any m_1 and any strategy $\sigma^1 \in \Sigma_t^1(m_1)$ and $\varepsilon > 0$ there exists a strategy $\sigma^2 \in \Sigma_t^2(m)$ such that

$$r_\infty^1(\sigma^1, \sigma^2) \leq \sup \inf r_\infty^1(\sigma, \tau) + \varepsilon$$

where the sup ranges over all stationary strategies of player 1 and the inf ranges over all strategies of player 2.

References

Amitai, M. (1989). "Stochastic Games with Automata", (Hebrew) MSc. Thesis, Hebrew University, Jerusalem.

Aumann, R. J. (1981). "Survey of Repeated Games" in *Essays in Game Theory and Mathematical Economics in Honor of Oskar Morgenstern*, 11-42. Mannheim/Vienna/Zurich:Bibliographisches Institute.

Aumann, R. J., and S. Sorin (1990). "Cooperation and Bounded Recall," *Games and Economic Behavior* **1** 5-39.

Ben-Porath, E., (1993). "Repeated games with finite automata," *Journal of Economic Theory* **59** 17-32.

Ben-Porath, E., and B. Peleg (1987). "On the Folk Theorem and Finite Automata." The Hebrew University (discussion paper).

Benoit, J.-P. and V. Krishna (1985), "Finitely Repeated Games," *Econometrica* 53, 905-922.

Benoit, J.-P. and V. Krishna (1987), "Nash Equilibria of Finitely Repeated Games," *International Journal of Game Theory* 16, 197-204.

Blackwell, D. and T. S. Ferguson (1968). "The Big Match," *Ann. Math. Stat.*, **39** 159-163.

Cover, T. M. (1969). "Hypothesis Testing with Finite Statistics," *Annals of Mathematical Statistics* **40** 828-835.

Forges, F., Mertens J.-F., and A. Neyman (1986), "A Counter Example to the Folk Theorem with Discounting," *Economics Letters* **20** 7.

Hopcroft, J., and J. Ullman (1979). "Introduction to Automata Theory, Languages, and Computation." Reading, Massachusetts: Addison-Wesley.

Kalai, E. (1990). "Bounded Rationality and Strategic Complexity in Repeated Games," *Game Theory and Applications*, edited by T. Ichiishi, A. Neyman and Y. Tauman, Academic Press, 131-157.

Kalai, E., and W. Stanford (1988). "Finite Rationality and Interpersonal Complexity in Repeated Games," *Econometrica* **56** 397-410.

Lehrer, E. (1988). "Repeated Games with Stationary Bounded Recall Strategies," *Journal of Economic Theory* **46** 130-144.

Lehrer, E. (1994). "Many Players with Bounded Recall in Infinite Repeated Games." *Games and Economic Behavior* **7** 390-405.

Megiddo, N., and A. Wigderson (1986). "On Play by Means of Computing Machines," in *Proc. 1st Cong. on Theoretical Aspects of Reasoning and Knowledge"* (J. Halpern, ed.), 259-274. Los Altos, California, Morgan Kaufman.

Mertens, J. F., and A. Neyman (1981). "Stochastic games," *International Journal of Game Theory.* 53-66.

Neyman, A. (1985). "Bounded Complexity Justifies Cooperation in the Finitely Repeated Prisoner's Dilemma," *Economics Letters.* **19**, 227-229.

Neyman, A. (1995). "Finitely Repeated Games with Finite Automata", Discussion Paper no. 69, Center for Rationality and Interactive Decision Theory. Submitted to *Mathematics of Operations Research.*

Papadimitriou, C. H., and M. Yannakakis (1994). "On Complexity as Bounded Rationality" (extended abstract). STOC - 94 , 726-733.

Papadimitriou,C. H., and M. Yannakakis (1995). "On Bounded Rationality and Computational Complexity" Mimeo.

Zemel, E. (1989). "Small Talk and Cooperation: A Note on Bounded Rationality," *Journal of Economic Theory* **49**, 1-9.

Learning in Games: Fictitious Play Dynamics

Vijay Krishna[1] and Tomas Sjöström[2]

[1]Department of Economics, Pennsylvania State University, University Park, PA 16802, USA
[2]Department of Economics, Harvard University, Cambridge, MA 02138, USA

Abstract. We present a selective survey of recent work on the Brown-Robinson learning process known as "fictitious play." We study the continuous time version of the process and report convergence results for specific classes of games.

Keywords. Equilibrium, learning, fictitious play, dynamical systems.

1. Introduction

The notion of a Nash equilibrium plays a key role in modern economics. Indeed, it is only a slight exaggeration to say that it has become the sole basis of positive theory. Despite its central role, it has become clear that there are few reasons to warrant the use of equilibrium as a predictive tool.

Theories of equilibrium may be divided into two broad categories. The first category consists of *epistemic* (or knowledge based) theories and the second of *dynamic* theories.[1]

This is intended to be the first of a two-part selective survey of some recent developments in the study of dynamical systems associated with games. In this paper we study a simple model of learning and in a companion paper we study a simple model of evolutionary dynamics.

1.1 Epistemic Versus Dynamic Theories

Epistemic theory explores the hypothesis that common knowledge of various aspects of the game will lead rational players, via a process of introspection, to play an equilibrium. However, the informational and rationality requirements needed to arrive at an equilibrium are extreme, and thus it is difficult to argue that such a theory provides a justification of the equilibrium idea for positive purposes. In particular, for players to reach an equilibrium requires not only that the rationality of the players and the payoffs be common knowledge, but also that the beliefs they hold about each other's behavior be commonly known.[2]

---

[1]Binmore (1988) calls these the *eductive* and *evolutive* approaches, respectively.
[2]Aumann and Brandenburger (1995) present a "state of the art" account of the epistemic theory.

Dynamic theories explore the hypothesis that equilibrium is reached via a process of gradual adjustment by boundedly rational players who encounter each other in a repeated setting. In contrast to the epistemic theory, the informational and rationality requirements of the dynamic theories are usually minimal. If anything, the dynamic theories are to be faulted for postulating behavior that is too naive. Nevertheless, Nash equilibria are rest points of most dynamic processes associated with games. The key question then is whether the particular dynamic process will converge to an equilibrium. Typical examples are the fictitious play learning process and the replicator dynamics from evolutionary biology. The former is the main subject of this survey and the latter the subject of an accompanying survey.

1.2 General Versus Special Theories

What can we learn from the two theories? At the most general level, the two theories make identical and rather weak predictions. Under the most plausible assumptions (common knowledge of rationality and payoffs), the epistemic theory implies that players will choose from the set of rationalizable outcomes.[3] For general games, most dynamic theories also make exactly the same prediction: in the long run the choices of the players will be rationalizable.

Is one approach superior to the other as the basis of Nash equilibrium? At a *general* level, the answer is no, and there is little to choose between the two theories. However, the dynamic theories can make sharper predictions in cases where the underlying processes can be shown to converge. Although certain well-known examples demonstrate that the convergence of these processes cannot be established in general, there are large classes of games with special structures for which convergence to an equilibrium can be guaranteed. Thus dynamic processes can form the basis of *special* theories. This is in marked contrast to the epistemic approach. There do not appear to exist interesting classes of games for which the predictions of the epistemic theory will be sharper than in general.[4]

In this paper we study the special learning process known as *fictitious play*. This process is chosen for its inherent simplicity, intuitive appeal, and historical interest. Moreover, it appears that fictitious play shares many important structural features with more complicated learning processes. It seems quite likely that results obtained for the fictitious play process will hold for a more general class of processes.

1.3 Experimental Evidence

Before proceeding further, we wish to draw attention to some of the lessons that seem to be emerging from the now large accumulation of experimental findings on the behavior of subjects in games. Smith (1990) suggests that (i) in a one-shot game, behavior is not well predicted by equilibrium; (ii) in a repeated, *complete* information setting players tend to "cooperate" and thus repeated game effects emerge, though they are hard to predict; and (iii) in a repeated setting with *incomplete* information, where players have

---

[3] See Bernheim (1984) and Pearce (1984).

[4] Unless, of course, the set of rationalizable strategy combinations is a singleton.

knowledge of their own payoffs only, the best predictors of long-run behavior are the equilibria of the one-shot game with complete information.

Since the environment in (iii) is exactly the environment that most learning processes postulate (see the next section), this emergent finding lends additional credence to the view that such processes are indeed worthy of study as the basis of a positive theory for games.[5]

2. Preliminaries

Let $G = (A, B)$ be a two-player *game* where A and B are $I \times J$ matrices. We will refer to $I = \{1, 2, \ldots, I\}$ and $J = \{1, 2, \ldots, J\}$ as the sets of pure strategies available to players 1 and 2, respectively. As usual, if player 1 chooses strategy i and player 2 chooses strategy j, the payoff to player 1 is a_{ij}, and the payoff to player 2 is b_{ij}. The sets of mixed strategies are denoted by $\Delta(I)$ and $\Delta(J)$, respectively. Let $\delta_i \in \Delta(I)$ be the mixed strategy that assigns weight 1 to i. We will identify i with δ_i and write $i \in \Delta(I)$ instead of $\delta_i \in \Delta(I)$.

For all $q \in \Delta(J)$, let $BR(q)$ be the set of *pure strategy best responses* for player 1. The strategy pair (p^*, q^*) is a *Nash equilibrium* if $\{i : p_i^* > 0\} \subseteq BR(q^*)$ and $\{j : q_j^* > 0\} \subseteq BR(p^*)$.

Let B be a selection from BR; that is, for all q, $B(q) \in BR(q)$. We assume that $BR(q) = BR(q')$ implies $B(q) = B(q')$. Similarly, for all $p \in \Delta(I)$, let $B(p) \in BR(p)$.

2.1 Fictitious Play

We consider the following dynamic process: At time $t = 0$, players randomly select pure strategies, $i(0), j(0)$. At time $t = 1$, player i chooses a strategy that is the best response to player j's strategy in the previous period; thus, $i(1) = B(j(0))$ and $j(1) = B(i(0))$. For all subsequent periods, player i chooses a strategy that is the best response to the history of strategies employed by player j, treating the history as outcomes arising from an underlying mixed strategy employed by j. This is the "fictitious play" process first defined by Brown (1951).

Definition 1 *For $t = 0, 1, 2, \ldots$, the sequence $(p(t), q(t))$ is a* discrete time fictitious play process (DFP) *if*

$$(p(0), q(0)) \in \Delta(I) \times \Delta(J);$$

and, for all $t \geq 0$,

$$p(t+1) = \frac{tp(t) + B(q(t))}{t+1}, \quad q(t+1) = \frac{tq(t) + B(p(t))}{t+1}.$$

[5] See Boylan and El-Gamal (1993) on some experimental results on fictitious play and Roth and Erev (1995) for a survey of experimental work on learning in games.

Thus $p(t+1)$ is a weighted average of $p(t)$ and $B(q(t))$ where the weights are $\frac{t}{t+1}$ and $\frac{1}{t+1}$. New strategies are chosen each "period". Now suppose $\delta > 0$ is the time between adjustments; replacing the weights by $\frac{t}{t+\delta}$ and $\frac{\delta}{t+\delta}$, we get:

$$p(t+\delta) = \frac{tp(t) + \delta B(q(t))}{t+\delta} \tag{1}$$

or, equivalently:

$$\frac{p(t+\delta) - p(t)}{\delta} = \frac{B(q(t)) - p(t+\delta)}{t}$$

As $\delta \to 0$, we obtain that the right derivative of $p(t)$:

$$\left.\frac{dp(t)}{dt}\right|_{+} = \frac{B(q(t)) - p(t)}{t},$$

This is not defined for $t = 0$, so the continuous time version should start at some $t_0 > 0$, say $t_0 = 1$. This leads to the following definition:

Definition 2 *For* $t \geq 1$, *the path* $(p(t), q(t))$ *is a* continuous time fictitious play process (CFP) *if*

$$(p(1), q(1)) \in \Delta(I) \times \Delta(J);$$

and

$$\left.\frac{dp(t)}{dt}\right|_{+} = \frac{B(q(t)) - p(t)}{t}, \quad \left.\frac{dq(t)}{dt}\right|_{+} = \frac{B(p(t)) - q(t)}{t}. \tag{2}$$

The system in (2) may be compactly written as:

$$\dot{p} = \frac{1}{t}\left[B(q) - p\right], \; \dot{q} = \frac{1}{t}\left[B(p) - q\right] \tag{3}$$

Proposition 1 *Suppose* $(p, q) \to (p^*, q^*)$. *Then* (p^*, q^*) *is a Nash equilibrium of* G.

Proof. Suppose not. Then there exists a player, say 1, and a pure strategy i for player 1 such that $p_i^* > 0$, but $i \notin BR(q^*)$. Since $q(t) \to q^*$ there exists a T such that for all $t > T$, $i \notin BR(q(t))$, and thus $i \neq B(q(t))$. But this implies that $\lim p_i(t) = p_i^* = 0$, which is a contradiction. ■

2.2 Best Response Dynamics

Definition 3 *For* $s \geq 0$, *the path* $(x(s), y(s))$ *is generated by* best response dynamics (BRD) *if*

$$(x(0), y(0)) \in \Delta(I) \times \Delta(J);$$

and

$$\frac{dx(s)}{ds} = B(y(s)) - x(s), \quad \frac{dy(s)}{ds} = B(x(s)) - y(s). \tag{4}$$

(See Hofbauer (1994) and references therein.)

We now argue that the best response dynamics BRD is "equivalent" to the CFP in the sense that there is a one to one correspondence between the trajectories generated by BRD and the trajectories generated by CFP and one system is convergent if and only if the other is.

Suppose that for all $t \geq 1, (p(t), q(t))$ is a CFP. For all $s \geq 0$, define $(x(s), y(s))$ by $x(s) = p(e^s)$ and $y(s) = q(e^s)$. Thus we are using the transformation $t = e^s$. We now obtain:

$$\begin{aligned}
\frac{dx(s)}{ds} &= \frac{dp}{dt}(e^s) \times \frac{dt}{ds} \\
&= e^s \frac{dp}{dt}(e^s) \\
&= e^s \frac{1}{e^s}\left[B(q(e^s)) - p(e^s)\right] \\
&= B(y(s)) - x(s).
\end{aligned}$$

A similar argument applies to $y(s)$ so that $(x(s), y(s))$ is a BRD.

Conversely, if for all $s \geq 0, (x(s), y(s))$ is a BRD then define for all $t \geq 1$, $(p(t), q(t))$ by $p(t) = x(\log t)$ and $q(t) = y(\log t)$. It is routine to verify that $(p(t), q(t))$ is a CFP.

This shows that the trajectory generated by a BRD with some initial condition $(x(0), y(0))$ is isomorphic to the trajectory generated by a CFP with the same initial condition, that is, $(p(1), q(1)) = (x(0), y(0))$.

2.3 Equivalent Games

Definition 4 *Let $G = (A, B)$ and $G' = (A', B')$ be two games with the same number of pure strategies for each player. Let BR and BR' be the pure strategy best response correspondences for G and G' respectively. The games G and G' are said to be* BR-equivalent, *written as $G \sim G'$, if $BR = BR'$.*

For our purposes, the importance of this definition lies in the fact the fictitious play process depends only on the best response correspondence of the game. Formally, let $p(t)$ be a CFP for the game G and let $p'(t)$ be a CFP for the game G'. If $G \sim G'$ and $p(1) = p'(1)$, then, for all t, $p(t) = p'(t)$. The same is true for DFP also.

Proposition 2 *Suppose G are G' are two games satisfying: there exist $\alpha > 0$ and $\beta > 0$ such that for all i, i', j, j':*

$$\begin{aligned}
a'_{ij} - a'_{i'j} &= \alpha\left(a_{ij} - a_{i'j}\right) \\
b'_{ij} - b'_{ij'} &= \beta\left(b_{ij} - b_{ij'}\right).
\end{aligned}$$

Then $G \sim G'$.

Proof. Let $q \in \Delta(J)$. For all i and i' we have that:

$$\sum_{j=1}^{J}\left(a_{ij} - a_{i'j}\right) q_j \geq 0 \text{ if and only if } \sum_{j=1}^{J}\left(a'_{ij} - a'_{i'j}\right) q_j \geq 0$$

and thus $BR_1 = BR'_1$. Similarly, $BR_2 = BR'_2$. ■

We now establish that every 2×2 game without weakly dominant strategies is BR-equivalent to either a zero-sum game or to a game with identical payoffs.

Proposition 3 *Suppose $G = (A, B)$ is a 2×2 game without any weakly dominated strategies. Then there exists a 2×2 matrix C such that either (i) $G \sim (C, -C)$; or (ii) $G \sim (C, C)$.*

Proof. Without loss of generality, let $a_{11} > a_{21}$ and $a_{12} < a_{22}$. Again, without loss of generality, let $a_{11} = b_{11} = 0$.
Case (i) : $0 = b_{11} < b_{12}$ and $b_{21} > b_{22}$.

By assumption $(a_{22} - a_{12}) - a_{21} > 0$ and $(b_{22} - b_{21}) - b_{12} < 0$. Thus there exists a $k < 0$ such that $(a_{22} - a_{12}) - a_{21} = k\left[(b_{22} - b_{21}) - b_{12}\right]$. Now again without loss of generality, we can assume that $k = -1$. Define C to be the matrix:

$$C = \begin{bmatrix} 0 & -b_{12} \\ a_{21} & (a_{22} - a_{12}) - b_{12} \end{bmatrix} = \begin{bmatrix} 0 & -b_{12} \\ a_{21} & a_{21} - (b_{22} - b_{21}) \end{bmatrix}$$

Using Proposition 2 it is easy to verify that $(A, B) \sim (C, -C)$.
Case (ii) : $0 = b_{11} > b_{12}$ and $b_{21} < b_{22}$.

By assumption $(a_{22} - a_{12}) - a_{21} > 0$ and $(b_{22} - b_{21}) - b_{12} > 0$. Thus there exists a $k > 0$ such that $(a_{22} - a_{12}) - a_{21} = k\left[(b_{22} - b_{21}) - b_{12}\right]$. Now again without loss of generality, we can assume that $k = 1$. Define C to be the matrix:

$$C = \begin{bmatrix} 0 & b_{12} \\ a_{21} & (a_{22} - a_{12}) + b_{12} \end{bmatrix} = \begin{bmatrix} 0 & b_{12} \\ a_{21} & a_{21} + (b_{22} - b_{21}) \end{bmatrix}$$

Again, using Proposition 2 it is easy to verify that $(A, B) \sim (C, C)$. ■

3. Lyapunov Functions

Before proceeding, some results on asymptotic stability and Lyapunov functions will prove helpful.

Definition 5 *Suppose x^* is an equilibrium of the differential equation*

$$\dot{x} = f(x), \tag{5}$$

where f is a C^1 map. Then x^* is an asymptotically stable *equilibrium if, (i) for every neighborhood U of x^*, there is a neighborhood U_1 of x^* in U such that every solution $x(t)$, with $x(0)$ in U_1, is defined and in U for all $t > 0$; and (ii) $\lim_{t \to \infty} x(t) = x^*$.*

Lemma 1 *Let x^* be an equilibrium for (5). Let $L : U \to \Re$ be a continuous function defined on a neighborhood U of x^*, differentiable on $U - x^*$, such that*

$$L(x^*) = 0 \text{ and } L(x) > 0 \text{ if } x \neq x^*;$$
$$\dot{L}(x) < 0 \text{ in } U - x^*$$

then x^* is asymptotically stable.

Proof. See Hirsch and Smale (1974). ■

4. Convergence Results

In this section we study some special classes of games for which the CFP can be shown to converge.

4.1 Zero-Sum Games

Definition 6 *The game* $G = (A, B)$ *is a* zero-sum *game if* $B = -A$.

The convergence of the DFP in zero-sum games was established by Robinson (1951). Robinson's proof is rather deep and difficult. However, a rather simple argument shows the convergence of the CFP in zero-sum games. [6]

Theorem 1 *Suppose* G *is a zero-sum game. Then every CFP converges.*

Proof. We show that every BRD process (defined in (4)) converges. This will then imply that every CFP converges (see section 2.2).

For $(x, y) \in \Delta(I) \times \Delta(J)$ define

$$L(x,y) = \max_{\overline{x}} \overline{x}Ay - \min_{\overline{y}} yA\overline{y} = B(y)Ay - xAB(x)$$

Observe that for all (x, y), $L(x, y) \geq 0$. Furthermore, $L(x, y) = 0$ if and only if (x, y) is an equilibrium of the game. We will argue that L is a Lyapunov function for the BRD process.

Observe that by the Envelope Theorem, we have $\frac{d}{dt}\left(\max_x xAy\right) = B(x)A\dot{y}$.

Differentiating with respect to t we obtain:

$$\begin{aligned}
\dot{L} &= \tfrac{d}{dt}\left(\max_x xAy\right) - \tfrac{d}{dt}\left(\min_y xAy\right) \\
&= B(y)A\dot{y} - \dot{x}AB(x) \\
&= B(y)A\left[B(x) - y\right] - \left[B(y) - x\right]AB(x) \\
&= -B(y)Ay + xAB(x) \\
&= -L(x,y) \\
&< 0.
\end{aligned}$$

[6] The simple proof given below was first shown to us by Christopher Harris. Brown (1951) seems to be aware of this proof. See Hofbauer (1994) also.

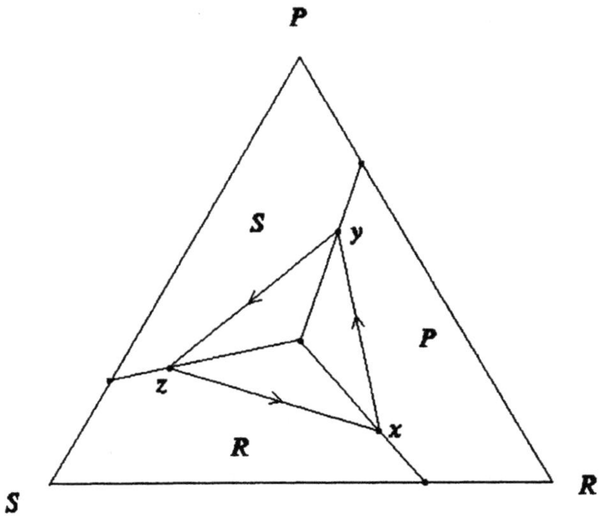

Figure 1: An example of non-convergence

as long as (x, y) is not an equilibrium. Thus BRD is asymptotically stable by Lemma 1. This completes the proof.[7] ■

Corollary 1 *The rate of convergence of CFP in zero-sum games is* $O(1/t)$.

4.2 An Example of Non-convergence

Shapley (1964) showed that the fictitious play process need not converge in general non-zero-sum games. We present a slight simplification of Shapley's example.

Consider the following game (also known as "Rock-Paper-Scissors").

	R	P	S
R	$0, 0$	$-1, \alpha$	$\alpha, -1$
P	$\alpha, -1$	$0, 0$	$-1, \alpha$
S	$-1, \alpha$	$\alpha, -1$	$0, 0$

where $\alpha > 0$. This game has a unique equilibrium at $p^* = q^* = \left(\frac{1}{3}, \frac{1}{3}, \frac{1}{3}\right)$.

Suppose $\alpha = \frac{1}{2}$. The best response correspondence for this game is depicted in the figure above. For all p in the region marked P (resp. S, R), it is a best-response to play P (resp. S, R). Suppose that both players start with the same initial beliefs about each

[7] Clearly we are glossing over some details since there are points where the use of the envelope theorem is invalid and L is not differentiable. However, these complications can be taken care of without too much difficulty. See Hofbauer (1994) or Harris (1995).

other, that is, $p(0) = q(0)$. Then, by the symmetry of the game, we will have that for all t, $p(t) = q(t)$. Let $x = (\frac{4}{7}, \frac{2}{7}, \frac{1}{7})$, $y = (\frac{1}{7}, \frac{4}{7}, \frac{2}{7})$ and $z = (\frac{2}{7}, \frac{1}{7}, \frac{4}{7})$. If $p(0) = x$, then as drawn, the CFP trajectory will cycle in the following manner: $x \to y \to z \to x$ and will not converge to the unique equilibrium.[8]

We now turn to two special classes of non-zero sum games for which the fictitious play process can be shown to converge.

4.3 Potential Games

Definition 7 *A game $G = (A, B)$ is a game with* identical payoffs *if $B = A$.*

Monderer and Shapley (1996a) and (1996b) have introduced the following class of games.

Definition 8 *A game $G = (A, B)$ is a* potential game *if there exists a matrix $P = (p_{ij})$ such that for all i, i', j, j':*

$$\begin{aligned} a_{i'j} - a_{ij} &= p_{i'j} - p_{ij} \\ b_{ij'} - b_{ij} &= p_{ij'} - p_{ij}. \end{aligned}$$

G is a weighted potential game *if there exist $\alpha > 0$, $\beta > 0$ and a matrix $P = (p_{ij})$ such that for all i, i', j, j':*

$$\begin{aligned} a_{i'j} - a_{ij} &= \alpha \left(p_{i'j} - p_{ij}\right) \\ b_{ij'} - b_{ij} &= \beta \left(p_{ij'} - p_{ij}\right). \end{aligned}$$

P is called the *potential* (resp. *weighted potential*) for the game G. By Proposition 2, every weighted potential game G is BR-equivalent to a game with identical payoffs $G' = (P, P)$.

Theorem 2 *Suppose G is a game with identical payoffs. Then every CFP converges.*

Proof. Once again, we argue that the BRD process (defined in (4)) converges. This will then imply that every CFP converges (see section 2.2).

For $(x, y) \in \Delta(I) \times \Delta(J)$ define

$$M(x, y) = xAy.$$

---

[8]Foster and Young (1995) have recently contructed an ingenious example of a game of coordination in which the FP process also cycles but fails to converge.

Differentiating with respect to t we obtain:

$$\begin{aligned}\dot{M} &= \tfrac{d}{dt}(xAy)\\ &= xA\dot{y} + \dot{x}Ay\\ &= xA\left[B(x) - y\right] + \left[B(y) - x\right]Ay\\ &\geq 0\end{aligned}$$

and $\dot{M}$ is strictly positive unless (x, y) is an equilibrium.

Define $M^* = \lim_{t\to\infty} M(x, y)$ starting from some initial condition, say $(x(0), y(0))$. If $L(x, y) = M^* - M(x, y)$, then L is a Lyapunov function.

This completes the proof. ■

From Theorem 2 and Proposition 2 we immediately obtain:

Theorem 3 *Suppose G is a weighted potential game. Then every CFP converges.*

4.4 2×2 Games

Recall, from Proposition 3, that every 2×2 game is BR-equivalent to either a zero-sum game or a game with identical payoffs. Now from Theorems 1 and 2 we obtain:

Theorem 4 *Suppose G is 2×2 game. Then every CFP converges.*

The convergence of the DFP in 2×2 games was established by Miyasawa (1961). Recently, Metrick and Polak (1994) have provided a geometric proof of Miyasawa's result.

4.5 Games with Strategic Complementarities

Definition 9 *The game $G = (A, B)$ satisfies* strategic complementarities *(SC) if, for all $i < i'$ and $j < j'$:*

$$(a_{i'j'} - a_{ij'}) > (a_{i'j} - a_{ij}) \text{ and } (b_{i'j'} - b_{i'j}) > (b_{ij'} - b_{ij}).$$

This class of games was introduced by Topkis (1979) and the properties of such games have been studied by Vives (1990) and Milgrom and Roberts (1990). In particular, such games always have a pure strategy equilibrium. Furthermore, if there is a unique equilibrium then the game is dominance solvable, that is, iterated removal of strongly dominated strategies will result in the equilibrium configuration. Thus if a game satisfying strategic complementarities has a unique equilibrium every CFP will converge. We now examine the behavior of CFP with multiple equilibria.

Denote by $B(q)$ the largest element of $BR(q)$. The set $BR^{-1}(q) = \{q : i \in BR(q)\}$ is convex and notice that $B^{-1}(i) = \{q : i = B(q)\}$ is also convex. Let $\succ$ denote the partial order on $\Delta(I)$ signifying first-order stochastic dominance, that is, $p'' \succ p'$ if for all i',

$$\sum_{i \leq i'} p_i'' \leq \sum_{i \leq i'} p_i'.$$

The same symbol will denote the partial order on $\Delta(J)$.

Lemma 2 *Suppose that G satisfies strategic complementarities. If $q'' \succ q'$, then $B(q'') \geq B(q')$.*

Proof. Let $i' = B(q')$ and $i'' = B(q'')$.

For all $i < i'$, $\sum_j (a_{i'j} - a_{ij})\, q_j' \geq 0$.

Since for all $i < i'$, $(a_{i'j} - a_{ij})$ is a non-decreasing function of j, and $q'' \succ q'$, for all $i < i'$, $\sum_j (a_{i'j} - a_{ij})\, q_j'' \geq 0$.

Thus, for all $i < i'$, $\sum_j a_{i'j} q_j'' \geq \sum_j a_{ij} q_j''$.

Hence, $B(q'') = i'' \geq i' = B(q')$. ■

Consider two initial conditions, $(p(1), q(1))$ and $(p'(1), q'(1))$ and the corresponding CFP trajectories $(p(t), q(t))$ and $(p'(t), q'(t))$. By Lemma 2 if $(p(1), q(1)) \succ (p'(1), q'(1))$ then for all t, $(p(t), q(t)) \succ (p'(t), q'(t))$. Thus in games with strategic complementarities, CFP is a *monotone* system as defined by Hirsch (1988).

Definition 10 *The game $G = (A, B)$ satisfies* diminishing marginal returns *(DMR) if, for all i, j:*

$$(a_{i+1,j} - a_{ij}) < (a_{ij} - a_{i-1,j}) \text{ and } (b_{i,j+1} - b_{ij}) < (b_{ij} - b_{i,j-1}).$$

The condition of diminishing marginal returns implies that for all q, $BR^{-1}(q)$ can consist of at most two elements and that these two elements must be consecutively numbered strategies for player 1. That is, if $BR^{-1}(q)$ is not a singleton then $BR^{-1}(q) = \{i, i+1\}$ for some $i \in I$.

We now present a convergence result for games with strategic complementarities and diminishing returns due to Krishna (1992).

Theorem 5 *Suppose G satisfies strategic complementarities and diminishing marginal returns. Then every CFP converges.*

Proof. (Sketch) Let Q be the set of limit points of $q(t)$. Let $B(Q) = \{i : i = B(q),\ q \in Q\}$ and $i_1 = \min B(Q)$.

Case 1. For all $i < i_1$, $\limsup p_i(t) = 0$ (and hence $\lim p_i(t) = 0$).

If $B(Q) = \{i_1\}$, then $\lim p_{i_1}(t) = 1$ and $p(t)$ converges. Hence $q(t)$ also converges.

So suppose that there exists an $i \in B(Q)$ such that $i > i_1$. Then it is the case that the trajectory $q(t)$ traverses the boundary between the sets $B^{-1}(i_1 + 1)$ and $B^{-1}(i_1)$

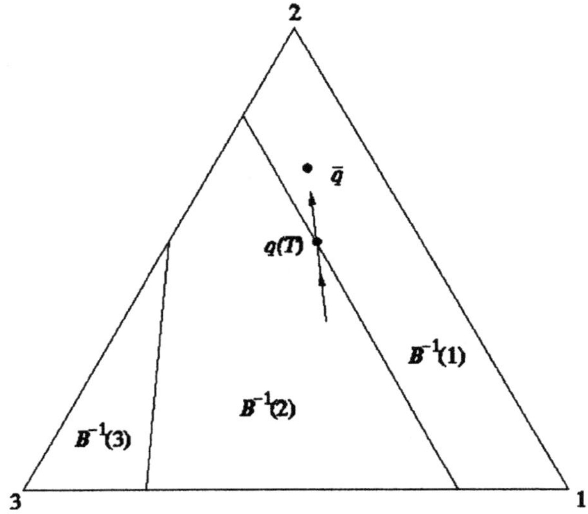

Figure 2: Games with SC and DMR

infinitely often. Consider a time T such that the trajectory $q(t)$ makes the transition from $B^{-1}(i_1+1)$ to $B^{-1}(i_1)$.

Suppose, as a simplification, that $\min B(Q) = i_1 = 1$ (as depicted in the accompanying figure). For all t in some interval $(T, T+\epsilon)$ $q(t) \in B^{-1}(1)$ and thus $i(t) = 1$. Thus, for all $t \in (T, T+\epsilon)$, $p(t)$ is a declining sequence, that is, for all $t', t'' \in (T, T+\epsilon)$, $t' < t''$ we have that $p(T) \succ p(t') \succ p(t'')$. Lemma 2 implies that for all $t \in (T, T+\epsilon)$, $j(t)$ is also a declining sequence, that is, for all $t', t'' \in (T, T+\epsilon)$, $t' < t''$ we have that $j(T) \geq j(t') \geq j(t'')$. (In the figure, $j(T) = 2$).

It can be established that because of diminishing returns, for all $t \in (T, T+\epsilon)$, $B(j(t)) \leq B(q(t))$. Thus the trajectory $q(t)$ cannot leave $B^{-1}(1)$ after T, contradicting the assumption that there is an $i > 1$ such that $i \in B(Q)$. l

If $i_1 > 1$ then it is no longer possible to argue that $p(T) \succ p(t') \succ p(t'')$. However, an indirect argument shows that we still have $j(T) \geq j(t') \geq j(t'')$ (Krishna 1992).

Case 2. $\limsup p_{i_1-1}(t) > 0$.

Since $i_1 = \min B(Q)$, it is certainly the case that for all $i < i'$, $\limsup p_i(t) = 0$. Since the transition from $B^{-1}(i_1)$ to $B^{-1}(i_1-1)$ occurs infinitely often, then using the same argument as in case 1 it can be argued that $B^{-1}(i_1-1)$ will absorb the trajectory $q(t)$. ■

We know of no examples with strategic complementarities for which CFP does not converge. Thus we make the following conjecture.

Conjecture 1 *Suppose G satisfies strategic complementarities. Then every CFP converges.*

5. Mixed Strategies

We have identified two classes of non-zero sum games for which CFP processes are convergent: potential games and games satisfying SC and DMR. Games in these classes share the important feature that CFP typically converges to a pure strategy equilibrium. Are there classes of (non-zero sum) games for which CFP can be shown to converge to mixed strategy equilibrium?

A CFP is said to be *cyclical* if there is a finite sequence of K pure strategy combinations, say $(i_1, j_1), (i_2, j_2), (i_3, j_3), ..., (i_K, j_K)$, such that the pure strategies played along the CFP follow this sequence over and over. We refer to this sequence as a *κ-cycle* if each player uses exactly κ distinct pure strategies. A κ-cycle is said to be *robust* if there is an open set of initial conditions $(p(1), q(1))$ such that the resulting CFP follows this cycle.

Krishna and Sjöström (1995) have shown the following.

Theorem 6 *For almost all games, if a robust κ-cycle converges then $\kappa \leq 2$.*

The result shows that cyclical convergence of CFP to mixed strategy equilibria is a rare event. (The 2×2 exception is a consequence of Proposition 3.) Whether there are non-cyclical CFPs remains an open question.

6. Convergence of Payoffs

The notion of convergence we have used throughout is that the beliefs generated by the CFP converge to equilibrium beliefs. Consider the following simple game of pure coordination.

	H	T
H	1,1	0,0
T	0,0	1,1

Let $p = p_H$ and $q = q_H$. This game has three equilibria: $(0,0)$, $(1,1)$ or $\left(\frac{1}{2}, \frac{1}{2}\right)$. Consider the discrete time process DFP when the weights on $p(t)$ and $B(q(t))$ are $\frac{t}{t+\delta}$ and $\frac{\delta}{t+\delta}$ respectively as defined in (1). If $(p(0), q(0)) = (1,0)$ then $(p(t), q(t))$ converges to $\left(\frac{1}{2}, \frac{1}{2}\right)$. But for all t, $(i(t), j(t))$ is either (T, H) or (H, T) and so the actual payoffs in each period are 0 to both players while the equilibrium payoffs are $\frac{1}{2}$.

Now consider the CFP. If $(p(1), q(1)) = (1,0)$ then for all $t \in [1,2)$ we have that $(i(t), j(t)) = (T, H)$ and thus $(p(2), q(2)) = \left(\frac{1}{2}, \frac{1}{2}\right)$. However, the CFP is not well defined at the point $\left(\frac{1}{2}, \frac{1}{2}\right)$ although it is natural to postulate that the CFP is a limit of the DFP as δ approaches 0. Thus in both the DFP and the CFP, while the beliefs converge to the equilibrium beliefs, the average of the actual payoffs is not the same as the expected payoffs in equilibrium. Of course, this can only happen when the convergence is to a mixed strategy equilibrium.

Consider the following result due to Monderer, Samet and Sela (1994).

Theorem 7 *Suppose that the CFP* $(p, q) \to (p^*, q^*)$ *and for all* t, $(p, q) \neq (p^*, q^*)$. *Then the limit of the average of the accumulated payoffs from the path* $(i(t), j(t))$ *is the same as the expected payoff in equilibrium.*

Proof. Let $\alpha(t) = \max_p pAq(t) = e_{i(t)}Aq(t)$ denote player 1's expected payoff in period t, where $e_{i(t)}$ is the $i(t)$th unit vector. By definition, the actual payoff obtained in period t is $e_{i(t)}Ae_{j(t)} = a_{i(t),j(t)}$. Then:

$$\begin{aligned} \dot{\alpha} &= e_i A\dot{q} \\ &= \tfrac{1}{t}e_i A\left[B(p) - q\right] \\ &= \tfrac{1}{t}e_i A\left[e_j - q\right] \\ &= \tfrac{1}{t}\left[e_i Ae_j - e_i Aq\right] \end{aligned}$$

and thus:

$$t\dot{\alpha} = e_i Ae_j - \alpha$$

which implies that:

$$\frac{d}{dt}\left[t\alpha\right] = e_i Ae_j = a_{ij}.$$

Thus, we have that:

$$t\alpha(t) = c + \int_1^t e_{i(s)}Ae_{j(s)}ds$$

or that:

$$\alpha(t) = \frac{c}{t} + \frac{1}{t}\int_1^t a_{i(s),j(s)}ds$$

Taking limits we obtain that:

$$\begin{aligned} p^*Aq^* &= \max_p pAq^* \\ &= \lim_{t\to\infty}\left(\max_p pAq(t)\right) \\ &= \lim_{t\to\infty}\alpha(t) \\ &= \lim_{t\to\infty}\tfrac{1}{t}\int_1^t a_{i(s),j(s)}ds. \end{aligned}$$

which completes the proof. ■

It can be shown that for zero-sum games the actual payoff always converges to the expected payoff. This is true even for the DFP (Riviere (1993), Monderer, Samet and Sela (1994)).

7. Continuous Versus Discrete Time

It should be apparent that there is much to be gained from considering the continuous time version of fictitious play (CFP). It leads to the development of strong convergence results whose proofs are simpler and more direct than those of their counterparts for the discrete process (DFP).[9]

We propose the following.

Conjecture 2 *Suppose that, for some game G, every CFP is convergent. Then every DFP is convergent.*

8. Conclusion

While salient in many ways, fictitious play is hardly the only interesting learning or evolutionary mechanism. A small and very incomplete sample of other proposed schemes that have been studied may be found in Fudenberg and Kreps (1988), Kandori, Mailath and Rob (1993) and Young (1993).

Throughout this survey the questions have been tightly focussed around the fictitious play process and around special classes of games. We believe that this focus affords two advantages. First, it allows many questions to be posed in a precise manner. Second, fictitious play seems to capture many important features associated with learning. It is hoped that the results will be generalizable to other classes of games and to other learning schemes. Thus, the approach espoused in this survey has been to concentrate on specific classes of games with a view to generating strong results.

Expressing similar sentiments, a mathematician (John Casti quoted by M. Hirsch) has observed that:

> "All current indications point toward the conclusion that seeking a completely general theory of nonlinear systems is somewhat akin to the search for the Holy Grail: a relatively harmless activity full of many pleasant surprises and mild disappointments, but ultimately unrewarding. A far more profitable path to follow is to concentrate upon special classes of nonlinear problems, usually motivated by applications, and to use the structure inherent in these classes as a guide to useful (i.e., applicable) results."

Acknowledgments. We are grateful to Christopher Harris, Dov Monderer and Ben Polak for insightful discussions and to John Morgan for able research assistance.

[9] Andreu Mas-Colell first suggested to us that the continuous time system may be simpler to work with than the discrete time system.

References

Aumann, R. and Brandenburger, A. (1995) Epistemic conditions for Nash equilibrium. Econometrica, 63, 1161-1180.

Bernheim, B. D. (1984) Rationalizable strategic behavior. Econometrica, 52, 1007-1028.

Binmore, K. (1988) Modelling rational players: part I. Economics and Philosophy, 3, 179-214.

Boylan, R. T. and El-Gamal, M. A. (1993). Fictitious play: a statistical study of multiple economic experiments. Games and Economic Behavior, 5, 205-222.

Brown, G. W. (1951) Iterative solution of games by fictitious play. In: Activity analysis of production and allocation. New York: Wiley.

Foster, D. and Young, H. P. (1995) On the nonconvergence of fictitious play in coordination games. mimeo. University of Pennsylvania and Johns Hopkins University.

Fudenberg, D. and Kreps, D. (1988) Learning and experimentation in games. mimeo. Department of economics, MIT, Cambridge, MA.

Harris, C. (1995) On the rate of convergence of continuous time fictitious play, mimeo, Department of economics, University of Cambridge, Cambridge, UK.

Hirsch, M. (1988) Stability and convergence in strongly monotone dynamical systems. Journal für die Reine und Angewandte Mathematik, 383, 1-53.

Hirsch, M. and Smale, S. (1974) Differential equations, dynamical systems and linear algebra. New York: Academic Press.

Hofbauer, J. (1994) Stability for the best response dynamics. mimeo. University of Vienna, Vienna, Austria.

Kandori, M., Mailath, G. and Rob, R. (1992) Learning, mutation and long run equilibria in games. Econometrica, 61, 29-56.

Krishna, V. (1992) Learning in games with strategic complementarities. Harvard Business School working paper no. 92-073. Boston.

Krishna, V. and Sjöström, T. (1995) On the convergence of fictitious play. mimeo. Penn State University and Harvard University.

Metrick, A. and Polak, B. (1994) Fictitious play in 2×2 games: a geometric proof of convergence. Economic Theory, 4, 923-933.

Milgrom, P., and Roberts, J. (1990) Rationalizability, learning and equilibrium in games with strategic complementarities. Econometrica, 58, 1255-1277.

Miyasawa, K. (1961) On the convergence of the learning process in a 2×2 non-zero-sum Game. Princeton University econometric research program, research memorandum no. 33. Princeton.

Monderer, D., Samet, D. and Sela, A. (1994) Belief affirming in learning processes. mimeo. The Technion, Haifa, Israel.

Monderer, D. and Shapley, L. (1996a) Fictitious play property for games with identical interests. Journal of Economic Theory, 68, 258-265.

Monderer, D. and Shapley, L. (1996b) Potential games. Games and Economic Behavior, forthcoming.

Pearce, D. (1984) Rationalizable strategic behavior and the problem of perfection. Econometrica, 52, 1029-1050.

Riviere, P. (1993) Private communication.

Robinson, J. (1951) An iterative method of solving a game. Annals of Mathematics, 54, 296-301.

Roth, A. E. and Erev, I. (1995) Learning in extensive form games: experimental data and simple dynamic models in the intermediate term. Games and Economic Behavior, 8, 164-212.

Shapley, L. (1964) Some topics in two-person games. In: Advances in Game Theory, Annals of Mathematical Studies 5, 1-28.

Smith, V. L. (1990) Experimental economics: behavioral lessons for microeconomic theory and applications. 1990 Nancy L. Schwartz Memorial Lecture, Kellogg Graduate School of Management, Northwestern University, Evanston.

Topkis, D. (1979) Equilibrium points in non-zero sum n-person submodular games. SIAM Journal on Control and Optimization, 17, 773-787.

Vives, X. (1990). Nash equilibrium with strategic complementarities. Journal of Mathematical Economics, 19, 305-321.

Young, H. P. (1992) The evolution of conventions. Econometrica, 61, 57-84.

Evolution and Games: Replicator Dynamics

Vijay Krishna[1] and Tomas Sjöström[2]

[1]Department of Economics, Pennsylvania State University, University Park, PA 16802, USA
[2]Department of Economics, Harvard University, Cambridge, MA 02138, USA

Abstract. We present a selective survey of work on evolutionary models of dynamics in games. We focus on the continuous time replicator dynamics and report convergence results for specific classes of games.

Keywords. Equilibrium, evolution, replicator dynamics, dynamical systems.

1. Introduction

This paper is a selective survey of work on a dynamic evolutionary model for games: the replicator dynamics introduced by Taylor and Jonker (1978). It is intended to be the second of a two-part survey on dynamical systems associated with games. Our main focus is on convergence results for specific classes of games. A companion piece surveys work on the Brown-Robinson learning process known as fictitious play, and in this paper our purpose is to compare the results obtained for fictitious play with those for the replicator dynamics.

Detailed accounts of evolutionary dynamical systems, including replicator dynamics, may be found in Hofbauer and Sigmund (1988) and Weibull (1995).

2. Preliminaries

Suppose that there is a single, infinite population of animals with types $i = 1, ..., I$. An individual animal's type i is simply a pure strategy which is genetically encoded and that the individual always plays. A single play of the game is viewed as a pairwise encounter between, say, animals with types i and j. The expected number of offspring of the animal of type i is then a_{ij}; and the expected number of offspring of the animal of type j is a_{ji}. Let $A = (a_{ij})$ be a *fitness* matrix of size $I \times I$. In game theoretic terms, there is an underlying *symmetric* game $G = (A, B)$, where A is a square matrix and $B = A^T$. Animals are matched randomly in each encounter and thus if in the current population the proportion of types is given by the vector $p = (p_1, p_2, ..., p_I) \in \Delta(I)$, the overall expected fitness of type i is $e_i Ap$ where e_i is the ith unit vector. Behavior which increases the animal's fitness will be selected and behavior is assumed to be inherited via asexual reproduction. Thus if type i does well relative to other types, in the next generation the proportion of type i animals will be larger.

2.1 Replicator Dynamics

We consider the following dynamic process: Let $N(t)$ denote the total number of animals in the population at time t. Let $N_i(t)$ denote the number of animals of type i in the population at time t. If μ represents the per period death rate, then

$$N_i(t+1) = N_i(t)\,[1 - \mu + e_i Ap(t)]$$

where $e_i Ap(t)$ is the expected number of offspring with type i, and $p(t)$ is a $k \times 1$ vector with typical element $p_i(t) = \frac{N_i(t)}{N(t)}$. Thus,

$$N(t+1) = N(t)\,[1 - \mu + p(t)Ap(t)]\,.$$

We may then write:

$$p_i(t+1) = \frac{N_i(t+1)}{N(t+1)} = p_i(t)\frac{1 - \mu + e_i Ap(t)}{1 - \mu + p(t)Ap(t)}$$

which yields:

$$p_i(t+1) - p_i(t) = p_i(t)\frac{e_i Ap(t) - p(t)Ap(t)}{1 - \mu + p(t)Ap(t)}$$

If we let the distance between time periods be denoted by δ and divide by δ we obtain:

$$\frac{p_i(t+\delta) - p_i(t)}{\delta} = p_i(t)\frac{e_i Ap(t) - p(t)Ap(t)}{1 - \delta\mu + p(t)hAp(t)}$$

Taking the limit as $\delta \to 0$ yields:

$$\dot{p}_i(t) = p_i(t)\,[e_i Ap(t) - p(t)Ap(t)] \tag{1}$$

(1) may be written more compactly as:

$$\dot{p}_i = p_i\,[e_i Ap - pAp)]$$

Given an initial condition $p(0) \gg 0$, (1) generates continuous time replicator dynamics (CRD) for the population. Observe that we may decompose (1) into $e_i Ap(t)$, which represents the fitness of type i, and $p(t)Ap(t)$, which represents the average fitness of the population. Thus, in CRD, all strategies continue to exist forever. This differs from the continuous time fictitious play (CFP) dynamics where only strategies that are best responses are present as $t \to \infty$.

The following property of (1) is called the *quotient rule:*

$$\frac{d}{dt}\left(\frac{p_i}{p_j}\right) = \frac{p_i}{p_j}\,[e_i Ap - e_j Ap]$$

which says that strategy i grows faster than j if and only if it has a higher expected payoff.

It is easy to argue that, like the CFP, if CRD converges then its limit must be an equilibrium.

Theorem 1 *If $p(0) \gg 0$ and $p(t) \to p^*$ then p^* is an equilibrium.*

Proof. Suppose p^* is not an equilibrium. Then there exists a pure strategy i such that $e_i A p^* - p^* A p^* = 2\delta > 0$. By continuity, there is a neighborhood P of p^* such that $e_i A q - q A q > \delta$ for all $q \in P$. Since $p(t)$ converges to p^*, $p(t) \in P$ for all $t \geq t_0$. Furthermore, since $p_i(0) > 0$, for all t, $p_i(t) > 0$. In fact,

$$\frac{\dot{p}_i(t)}{p_i(t)} = e_i A p(t) - p(t) A p(t) > \delta, \text{ for all } t$$

Therefore, $p_i(t) > p_i(0)e^{\delta t} \to \infty$, contradicting $p(t) \in P$. ■

3. Evolutionary Stable Strategies

Maynard Smith (1982) introduced the following definition.

Definition 1 *A mixed strategy (set of types) p^* is an* Evolutionarily Stable Strategy (ESS) *if for any $p \neq p^*$, there exists $\varepsilon' > 0$ such that for all $\varepsilon \in (0, \varepsilon')$: $p^* A((1-\varepsilon)p^* + \varepsilon p) > pA((1-\varepsilon)p^* + \varepsilon p)$.*

The notion of an ESS is intended to capture the idea that the strategy p^* is immune to small invasion by another strategy p. It is easy to verify that p^* is an ESS if and only if for all $p \neq p^*$:

$$\begin{array}{ll} \text{(i)} & p^* A p^* \geq pAp^* \\ \text{(ii)} & \text{if } p^* A p^* = pAp^* \text{ then } p^* A p > pAp. \end{array} \tag{2}$$

Thus for symmetric games, an ESS is a refinement of the set of Nash equilibria. In particular every ESS is an undominated equilibrium. Not every game has an ESS.

Theorem 2 *Suppose $p^* \gg 0$ is an ESS. Then every CRD converges to p^*.*

Proof. First, observe that since $p^* \gg 0$, by the definition of an equilibrium we have that for all p, $pAp^* = p^* A p^*$ and hence by (2), for all $p \neq p^*$:

$$p^* A p > pAp \tag{3}$$

Next, define the function $Z : \Delta^{I-1} \to \mathbf{R}$ by:

$$Z(p) = \sum_i p_i^* \ln p_i$$

For all $p \neq p^*$, we have:

$$\begin{aligned} Z(p) - Z(p^*) &= \sum_i p_i^* (\ln p_i - \ln p_i^*) \\ &= \sum_i p_i^* \ln \frac{p_i}{p_i^*} \\ &< \ln\left(\sum_i p_i^* \frac{p_i}{p_i^*}\right) \end{aligned}$$

$$
\begin{aligned}
&= \ln \sum_i p_i \\
&= 0
\end{aligned}
$$

where we have used Jensen's inequality. Thus Z has a strict maximum at p^*.

Now observe that:

$$
\begin{aligned}
\dot{Z}(p) &= \sum_i p_i^* \frac{\dot{p}_i}{p_i} \\
&= \sum_i p_i^* \left[e_i Ap - pAp \right] \\
&= p^* Ap - pAp \\
&> 0
\end{aligned}
$$

as long as $p \neq p^*$, by using (3). Let $L(p) = Z(p^*) - Z(p)$, then L is a Lyapunov function; hence p^* is an asymptotically stable equilibrium. ■

3.1 Equivalent Games

Suppose that A and A' are two fitness matrices such that for all i, i', j :

$$a'_{i'j} - a'_{ij} = a_{i'j} - a_{ij}.$$

Then the replicator dynamics for A' is the same as the replicator dynamics for A. To see this, write $a'_{ij} = a_{ij} + c_j$. The replicator dynamics for the fitness matrix A' is:

$$
\begin{aligned}
\dot{p}_i &= p_i \left[e_i A'p - pA'p) \right] \\
&= p_i \left[\sum_j a'_{ij} p_j - \sum_k \sum_j a'_{kj} p_k p_j \right] \\
&= p_i \left[\sum_j (a_{ij} + c_j) \, p_j - \sum_k \sum_j (a_{ij} + c_j) \, p_k p_j \right] \\
&= p_i \left[\sum_j a_{ij} p_j - \sum_k \sum_j a_{kj} p_k p_j + \sum_j c_j p_j - \sum_k p_k \sum_j c_j p_j \right] \\
&= p_i \left[\sum_j a_{ij} p_j - \sum_k \sum_j a_{kj} p_k p_j \right] \\
&= p_i \left[e_i Ap - pAp) \right]
\end{aligned}
$$

and this is the same as the replicator dynamics for the fitness matrix A.

4. Convergence Results

In this section we examine the convergence properties of the replicator dynamics. It is useful to begin by considering an example.

4.1 An Example

Consider the following game, also known as "Rock-Scissors-Paper":

	Rock	*Scissors*	*Paper*
Rock	1	α	0
Scissors	0	1	α
Paper	α	0	1

where $\alpha > 1$. This game has a unique equilibrium $p^* = (\frac{1}{3}, \frac{1}{3}, \frac{1}{3})$. According to Theorem 1, if the orbits converge, they converge to the unique symmetric equilibrium p^*. For any p,

$$pAp = |p|^2 + \alpha\,(p_1p_2 + p_2p_3 + p_1p_3) = |p|^2 + \alpha\frac{1 - |p|^2}{2}$$

The replicator dynamics are:

$$\begin{aligned}
\dot{p}_1 &= p_1\,[p_1 + \alpha p_2 - pAp] \\
\dot{p}_2 &= p_2\,[p_2 + \alpha p_3 - pAp] \\
\dot{p}_3 &= p_3\,[\alpha p_1 + p_3 - pAp]
\end{aligned}$$

Consider the function $V(p) = p_1p_2p_3$ defined on the simplex. $V = 0$ on the boundary of the simplex, and V attains its maximum at p^*. We compute the time derivative of $\ln V$:

$$\begin{aligned}
\tfrac{d}{dt}\ln V &= \tfrac{\dot{p}_1}{p_1} + \tfrac{\dot{p}_2}{p_2} + \tfrac{\dot{p}_3}{p_3} \\
&= [p_1 + \alpha p_2 - pAp] + [p_2 + \alpha p_3 - pAp] + [\alpha p_1 + p_3 - pAp] \\
&= 1 + \alpha - 3pAp \\
&= \left(\tfrac{\alpha}{2} - 1\right)\left[3\,|p|^2 - 1\right]
\end{aligned}$$

Since $3\,|p|^2 > 1$ for $p \neq p^*$, we have:

$$\dot{V}\begin{cases} > 0 \text{ if } \alpha > 2 \\ = 0 \text{ if } \alpha = 2 \\ < 0 \text{ if } \alpha < 2 \end{cases}$$

Thus, we have that if $\alpha \leq 2$, CRD does not converge.

4.2 Zero-Sum Games

Definition 2 *The game $G = (A, B)$ is a* symmetric zero-sum *game if $B = -A = A^T$.*

We observe that the CRD does not converge in general for symmetric zero-sum games. More formally:

Theorem 3 *Suppose that G is a (symmetric) zero-sum game with a completely mixed equilibrium p^*. Then every CRD starting from $p(0) \gg 0$, $p(0) \neq p^*$ does not converge to p^*.*

Proof. Suppose p^* is a completely mixed equilibrium. Since the matrix A is skew-symmetric, that is, $A^T = -A$, its value is 0 and by definition, we have that for all i :

$$0 = p^* A p^* = e_i A p^* = p^* A^T e_i = -p^* A e_i.$$

Hence we have that for all i, $p^* A e_i = 0$ and thus for all p :

$$p^* A p = 0$$

We also have that $pAp = (pAp)^T = pA^T p = -pAp$ and thus for all p :

$$pAp = 0$$

Now define the function:

$$V(p) = \sum_i p_i^* \ln p_i$$

Suppose $p \gg 0$. Then

$$\begin{aligned} \dot{V}(p) &= \textstyle\sum_i p_i^* \frac{\dot{p}_i}{p_i} \\ &= \textstyle\sum_i p_i^* \left[e_i A p - pAp\right] \\ &= p^* A p - pAp \\ &= 0 \end{aligned}$$

Therefore the orbits of CRD lie on the level curves of V and hence do not converge to p^*. ■

4.3 Potential Games

Ronald Fisher first considered games in which the two players have identical payoffs, that is $A^T = A$, and such games play an important role in evolutionary biology. It is useful to think of a game as being played by bits of DNA, called *alleles*, which are competing for a place on some gene locus. Alleles may be thought of as "candidates for genes." Each allele is identified as a strategy and an animal consists of a pair (i, j) of alleles, which is known as its *genotype.* Since the two alleles belong to one animal, the *phenotype*, their survival and reproduction must be the same. Thus a_{ij} represents the joint fitness of the pair (i, j) and we have that $a_{ij} = a_{ji}$.

Theorem 4 *Suppose that G is a (symmetric) game with identical payoffs. Then every CRD converges.*

Proof. Define $V(p) = pAp$. We have:

$$\begin{aligned} \dot{V}(p) &= \dot{p}Ap + pA\dot{p} \\ &= 2\dot{p}Ap \end{aligned}$$

since A is symmetric.

Now:

$$\begin{aligned} \tfrac{1}{2}\dot{V}(p) &= \dot{p}Ap \\ &= \textstyle\sum_i \dot{p}_i\,(e_iAp) \\ &= \textstyle\sum_i p_i\,[e_iAp - pAp]\,(e_iAp) \\ &= \textstyle\sum_i p_i\,(e_iAp)^2 - \left(\sum_i p_i\,(e_iAp)\right)^2 \\ &= \textstyle\sum_i p_i\,[e_iAp - pAp]^2 \\ &\geq 0 \end{aligned}$$

and is strictly positive unless p is an equilibrium. Define $V^* = \lim_{t\to\infty} V(p)$ starting from some initial condition $p(0)$. Let $L(p) = V^* - V(p)$, then L is a Lyapunov function; hence $\lim_{t\to\infty} p(t)$ is an asymptotically stable equilibrium. ■

The following result is called the "Fundamental Theorem of Natural Selection" and is attributed to R. Fisher.

Corollary 1 (The Fundamental Theorem of Natural Selection) *Average fitness is increasing.*

4.3.1 Gradient Fields

In this subsection we establish "Kimura's Maximum Principle": The change in gene frequencies occurs in such a way that the increase in average fitness is maximal.

The average fitness in the gene pool is pAp and the increase in average fitness is maximal is gene frequencies move in the direction of the gradient of pAp, that is, $2Ap$. Thus the dynamical system that maximizes the increase in average fitness is:

$$\dot{p} = 2Ap \tag{4}$$

But the CRD is not the same as (4).

A formal treatment of Kimura's principle requires the introduction of a new metric on Δ^{I-1} that is different from the Euclidean metric.

For any $p \in \Delta^{I-1}$ consider the inner product $\langle \cdot, \cdot \rangle_p$ on the tangent space at p defined by[1]:

$$\langle y, z \rangle_p = \sum_i \frac{1}{p_i} y_i z_i$$

Given a function $V(p)$, there exists a unique vector $\operatorname{grad} V(p)$ in the tangent space at p called the *gradient of* V *at* p such that for all z in the tangent space at p:

$$\langle \operatorname{grad} V(p), z \rangle_p = \sum_i \frac{\partial V(p)}{\partial p_i} z_i$$

Lemma 1 *The gradient corresponding to the S-inner product is:*

$$(\operatorname{grad} V(p))_i = p_i \left(\frac{\partial V}{\partial p_i} - \sum_j p_j \frac{\partial V}{\partial p_j} \right)$$

Proof. Verify that $\sum_i (\operatorname{grad} V(p))_i = 0$.
Now:

$$\begin{aligned} \langle \operatorname{grad} V(p), z \rangle_p &= \sum_i \tfrac{1}{p_i} (\operatorname{grad} V(p))_i z_i \\ &= \sum_i \left(\tfrac{\partial V}{\partial p_i} - \sum_j p_j \tfrac{\partial V}{\partial p_j} \right) z_i \\ &= \sum_i \tfrac{\partial V}{\partial p_i} z_i - \sum_j p_j \tfrac{\partial V}{\partial p_j} \left(\sum_i z_i \right) \\ &= \sum_i \tfrac{\partial V}{\partial p_i} z_i \quad \blacksquare \end{aligned}$$

Definition 3 *A dynamical system* $\dot{p} = f(p)$ *is said to be a* gradient field *with* potential function $V(p)$ *if* $f(p) = \operatorname{grad} V(p)$.

Since $\operatorname{grad} V(p)$ is the direction of steepest increase of V, Kimura's maximum principle is a consequence of the following result.

Theorem 5 *The CRD is a S-gradient vector field with potential function* $\frac{1}{2} pAp$.

Proof. Consider $V(p) = \frac{1}{2} pAp$. Then $\frac{\partial V}{\partial p_i} = e_i Ap$ and $\sum_j p_j \frac{\partial V}{\partial p_j} = pAp$. The replicator dynamics is:

$$\begin{aligned} \dot{p}_i &= p_i (e_i Ap - pAp) \\ &= p_i \left(\tfrac{\partial V}{\partial p_i} - \sum_j p_j \tfrac{\partial V}{\partial p_j} \right) \end{aligned}$$

Thus from the previous lemma the CRD can be rewritten as:

$$\dot{p} = \operatorname{grad} V(p). \quad \blacksquare$$

---

[1] The tangent space at p is the set of vectors $v \in \mathbf{R}^I$ such that $\sum_i v_i = 0$.

4.4 2 × 2 Games

For 2×2 games we prove the result that the orbits converge if they start in the interior.

Theorem 6 *In non-degenerate* 2×2 *games, if* $p(0) \gg 0$ *then* $p(t)$ *converges to an equilibrium.*

Proof. Without loss of generality, assume $a_{12} = a_{21} = 0$, and $a_{11}a_{22} \neq 0$ (non-degeneracy). We need to consider only strategy one; thus, we have:

$$\begin{aligned}\dot{p}_1 &= p_1\left(p_1a_{11} - \left(p_1^2a_{11} + (1-p_1)^2a_{22}\right)\right) \\ &= p_1(1-p_1)\left(p_1a_{11} + (1-p_1)(-a_{22})\right)\end{aligned}$$

There are four cases to consider:
Case 1. $a_{11} > 0 > a_{22}$. Then $\dot{p}_1(t) > 0$ for all t and $\lim p_1(t) = 1$.
Case 2. $a_{11} < 0 < a_{22}$. Then $\dot{p}_1(t) < 0$ for all t and $\lim p_1(t) = 0$.
Case 3. $a_{11} > 0, a_{22} > 0$. Then

$$\dot{p}_1(t)\begin{cases} > 0 \text{ for all } t \geq 0 \text{ if } p_1(0) > \frac{a_{22}}{a_{11}+a_{22}} \\ < 0 \text{ for all } t \geq 0 \text{ if } p_1(0) < \frac{a_{22}}{a_{11}+a_{22}} \end{cases}$$

thus, $p_1(t) \to 1$ if $p_1(0) > \frac{a_{22}}{a_{11}+a_{22}}$ and $p_1(t) \to 0$ if $p_1(0) < \frac{a_{22}}{a_{11}+a_{22}}$.
Case 4. $a_{11} < 0, a_{22} < 0$. Then

$$\dot{p}_1(t)\begin{cases} > 0 \text{ for all } t \geq 0 \text{ if } p_1(0) < \frac{a_{22}}{a_{11}+a_{22}} \\ < 0 \text{ for all } t \geq 0 \text{ if } p_1(0) > \frac{a_{22}}{a_{11}+a_{22}} \end{cases}$$

thus, $p_1(t) \to \frac{a_{22}}{a_{11}+a_{22}}$. The limits are equilibria by Theorem 1. ■

4.5 Games with Strategic Complementarities

The behavior of CRD in symmetric games with strategic complementarities (see the previous chapter for a definition) remains an open problem.

5. Convergence of the Time Average

In many instances even if the trajectory $p(t)$ does not converge, it is possible to show convergence of the "time average" of $p(t)$.

Definition 4 *The* time average *of the trajectory* $p(t)$ *at time* T *is*

$$m(T) = \frac{1}{T}\int_0^T p(t)dt$$

Definition 5 *The strategy* i *is* permanent *if there exists a* $\varepsilon > 0$ *such that for all* t, $p_i(t) > \varepsilon$.[2]

[2]Our definition of permanent strategies is different from that in Hofbauer and Sigmund (1988).

Theorem 7 *If every strategy is permanent, then every limit point of the sequence of time averages of CRD, $m(T)$, is an equilibrium.*

Proof. Observe that

$$\begin{aligned}\tfrac{1}{T}\left[\ln p_i(T) - \ln p_i(0)\right] &= \tfrac{1}{T}\int_0^T d\ln p_i(t)\\ &= \tfrac{1}{T}\int_0^T \tfrac{\dot{p}_i(t)}{p_i(t)}\,dt\\ &= \tfrac{1}{T}\int_0^T (e_iAp - pAp)\,dt \quad \text{for all } i\end{aligned}$$

thus,

$$\begin{aligned}\tfrac{1}{T}\left[\ln p_i(T) - \ln p_i(0)\right] &= e_iAm(T) - \tfrac{1}{T}\int_0^T pAp\,dt\\ &= \tfrac{1}{T}e_iA\int_0^T p\,dt - \tfrac{1}{T}\int_0^T pAp\,dt\end{aligned}$$

Suppose that along a convergent subsequence $m(T^s)$ of $m(T)$, $m(T^s) \to \hat{p}$. Then, for all i :

$$\begin{aligned}e_iA\hat{p} &= \lim_{s\to\infty} e_iAm(T^s)\\ &= \lim_{s\to\infty}\left[\tfrac{1}{T^s}\left(\ln p_i(T^s) - \ln p_i(0)\right) + \tfrac{1}{T^s}\int_0^{T^s} pAp\,dt\right]\\ &= \lim_{s\to\infty}\tfrac{1}{T^s}\int_0^{T^s} pAp\,dt\end{aligned}$$

since, by assumption, $p_i(T^s)$ stays away from the boundary of the simplex. Observe that the final expression is independent of i; hence, $\hat{p}$ is an equilibrium. ■

Corollary 2 *In zero-sum games where $p^* \gg 0$, $\lim_{T\to\infty} m(T) = p^*$.*

Proof. By Theorem 3 CRD trajectory lies on the level curves of the function

$$V(p) = \sum_i p_i^* \ln p_i$$

Since no level curve of this function can intersect the boundary of Δ^{I-1} every pure strategy is permanent. Now the result follows immediately from Theorem 7. ■

6. Conclusions

We have compared the convergence properties of the continuous time fictitious play process (CFP) with those of the continuous time replicator dynamics (CRD). Our focus has been to identify specific classes of games in which convergence of the processes may be guaranteed. The results may be summarized in the following table.

Type of game	*CFP*	*CRD*
Zero-sum	Convergence	Non-convergence $m(T)$ converges
2×2	Convergence	Convergence
Potential	Convergence	Convergence
SC + DMR	Convergence	Open Problem

We end with an interesting conjecture due to Gaunersdorfer and Hofbauer (1994):

Conjecture 1 *CFP converges if and only if the time average of CRD,* $m(T)$, *converges.*

References

Gaunersdorfer, A. and Hofbauer, J. (1994) Fictitious play, Shapley polygons and the replicator equation. Mimeo, University of Vienna, forthcoming in Games and Economic Behavior.

Hofbauer, J. and Sigmund, K. (1988) The theory of evolution and dynamical systems. Cambridge University Press, Cambridge.

Maynard Smith, J. (1982) Evolution and the theory of games. Cambridge University Press, Cambridge.

Taylor, P. and Jonker, L. (1978) Evolutionary stable strategies and game dynamics. Mathematical Biosciences, 40, 145-156.

Weibull, J. (1995) Evolutionary game theory. MIT Press, Cambridge, MA.

PART D

DESCRIPTIVE THEORY

Descriptive Approaches to Cooperation

Reinhard Selten[1]

Wirtschaftstheorie I, Universität Bonn, Adenauerallee 24-42, D-53113 Bonn, Germany

1. Introduction

There are three types of decision and game theory:

ideal-normative theory,
prescriptive theory, and
descriptive theory.

In *ideal normative game theory* one assumes fully rational players and often also common knowledge of full rationality. The point of interest of ideal-normative game theory is the strategic behavior under these conditions. The assumptions are not realistic, but nevertheless, ideal normative game theory is an important intellectual pursuit. The consequences of ideal normative rationality are of great philosophical significance.

Prescriptive game theory asks the question what a player should do if the participants in the game are not fully rational. For prescriptive theory one cannot take a completely Bayesian view because Bayesian decision making needs probabilities and utilities as inputs. If people are not able to come up with consistent utilities and probabilities then Bayesian methods are not directly applicable to their decision problems.

Descriptive game theory is not concerned with the question how players should act, but how they actually do act. This lecture will be concerned with descriptive game theory only.

Experimental observations strongly suggest that human players in games usually

do not optimize,
do not have utility functions, and
do not form probability distributions.

Thus the question arises: what do players do instead?

---

[1] This paper is based on a written account of lectures given at the *NATO Advanced Study Institute* on *Cooperation: Game Theoretic Approaches* 1994 in Stony Brook, prepared by Bettina Kuon.

Descriptive game theory is still in its infancy but nevertheless much more could be said about it than is possible in this lecture. We shall restrict ourselves to topics directly related to cooperation and even within this restriction no claim of completeness is made.

2. The equity principle

The equity principle is a very simple principle which is of great practical importance. Homans (1961) emphasized it in his writings, but probably it has been spelled out more or less clearly in the literature much earlier. The equity principle is applicable to situations where something has to distributed.

A player i receives a share x_i which is determined according to the weight w_i of player i. The shares are measured by a *standard of distribution*. A *standard of comparison* determines the weights of the players. This terminology has been introduced in a paper by Selten (1988). The *equity principle* proposes a distribution such that the shares are proportional to the weights. This means

$$\frac{x_1}{w_1} = \frac{x_2}{w_2} = \ldots = \frac{x_n}{w_n}.$$

A context in which the equity principle is often applied is cartel formation. Consider the fictitious example of a cartel of three firms which wants to restrict output to 600. The standard of distribution is production quantity. The firms have to decide on a quota up to which they are permitted to produce. If the standard of comparison is the capacity of the firms then the quotas shown in table 2.1. emerge.

Table 2.1. Cartel formation

Firm	Capacity	Quota
I	500	300
II	300	180
III	200	120

In actual praxis most of the bargaining is about the standards of comparison. However, there are some requirements these standards have to satisfy: the standards of distribution and comparison have to be *observable* and *relevant*. Observability means that everybody can clearly see the basis of computing shares and weights. Relevance means that the standards are not arbitrary but closely related to the nature of the problem.

In another fictitious example for the equity principle we look at the construction of a common river water purification plant by three towns. The total cost of the plant is 40 million \$. Suppose that the towns select the number of inhabitants as the standard of comparison. Table 2.2 shows the resulting costs for each town.

Table 2.2. River water purification plant

Town	Inhabitants	Million \$
A	90,000	18
B	70,000	14
C	40,000	8

In this example the standard of distribution is the cost to be carried. Instead of the number of inhabitants another standard of comparison could be used, e.g. water consumption in the last three years. Water consumption may be considered to be more relevant than the number of inhabitants and may therefore have a better chance to be adopted in the bargaining process.

3. Equity and coalitional bargaining in three-person characteristic function games

In a *three-person characteristic function game*, in this section often simply called a *game*, three players, named 1, 2, and 3, can form coalitions. A *coalition* C is a non-empty subset of the set of all players. Each coalition C has a *value* v(C) which the members can distribute among themselves. The value of the grand coalition including all three players will be denoted by $v(123)=g$. The coalition of players 1 and 2 has the value $v(12)=a$, the coalition of players 1 and 3 has the value $v(13)=b$ and the coalition of players 2 and 3 has the value $v(23)=c$. Without loss of generality assume a numbering of the players such that $a \geq b \geq c$. We shall restrict our attention to *zero-normalized games*, which means that a player i which is not member of a coalition receives $v(i)=d_i=0$ $(i=1, 2, 3)$. The function v which assigns values to coalitions is called the *characteristic function* of the game.

A coalition with more than one member is called a *genuine coalition*. We also shall look at games in which not all genuine coalitions are permissible. The set of all permissible genuine coalitions is denoted by Q. A three-person game is called *superadditive*, if all genuine coalitions are permissible which means $Q=\{12,13,23,123\}$ and if in addition to this we have

$$v(C \cup D) \geq v(C) + v(D) \quad \text{for } C \cap D = \varnothing ,$$

where C and D may be any two non-intersecting coalitions, not necessarily genuine ones.

A *coalition structure* is a partition of the set of players. In a three-person game at most one genuine coalition can be formed. Therefore, a coalition structure can be described by the genuine coalition which is formed or by the absence of any genuine coalition. There are five coalition structures: the grand coalition 123, the three two-person coalitions 12, 13, and 23, and the "null-structure" that no coalition is formed indicated by " – ".

A *payoff configuration* is a coalition structure together with a payoff distribution with the property that each coalition in the partition fully distributes its values among its members in such a way that everybody receives at least his one-person payoff d_i. This means the payoff configurations can have the following form

$(-;d_1,d_2,d_3)$ if no genuine coalition is formed

$(C;x_1,x_2,x_3)$ with $C \in Q$ and $\sum_{i \in C} x_i = v(C)$

and $x_i = d_i$ for $i \notin C$ and $x_i \geq d_i$ for $i \in C$.

A *grid game* is a game which involves a smallest money unit γ. The values of the characteristic function are integer multiples of γ. In grid games only payoff configurations are permitted which specify payoffs which are integer multiples of γ. This has the consequence that grid games have only a finite number of configurations. Usually, games played in experiments are grid games. In the following we shall consider only grid games, shortly referred to as *games*.

An *area theory* predicts a subset of the set of all configurations for every game in a specified class of games. Normative game theory has produced quite a number of area theories. The core and the Aumann-Maschler bargaining set (Aumann and Maschler 1965) seem to be the most important ones. Point theories like the *Shapley value* (Shapley 1953) or the *nucleolus* (Schmeidler 1969) are an alternative to area theories. However, in view of the great variance usually shown by the results of game experiments point theories do not seem to be appropriate for descriptive purposes. Ideally, one would want a theory which predicts the distribution of the results, but the test of such theories requires many more data than are available at the moment.

3.1 The bargaining set

The *bargaining set* is of special importance for this lecture since it has been proposed not only as a normative but also as a descriptive theory (Maschler 1978). In the following we shall not give a definition of the bargaining set. It will only be explained which payoff configurations are predicted by the bargaining set for the case of the superadditive zero-normalized three-person game. As we shall see certain numbers called *quotas* are of special significance for this theory. The quotas of players 1, 2, and 3 are the numbers q_1, q_2, and q_3, resp., which solve the following set of equations:

$$q_1 + q_2 = a$$
$$q_1 + q_3 = b$$
$$q_2 + q_3 = c\ .$$

This means:

$$q_1 = \frac{a+b-c}{2}$$

$$q_2 = \frac{a+c-b}{2}$$

$$q_3 = \frac{b+c-a}{2}\ .$$

A game is called a *quota game* if $q_i \geq 0$ for $i=1, 2, 3$, which is equivalent to $b+c \geq a$. For each of the five coalition structures table 3.1 shows which payoff configurations are predicted by the bargaining set.

Table 3.1. The bargaining set for zero-normalized three-person games

Coalition structure	$b+c \geq a$ *quota games*		$b+c<a$
	$2g<a+b+c$	$2g \geq a+b+c$ *non-empty core*	
—	$(-;0,0,0)$		
12	$(12;q_1,q_2,0)$		$(12;x_1,x_2,0)$ with $x_1 \geq b$ and $x_2 \geq c$
13	$(13;q_1,0,q_3)$		$(13;b,0,0)$
23	$(23;0,q_2,q_3)$		$(23;0,c,0)$
123	$(123;x_1,x_2,x_3)$ with $x_i = q_i-(q_1+q_2+q_3-g)/3$	$(123;x_1,x_2,x_3)$ with $x_1+x_2 \geq a$ and $x_1+x_3 \geq b$ and $x_2+x_3 \geq c$ *core*	

In table 3.1. a case distinction is made between quota games and other games. In addition to this, one has to distinguish between games with empty and non-empty core. The core is non-empty if and only if $2g \geq a+b+c$.

If the core is non-empty the bargaining set predicts the configurations in the core for the coalition structure 123. If the core is empty then the game must be a quota game and the bargaining set predicts equal distances of payoffs from the quotas for the coalition structure 123. In quota games configurations for two-person coalitions predicted by the bargaining set split the value according to the quotas. If the game is not a quota game then the bargaining set predicts that in coalition 12 player 1 receives at least b and player 2 receives at least c; in coalition 13 player 1 receives the whole coalition value b and in coalition 23 player 2 receives the whole coalition value c. The bargaining set does not exclude the possibility that no genuine coalition is formed. The only configuration for this coalition structure is also predicted. In games without the grand coalition, in which two-person coalitions are permissible but not the three-person coalition, the predictions for the two-person coalitions and the null-structure are the same.

Example
Consider the following zero-normalized three-person game:
$g=110$, $a=100$, $b=80$, and $c=70$.
This game has the quotas
$q_1=55$, $q_2=45$, and $q_3=25$.
The bargaining set in this example is
(−;0,0,0)
(12;55,45,0), (13;55,0,25), (23;0,45,25)
(123;50,40,20).

3.1.1 The united bargaining set

Maschler (1978) proposed to consider certain transformations of the game instead of the original one: the *power transformations*. These transformations can be interpreted as an application of equity considerations.

$$v_1(C) = v(C) + \frac{1}{2}[g-v(C)-v(N\setminus C)]$$

$$v_2(C) = v(C) + \frac{|C|}{3}[g-v(C)-v(N\setminus C)]$$

$$v_3(ij) = \max[v_1(ij),\frac{2}{3}g]$$

$$v_3(i) = g - v_3(jk) \qquad i,j,k\in\{1,2,3\}, \quad i\neq j\neq k\neq i.$$

Here, $|C|$ is the number of members of C.

$v_1(C)$ can be interpreted as the *equal split of the surplus* transformation. The surplus of the value g of the grand coalition over the sum of the values of C and its complement N\C is split evenly between C and its complement. The transformation $v_2(C)$ splits the same surplus proportionally to the numbers $|C|$ and $|N\setminus C|$ of members of both coalitions. We may say that v_1 and v_2 both use the same surplus as the standard of distribution but different standards of comparison. In the case of v_1 both coalitions receive the same weight, whereas in the case of v_2 the weight of a genuine coalition is the number of its members.

The transformation of v_3 is called the *Maschler power*. It can be interpreted based on the idea that a two-person coalition can either rely on v_1 or instead of this treat the game as if only the three-person coalition were permissible; this means the two players i and j declare that they refuse to enter any two-person coalition and insist on their equal shares in the three-person coalition.

A problem of the power transformation v_1 is that the one-person values can add up to more than g:

$v_1(1) + v_1(2) + v_1(3) > g$ for $g>a+b+c$.

A problem of the other two power transformations is that dummies in v fail to be dummies in v_2 and v_3. A *dummy* in a superadditive characteristic function game is a player i with

$v(C)-v(C\backslash i)=v(i)$ for every permissible genuine coalition with $i\in C$.

The power transformations allow to define the *power bargaining sets*. Let B be the ordinary bargaining set for v and let B_m be the set of all configurations in the bargaining set for v_m ($m=1,2,3$) which are also configurations for v. Maschler proposes not only to consider the configurations in B as predictions but also those in B_m, with $m=1,2,3$.

The power bargaining sets as well as the original bargaining set do not exclude the null-structure. However, the null-structure very rarely is observed in experiments. Therefore, the predictive success of bargaining set theory can be considerably improved by excluding the null-structure. Accordingly, we define the bargaining set B_0 and B_{0m} as the bargaining sets B and B_m, respectively, without the configuration involving the null-structure. B_0 is called the *bargaining set without the null-structure*; and the B_{0m} are called *power bargaining sets without the null-structure*.

Maschler (1978) observed that in his experiments players would have a tendency to agree on payoffs divisible by 5. Therefore, he proposed to neglect deviations smaller or equal to 5. As we shall see later the number 5 has the significance of a "prominence level" which determines the dividing line between reasonably round numbers and other numbers in the perception of the players. The prominence level is 5 in Maschler's experiments but in other experiments with coalition payoffs in a different range it may assume other values. In order to do justice to prominence in this sense we define the bargaining set $B_0[\Delta]$ as the set of all configurations $(C;x_1,x_2,x_3)$ such that a configuration $(C;y_1,y_2,y_3)\in B_0$ with $|y_i-x_i|\leq\Delta$ for $i=1, 2, 3$ can be found. Analogously, $B_{0m}[\Delta]$ is the set of all configurations $(C;x_1,x_2,x_3)$ such that a configuration $(C;y_1,y_2,y_3)\in B_{0m}$ with $|y_i-x_i|\leq\Delta$ for $i=1, 2, 3$ can be found. Usually, we will have $\Delta=5$. The sets $B_0[\Delta]$ and $B_0[\Delta]$, resp., will be called *bargaining sets* and *power bargaining sets without the null-structure and with deviations up to* Δ.

Now the *united bargaining set* can be defined as the union of the three bargaining sets without null-structure and with deviations up to Δ

$U[\Delta] = B_0[\Delta] \cup B_{01}[\Delta] \cup B_{02}[\Delta]$.

The Maschler power bargaining set $B_{03}[\Delta]$ is not considered by this definition since its inclusion does not improve predictive success.

3.2 The experiment of Murnighan and Roth

Murnighan and Roth (1977) conducted an experiment on the following zero-normalized three-person game:

g = 100, a = 100, b = 100, the coalition 23 is not permitted.

All other genuine coalitions are permitted. They observed 412 plays in which two-person coalitions were formed. Figure 3.1. shows the frequency distribution over the share of player 1. For the numbers 50, 55, ..., divisible by 5, the figure shows the number of cases in which player 1 received this amount. The cases with payoffs for player 1 between such two amounts are aggregated to a single category. Thus, the bar between the bars for 50 and 55 shows the frequency of payoffs for player 1 between 50 and 55.

A strong tendency to allocations divisible by 5 is observable.

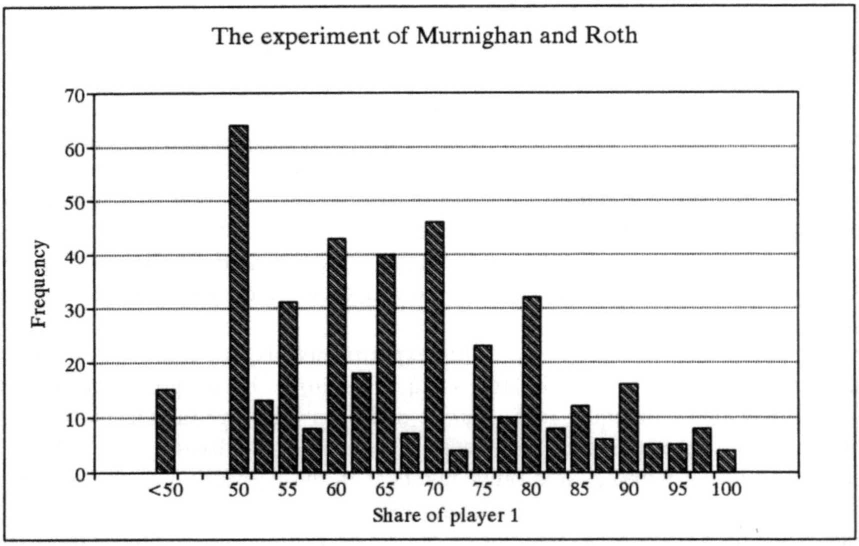

Figure 3.1. The experiment of Murnighan and Roth

The predictions of the different bargaining set theories are shown in table 3.2. In addition to this, the second last row shows the performance of the simple theory that player 1 gets at least 50. This theory is suggested by equity considerations. If both players in a permissible two-person coalition were equally strong both should receive 50. Since it is obvious that player 1 is stronger than the other player he should expect a payoff of at least 50. Since he is more powerful there are good reasons to suppose that he should get more than 50, but since it is very difficult to say how much more, 50 is the lower limit of what player 1 should receive.

Table 3.2. Bargaining set predictions for the game by Murnighan and Roth

Theory	Predicted range	Number of cases	Percentage of correct predictions
$B_0[5]$	$95 \leq x_1 \leq 100$	16	3.9
$B_1[5]$	$70 \leq x_1 \leq 80$	113	27.4
$B_2[5]=B_3[5]$	$61.67 \leq x_1 \leq 71.66$	103	25.0
U[5]	$95 \leq x_1 \leq 100$ or $61.67 \leq x_1 \leq 80$	188	45.6
	$50 \leq x_1 \leq 100$	399	96.8
	$0 \leq x_1 \leq 100$	412	100

The table shows that the predictions of the various bargaining sets are not very accurate. The united bargaining set U[5] predicts correctly in 45.6% of the cases only. The simple theory that player 1 receives at least 50 predicts correctly in 96.8% of the cases. Of course, this simple theory predicts a greater range than the united bargaining set and it is easier to achieve more correct predictions with a greater range. However, even if the right kind of correction for range size differences is made, the simple theory is much more successful than the united bargaining set or each of its components. The way in which range sizes should be taken into account in comparisons between area theories will be discussed in section 3.5.

It is very interesting to see how well the simple theory agrees with the data. The idea that the stronger player in a two-person coalition should get at least his equal share of the value seems to be a better predictor of behavior than the sophisticated bargaining set concept. The simple theory combines equity and power considerations. It does not say that the strong player must receive his equal share as payoff, but rather that, in view of his power position, the equal share is a lower bound for his payoff.

3.3 The equal division payoff bounds

This theory was introduced by Selten (1987) for three-person characteristic function games. The theory of equal division payoff bounds describes a hypothetical thought process. Unlike other area theories the theory of equal division payoff bounds is not based on a notion of stability. It does not exhibit the typical circularity of normative game theoretic concepts, but describes a finite sequence of reasoning steps leading to lower bounds for the players' payoffs. For the sake of simplicity the theory of equal division payoff bounds will only be presented for superadditive three-person games. It will however be clear how the theory has to be adjusted to the more general case.

A subject looking at the game structure immediately perceives power differences among the players which can be described by an *order of strength*.

Table 3.3. Order of strength for zero-normalized three-person games

inequality	$a > b > c$	$a > b = c$	$a = b > c$	$a = b = c$
order of strength	$1 \succ 2 \succ 3$	$1 \sim 2 \succ 3$	$1 \succ 2 \sim 3$	$1 \sim 2 \sim 3$

Table 3.3. shows the order of strength of the players 1, 2, and 3 for zero-normalized three-person games. The sign "$\succ$" means "stronger" and the sign "$\sim$" stands for "equally strong".

The order of strength is quite obvious even for naive subjects without any knowledge of game theory. Thus, if $b>c$ then 1 is stronger than 2 since player 1 can get b with player 3 whereas 2 can get only c with player 3.

The theory of equal division payoff bounds first focusses its attention on the strongest player and deduces what this player should expect to receive. Then it looks at the second strongest player, and so on. These considerations determine bounds (the *tentative bounds*), which limit the amount a player could expect.

The first tentative bound is the *coalition share* $v(C)/|C|$ for $i \in C$, if no other member of C is stronger than i. Thus, for $b>c$ in coalition 12 player 1 has the coalition share a/2, but player 2 does not have the coalition share a/2 in coalition 12 because player 1 is stronger.

For player 2 the *substitution share* is defined as $(a-b)/2$, which is half of the surplus of coalition 12 over coalition 13. The idea is that player 2 can replace player 3 in 13 and therefore can claim half of the surplus $a-b$. Neither for player 1 nor for player 3 a similar substitution share needs to be defined. Player 1 could replace either 2 or 3 in coalition 23 but the share in the surplus would be at most a/2, player 1's coalition share in 12. Player 3 cannot produce a surplus by replacing another player in a two-person coalition.

Player i's completion share $(g-v(jk))/3$ is based on the idea that player i should get at least one third of the surplus of the grand coalition over the two-person coalition of the other two players. Obviously, there is a connection to the proportional surplus power transformation v_2 (see 3.1.1).

The *highest tentative bounds for 1 and 2* are

$$t_1 = \max\left[\frac{a}{2}, \frac{g}{3}\right]$$

$$t_2 = \begin{cases} t_1 & \text{for } b=c \\ \max[\frac{c}{2}, \frac{a-b}{2}, \frac{g-b}{3}] & \text{for } b>c\ . \end{cases}$$

For player 3 a *competitive bound* w is defined. Player 3 is a very weak player. He has to make high offers to 1 and/or 2 in order to avoid that coalition 12 is formed. Therefore, player 3 is interested in the highest amounts player 1 and 2 can receive in coalition 12. These highest amounts are $h_1 = a - t_2$ for player 1 and $h_2 = a - t_1$ for player 2. The minimum of 3's surpluses over h_1 and h_2 is 3's competitive bound:

$$w = \min[b-h_1, c-h_2].$$

The *highest tentative bound for player 3* is the maximum of the completion share and the competitive bound

$$t_3 = \max[\frac{g-a}{3}, w]\ .$$

Various simple considerations of equity and power have lead to the definition of tentative bounds. For every player the highest tentative bound is a natural lower limit for his payoff aspirations. However, it may happen that in a game with $g>a$ the sum of the highest tentative bounds $t_1+t_2+t_3$ is greater than g. However, for $g>a$ the players feel a strong urge to form the three-person coalition, specifically if the game is run under favorable communication conditions with face to face interaction. In order to make it possible to form a three-person coalition in spite of $t_1+t_2+t_3>g$, at least one player must decrease his aspiration level to a value below his highest tentative bound. For $a>b$, which means that 2 is stronger than 3, player 3 as the weakest one has to yield by decreasing his aspiration level below t_3. In the case $a=b>c$, where 2 and 3 are equally strong, player 1 is the one who yields since otherwise two players instead of only one would have to decrease their aspiration levels. In the case $a=b=c$ the equal share g/3 in the grand coalition is the natural aspiration level for all three players.

These considerations lead us from the highest tentative bounds to what we call *preliminary bounds*. The bounds are preliminary only in as far as the influence of the prominence level still has to be considered. The preliminary bounds p_1, p_2, and p_3 are as follows:

For $t_1+t_2+t_3 \leq g$:

$$p_i = t_i \text{ for } i=1, 2, 3$$

For $t_1+t_2+t_3>g$:

$$p_1 = t_1 \quad p_2 = t_2 \quad p_3 = g-a \qquad \text{for } a>b\geq c$$

$$p_1 = \frac{g}{3} \quad p_2 = p_3 = \frac{1}{2}(g-\frac{a}{2}) \qquad \text{for } a=b>c$$

$$p_1 = p_2 = p_3 = \frac{g}{3} \qquad \text{for } a=b=c\ .$$

We always have $p_1+p_2+p_3\leq g$.

The *final bounds* emerge from the preliminary bounds by taking into account the prominence level Δ and the smallest money unit γ

$u_i = \max[\gamma, \Delta \mathrm{int}\frac{p_i}{\Delta}]$ where int x is the greatest integer not greater than x.

The final bounds are reached by rounding the preliminary bounds to the next lower integer multiple of the prominence level Δ. However, if this rounding should lead to zero the final bounds will be one smallest money unit γ.

The *predictions* of the equal division payoff bounds are that *a genuine coalition* C *with* $v(C) \geq \sum_{i\in C} u_i$ *is formed, if possible, and that* $x_i \geq u_i$ *for* $i\in C$, if C *is formed*. The words "if possible" indicate that no prediction is made for those extreme cases in which no configurations exist with the properties required above. For example this is the case for $a=b=c=g=\gamma$.

Table 3.4. gives an overview over the necessary calculations for the equal division payoff bounds.

Table 3.4. Equal division payoff bounds

Tentative bounds for 1 and 2

$$t_1 = \max[\frac{a}{2},\frac{g}{3}]$$

$$t_2 = \begin{cases} t_1 & \text{for } b=c \\ \max[\frac{c}{2},\frac{a-b}{2},\frac{g-b}{3}] & \text{for } b>c \end{cases}$$

Competitive bound w

$h_1=a-t_2$ and $h_2=a-t_1$

$w = \min[b-h_1, c-h_2]$.

Player 3's tentative bound

$$t_3 = \begin{cases} t_2 & \text{for } a=b \\ \max[\frac{g-a}{3},w] & \text{for } a>b \text{ and } t_1+t_2\leq a \\ \frac{g-a}{3} & \text{for } a>b \text{ and } t_1+t_2>a \end{cases}$$

Preliminary bounds

$p_i = t_i$ for $i=1, 2, 3$ if $t_1+t_2+t_3\leq g$ or $g=a$.

For $t_1+t_2+t_3>g$:

$p_1 = t_1 \quad p_2 = t_2 \quad p_3 = g-a \quad$ for $a>b\geq c$

$p_1 = \frac{g}{3} \quad p_2 = p_3 = \frac{1}{2}(g-\frac{a}{2}) \quad$ for $a=b>c$

$p_1 = p_2 = p_3 = \frac{g}{3} \quad$ for $a=b=c$

Final bounds

$u_i = \max[\gamma, \Delta \operatorname{int}\frac{p_i}{\Delta}]$ where int x is the greatest integer not greater than x

3.4 Prominence

Experiments have shown that players tend to select "round" numbers. In order to clarify this phenomenon a theory of prominence in the decimal system has been developed by Albers and Albers (1983). Let X be the set of all integer multiples of the smallest money unit γ. The *prominence level* $\Delta \in X$ is of the form

$\Delta = \mu 10^{\eta}\gamma$ where $\mu=1, 2, 5, 25$ and $\eta=0, 1, 2, \ldots$

The prominence level of x is the greatest prominence level such that x is an integer multiple of Δ.

The basic idea behind these definitions is a picture of the mental process which takes place if a person has to decide on a number such as a price to be set. As an illustrative example we shall look at the case of a person which has to guess the number of inhabitants of Islamabad. The first step in the process is the perception of a broad range in which the answer must lie, say, between 0 and 20 million. Then the person looks at the midpoint of this range and asks himself whether the number of inhabitants is greater or smaller than 10 million. The process stops and ends with the answer 10 million if the person feels that there is no reason to answer one way or the other. Suppose, that the person decides that the number of inhabitants is smaller than 10 million. This narrows the range to the numbers between 0 and 10 million. Again, the person will look at the midpoint of the range, 5 million and consider the question whether the number of inhabitants is greater or smaller. The process stops with the estimate 5 million if there are no good reasons to answer the question one way or the other. Suppose, that the person decides that the number of inhabitants is greater than 5 million. The midpoint of the remaining range between 5 and 10 million is 7.5 million. In this situation some persons may focus on 7.5 million but others on 7 or 8 million, since they perceive these numbers as "rounder". Whether 7.5 or 7 is considered as rounder may vary from person to person. Suppose, the subject focusses on 7 million and then decides that the number of inhabitants is smaller. He then may focus on 6 million and finally come to the conclusion that this is his estimate since he cannot find good reasons for deciding that the number to be guessed is greater or smaller.

Basically, the process just outlined successively divides the range still considered into two roughly equal parts and then focusses on the point separating the subintervals. This point may be different from the exact midpoint if otherwise one would obtain a too much "broken" number. Thus, an interval of the length $5 \cdot 10^{\eta}\gamma$ may be divided into two subintervals of lengths $2 \cdot 10^{\eta}\gamma$ and $3 \cdot 10^{\eta}\gamma$, rather than split evenly. The decimal system makes successive equal division of round intervals inconvenient since eventually a point will be reached where the result becomes too messy. Intervals of the length $2.5 \cdot 10^{\eta}\gamma$ may still be tolerated but a further equal subdivision to subintervals of the length $1.25 \cdot 10^{\eta}\gamma$ is too inconvenient. This has the consequence that at least one of the subintervals will always have a length of the form $\Delta = \mu 10^{\eta}\gamma$ where $\mu=1, 2, 5, 25$ and $\eta=0, 1, 2,\ldots$. This is the motivation of the definition of the prominence level given above.

My ideas about the role of decimal prominence in decision making may not be exactly the same ones as those of Albers and Albers (1983), but they are not more than an elaboration of their basic picture. Albers is now in the process of developing a quite different view which, however, cannot be described here.

For the purpose of the comparison of descriptive theories it is necessary to define a prominence level for a whole data set as an estimate of the dividing line separating sufficiently round numbers from other numbers. Table 3.5. illustrates the computation of the prominence level for the data set of Murnighan and Roth. In this experiment the smallest money unit has the value $\gamma = .01$.

Table 3.5. The prominence levels of player 1's share in the data of Murnighan and Roth

Δ	number of values $m(\Delta)$	number of observations $h(\Delta)$	$\frac{h(\Delta)}{m(\Delta)}$	$M(\Delta)$	$H(\Delta)$	$\frac{M(\Delta)}{M}$	$\frac{H(\Delta)}{H}$	$D(\Delta)$
100	1	3	3.00	1	3	.016	.007	−.009
50	1	64	64.00	2	67	.031	.163	.132
25	2	24	12.00	4	91	.062	.221	.159
20	3	79	26.33	7	170	.109	.413	.304
10	3	61	20.33	10	231	.156	.561	.405
5	5	90	18.00	15	321	.234	.779	.545
2.5	11	27	2.45	26	348	.406	.845	.439
2	11	24	2.18	37	372	.578	.903	.325
1	11	20	1.81	48	392	.750	.951	.201
.5	8	10	1.25	56	402	.875	.976	.101
.25	1	1	1.00	57	403	.891	.978	.087
.2	–	–	–	57	403	.891	.978	.087
.1	4	4	1.00	61	407	.953	.988	.035
.05	1	1	1.00	62	408	.969	.990	.021
.02	1	3	3.00	63	411	.984	.998	.014
.01	1	1	1.00	64	412	1.000	1.000	.000

The first column of table 3.5. shows the prominence levels which appeared in the distribution of the shares of player 1, shown in figure 3.1. The second column shows the number of values observed at the concerning prominence level. Thus, for the prominence level of 25 only two values were observed, namely 25 and 75. The third column shows the number of observations at the concerning prominence level. In the case of the prominence level 25 there are 24 observations. This means that 24 shares were either 25 or 75. For low levels of prominence only much fewer values are observed than there are numbers with this prominence level in the range from 0 to 100.

The fourth column shows the ratio $h(\Delta)/m(\Delta)$ of the number of observations $h(\Delta)$ to the number of values $m(\Delta)$. This ratio can be looked upon as the *relative occupation* of the values with the concerning prominence level. It can be seen that with some exceptions the relative occupation has a tendency to decrease with decreasing prominence level. Moreover, the relative occupation drastically moves down from $\Delta=5$ to $\Delta=2.5$.

The fifth column shows the cumulative number of values $M(\Delta)$ and the sixth column shows the cumulative number of observations $H(\Delta)$. The number of all values observed in the data set is denoted by M and the number of all observations is denoted by H. In table 3.5. we have $M=64$ and $H=412$. The next two columns show the relative cumulative number of values $M(\Delta)/M$ and the relative cumulative number of observations $H(\Delta)/H$. The last column shows the difference

$$D(\Delta) = \frac{H(\Delta)}{H} - \frac{M(\Delta)}{M}.$$

We call $D(\Delta)$ the *relative cumulative occupation surplus* or shortly the *occupation surplus*. The *prominence level* of a data set is defined as the greatest maximizer of the occupation surplus. In the case of table 3.5. the prominence level of the data set is 5. Already in figure 3.1. is was clearly visible that shares divisible by 5 have a much greater frequency than others. The method of computation confirms the visual impression that 5 is a reasonable value for the prominence level of the data set.

The interpretation of this way of computing the prominence level of the data set becomes clear if we look at the special case that the relative occupation $h(\Delta)/m(\Delta)$ is monotonically decreasing with decreasing prominence levels. In this case the maximum of $D(\Delta)$ is obtained at the prominence level Δ^* with the property that the relative occupation $h(\Delta)/m(\Delta)$ is greater than the mean occupation H/M if and only if $\Delta \geq \Delta^*$. This means that the prominence level of the data set separates the prominence levels with more than average occupation from those with less than average occupation.

As in table 3.5. the relative occupation $h(\Delta)/m(\Delta)$ is not always decreasing with decreasing prominence levels. This is partly due to random influences in the case of prominence levels with a small number of observations $h(\Delta)$. However, there are also other influences like the fact that often prominence levels of the form of $25 \cdot 10^{\eta}\gamma$ are less frequent than those of the form $20 \cdot 10^{\eta}\gamma$ (with the same η in both cases). An individual does not really need both types of prominence levels and may omit one in favor of the other. Thus, in table 3.5. the prominence level 25 has a relative occupation of 12, which is lower than the relative occupations of 20, 10, and 5. In spite of this lack of monotonicity the computation method for the determination of the prominence level of a data set still separates those prominence levels which on the whole have a higher relative occupation than average from those which on the whole have a lower one.

3.5 The difference measure of predictive success

Area theories differ with respect to the size of the predicted area. This has to be taken into account in the comparison of different area theories. We cannot simply look at the *hit rate* which is defined as the relative frequency of correct predictions. A correction for the relative size of the area has to be made. In the following the relative size of the predicted area within the set of all possible outcomes will simply be called the *area*.

Selten and Krischker (1983) introduced the difference between hit rate and area as a measure of predictive success. In the following the hit rate will be denoted by r and the area by a. The difference measure by Selten and Krischker is

$$m = r - a.$$

At first glance it might seem to be arbitrary to measure predictive success by this difference rather than by another function of r and a, e.g. r/a. It has been argued by Selten (1991) that the simple alternatives to the difference measure have bad properties. E.g. if there is a unique most frequent outcome the measure r/a favors the theory which predicts this single outcome only. All other theories have a lower ratio r/a. This means that a theory may be singled out by r/a even if it is almost always wrong.

In the paper by Selten (1991) an axiomatic justification of the difference measure has been given. Let m(r,a) be a measure based on hit rate r and area a. The following five axioms characterize the difference measure up to increasing monotonic transformations.

Axiom 1 *(r-monotonicity)*: $m(r,a) > m(r',a)$ for $r' < r$

Axiom 2 *(a-monotonicity)*: $m(r,a) > m(r,a')$ for $a' > a$

Axiom 3 *(continuity)*: $m(r,a)$ is continuous on $[0,1] \times [0,1]$

Axiom 4 *(independence)*: if $m(r'+r,a'+a) > m(r',a')$ then $m(r''+r,a''+a) > m(r'',a'')$

Axiom 5 *(indifference between trivial theories)*: $m(0,0) = m(1,1)$

Axioms 1 to 3 hardly need any comment. Obviously, r-monotonicity and a-monotonicity should be satisfied and continuity is a reasonable requirement. Axiom 4 can be interpreted as follows. Suppose there are three theories T, T', and T'' such that the predicted area of T intersects neither that of T' nor that of T''. Let the hit rates and areas be r, r', r'', resp., and a, a', a'', respectively. Consider the *union* $T \cup T'$ of the theories T and T' in the sense that the union of both areas is predicted. Suppose that T∪T' has a higher measure than T'. Then T∪T'', understood in the same way, should have a higher measure than T''. This means that whether joining a theory T to another one with a non-intersecting predicted area is an improvement or not, depends only on T and not on the other theory. This seems to be a reasonable requirement.

In the paper presenting the axiomatization (Selten 1991) axiom 4 is expressed in another way, but it can be seen easily that the mathematical content is the same. The intuitive justification given here is a different one.

A theory which predicts nothing is very sharp but never accurate. A theory which predicts everything is always correct but absolutely undiscriminating. Obviously, both types of theories are equally useless. It seems to be reasonable to give the same measure to both. This is expressed by axiom 5.

In the application of the measure to characteristic function games the problem arises how the relative size of the area should be computed. It is not adequate simply to determine the number of predicted configurations and to divide them by the number of all configurations. Doing this would put a too great emphasis on the grand coalition. There are 101 ways of dividing 100 smallest money units among two players but 5151 ways of dividing 100 smallest money units among three players. It seems to be adequate to give equal weights to all permissible coalition structures and within each structure equal weights to all configurations. This means that in the case of the zero-normalized three-person game with all genuine coalitions permissible each of the five coalition structures gets the weight 1/5. In a two-person coalition with the value 100 each configuration then receives the weight 1/505 and in a three-person coalition with a value of 100 the weight of a configuration is 1/25755. The area is computed as the sum of all weights of predicted configurations.

Table 3.6. shows hit rates, areas, and success measures for four data sets and the theories $B_0[5]$, U[5], and the equal division payoff bounds with $\Delta = 5$ (denoted by E_5). The table also shows significance levels for comparisons between theories according to the Wilcoxon signed-pairs matched-ranks test applied to success measures of independent subject groups.

For the purpose of comparing U[5] and $B_0[5]$ the last two data sets have been combined, since for each of them alone one obtains no significance.

For all four data sets the equal division payoff bounds have the highest success measure and it is significantly better than the other two theories.

Table 3.6. Comparison of predictive success

<table>
<tr><th>Theory</th><th>$B_0[5]$</th><th>$U[5]$</th><th>E_5</th><th>Experiment</th><th colspan="2">Significance</th></tr>
<tr><td>hit rate</td><td>.59</td><td>.89</td><td>.89</td><td rowspan="3">Maschler's 27 plays with grand coalition</td><td>$U[5]$</td><td>E_5</td></tr>
<tr><td>area</td><td>.19</td><td>.20</td><td>.13</td><td colspan="2">better than</td></tr>
<tr><td>success</td><td>.40</td><td>.69</td><td>.76</td><td>$B_0[5]$</td><td>$U[5]$</td></tr>
<tr><td>hit rate</td><td>.04</td><td>.44</td><td>.92</td><td rowspan="3">Murnighan and Roth 412 plays</td><td rowspan="3">.0001</td><td rowspan="3">.0001</td></tr>
<tr><td>area</td><td>.03</td><td>.12</td><td>.31</td></tr>
<tr><td>success</td><td>.01</td><td>.32</td><td>.61</td></tr>
<tr><td>hit rate</td><td>.51</td><td>.55</td><td>.92</td><td rowspan="3">Rapoport and Kahan 160 plays 8 quartets</td><td rowspan="6">.05</td><td rowspan="3">.01</td></tr>
<tr><td>area</td><td>.08</td><td>.08</td><td>.18</td></tr>
<tr><td>success</td><td>.43</td><td>.47</td><td>.74</td></tr>
<tr><td>hit rate</td><td>.68</td><td>.72</td><td>.93</td><td rowspan="3">Medlin 160 plays 8 quartets</td><td rowspan="3">.05</td></tr>
<tr><td>area</td><td>.09</td><td>.09</td><td>.20</td></tr>
<tr><td>success</td><td>.59</td><td>.63</td><td>.73</td></tr>
</table>

E_5 denotes the equal division payoff bounds with $\Delta=5$

Uhlich (1989) compared the bargaining set and the equal division payoff bounds in 25 independent data sets and found a higher success measure for the equal division payoff bounds. Table 3.7. shows this result.

Table 3.7. Uhlich's 25 independent data sets

Theory	Success measure
$B_0[\Delta]$	.1899
E_Δ	.6357

Δ is the prominence level of the data set

Uhlich (1989) developed a theory of *proportional payoff bounds* P_Δ, which is a modification of E_Δ. It is also applicable to three-person games which are not zero-normalized. For zero-normalized games E_Δ is slightly better than P_Δ.

4. The negotiation agreement area

The *negotiation agreement area* was first introduced in Uhlich (1989) and further tested and refined in Kuon and Uhlich (1993). It is a descriptive area theory for two-person games in which the conflict payoffs may be different from zero.

Two players bargain over a constant sum of money v(12). The players alternate in proposing until they either agree or one player decides to terminate the negotiations. In case of conflict player i receives v(i). Without loss of generality, number the players such that $v(1) \geq v(2)$, so that player 1 is the stronger player. The negotiation agreement area assumes that the strong player will start the bargaining with a high demand close to the surplus.

$A_1^{max} = v(12) - v(2).$

The weak player, however, will start with a lower demand in the area between the equal split of the surplus in addition to v(2) and the whole surplus.

$A_2^{max} = v(12) - v(1)$ and $A_2^{min} = v(2) + ½(v(12) - v(1) - v(2)).$

The negotiation agreement area assumes that the final agreement is reached by equal relative concessions from the initial demands. This assumption defines an area which specifies lower bounds x_i for the two players:

$$x_1 = \frac{A_1^{max}}{A_1^{max} + A_2^{max}} v(12) \qquad x_2 = \frac{A_2^{min}}{A_1^{max} + A_2^{min}} v(12).$$

An adjustment of these bounds to the prominence level of the data set leads to the final bounds u_i

$u_i = \max[v(i) + \gamma, \Delta \text{int}(\frac{x_i}{\Delta})]$ where Δ is the prominence level and γ is the smallest money unit.

Figure 4.1 graphically displays the idea of the negotiation agreement area.

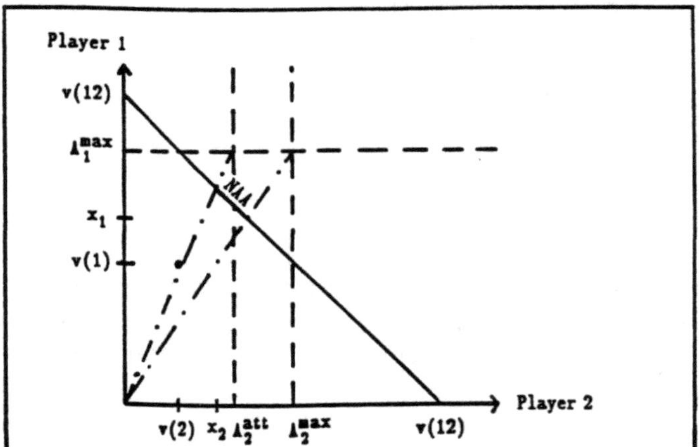

Figure 4.1. The negotiation agreement area

The negotiation agreement area was tested in 12 independent sessions of a laboratory experiment. It was compared to several other solution concepts. The best of the alternative theories was an interval around the equal split of the surplus in addition to the alternative values, bounded by the next prominent numbers (numbers having at least the prominence level of the data set). However, table 4.1. shows that the negotiation agreement area has a higher success measure than the equal split of the surplus interval. The difference between these theories is significant.

Table 4.1. The negotiation agreement area (NAA) and the equal split of the surplus interval (ES)

Theory	Success measure
NAA	.4873
ES	.2639

These results have been obtained by Kuon and Uhlich (1993). There, also games with negative conflict outcomes have been investigated. For such games the theory imposes the additional requirement that bargaining shares should not be negative. In the case of negative conflict outcomes the negotiation agreement area has a success measure of .7275 and is significantly better than the alternative theories.

5. The principle of balanced aspiration levels

The principle of balanced aspiration levels has been discovered in the context of a macroeconomic decision game, where groups of students had to play the roles of an employers' association, a workers' union, and a central bank. One of the tasks was the wage bargaining between the employers' association and the workers' union. Before the bargaining the players had to answer questionnaires on what they will try to achieve. This experiment was conducted by Tietz and Weber (1972) and Tietz (1973).

The employers' association as well as the workers' union had to specify the following numbers:

first demand

planned outcome (the outcome they plan to achieve)

at least attainable outcome (the outcome they expect to be able to push through at least)

conflict threat (at this point they threaten to strike or lockout)

conflict limit (at this point they would rather strike or lockout than accept the proposal)

These five levels form an ordinal scale of potential aspiration levels. The static *principle of balanced aspiration levels* predicts that *in terms of the ordinal scale the highest level achieved or surpassed by the final outcome will be the same one for both bargainers*. This level is the highest common level. The determination of the final outcome by the highest common level involves an interpersonal comparison between both players' aspiration scales. Apparently, bargainers are guided by such comparisons.

It turned out that the prediction of the static principle of balanced aspiration levels is very good. Tietz and Weber (1972) also proposed a theory, called the *planning difference theory*, which does not only predict the final outcome but also aspects of the bargaining process. In the next section we shall look at this semi-dynamic theory.

6. The planning difference theory

The following notation will be used for the five levels and for the expected first demand of the other player, which also had to be filled in in the preplay questionnaires:

first demand F
planned outcome P
at least attainable outcome A
conflict threat T
conflict limit L
expected first demand of the other player E

The lower indices 1 and 2 will indicate variables specified in the questionnaire of the workers' union and the employers' association, respectively.
Usually these numbers satisfy the inequalities

$$F_1 \geq P_1 \geq A_1 \geq T_1 \geq L_1$$

for player 1 (the workers' union), and

$$F_2 \leq P_2 \leq A_2 \leq T_2 \leq L_2$$

for player 2 (the employers' association).
The *planning difference* is defined as $P_1 - P_2$. If the planning difference is positive then the plans cannot be realized. The *concession reserve* is $F_1 - A_1$ for player 1 and $A_2 - F_2$ for player 2. The *tacit concession* is $E_2 - F_1$ for player 1 and $F_1 - E_2$ for player 2. If the other bargaining side expected a higher first demand of the proposer than this can be seen as if the proposer already made a concession.

We speak of an *easy bargaining situation* if $P_1 \leq P_2$. Then the player with the smaller tacit concession makes the first concession. On the other hand, we speak of a *tough bargaining situation* if $P_1 > P_2$. Then the player with the greater concession reserve makes the first concession.

Let V be the highest common level reachable by both bargaining partners. Then the bargaining result is that one of the values V_1 and V_2 which is more favorable for the first concession maker.

The planning difference theory involves a *strategic stability problem*. It seems to be advantageous to name a ridiculously high first demand. Here "high" must be understood in terms of the ordinal aspiration scale; for the workers' union a high demand is a high wage increase and for the employers' association a high demand is a low wage increase. The higher the first demand is in this sense, the lower is the tacit concession and the higher is the concession reserve. Therefore, the chance to be the first concession maker is increased by a more aggressive first demand. Since the first concession maker is favored by the final outcome, this has the consequence that for a given first demand of the other player it is always possible to obtain the more favorable outcome by a sufficiently aggressive first demand. If this were the case, then experience should drive players to more and more aggressive first demands. However, no such tendency is observed. Probably an excessive first demand becomes unbelievable and therefore does not have the predicted effect. The planning difference theory is unsatisfactory in this respect.

A second problem raised by the planning difference theory is an *information transmission problem*. How is the private information on aspiration scales transmitted by bargaining? If the way in which bargaining proceeds determines the final outcome it is advantageous for a player to behave as if he had aspiration levels leading to a better final result than honest information transmission. Players should be able to learn this by experience. Thereby, the validity of the theory would be destroyed.

In the experiment conducted by Tietz the players were represented by groups of students which interacted repeatedly for quite a number of periods. The situation had aspects of a supergame and the players may have taken advantage of this fact. Supergames facilitate cooperation and in the case at hand cooperation may take the form of not being too dishonest about the strength of one's motivation. In repeated interaction dishonesty could not be maintained without inducing strong suspicions on the other side. This kind of cooperation may explain how the transmission problem is solved.

Unfortunately, the answer to the transmission problem outlined above is probably not the whole story because the principle of balanced aspiration levels also seems to work in the studies with one shot bargaining situations by Scholz (1980) and Scholz, Fleischer, and Bentrup (1983). Maybe, the way in which people bargain is adapted to repeated bargaining situations, but not to one shot bargaining. The reasons for this could either lie in man's evolutionary past or in everyday experience. Maybe, many subjects have learned their bargaining behavior mainly in repetitive interactions with the same people.

In the next section the dynamic aspiration balance theory (Tietz 1976) will be described. Like the planning difference theory this more ambitious fully dynamic theory also faces the information transmission problem. However, simulations by Tietz, Daus, Lautsch, and Lotz (1988) indicate that the strategic stability problem does not arise in the dynamic aspiration balance theory.

7. The dynamic aspiration balance theory

The *dynamic aspiration balance theory* aims at the prediction of the whole bargaining process. It has been developed by Tietz (1976). The theory first refines the scale of aspiration levels by introducing midpoint levels.

F	level 8
½(F+P)	level 7
P	level 6
⋮	⋮ ⋮
L	level 0

The *secured level* is the highest level reached by the opponent's last offer. The *aspiration disadvantage* is the number of levels by which the opponent's secured level surpasses the own secured level. A *normal concession* is a concession by two level units. The theory determines who the first concession maker will be. This is done with the help of three successively applied criteria, called *filters*.

Filter 1: *Who has the higher secured level?*
Filter 2: *Who has the higher concession reserve* F_1-A_1 *or* A_2-F_2, *resp.?*
Filter 3: *Who has the lower tacit concession* E_2-F_1 *or* F_2-E_1, *resp.?*

The player which is selected by the highest filter is the first concession maker. If all filters are indecisive a random selection will determine the first concession maker. The *strength of the first concession maker* is

3 if filter 2 or filter 3 would select the other party,
2 if filter 3 is decisive,
1 else.

The strength determines the *size of the first concession*. The first concession is a normal concession unless (1), (2), or (3) is fulfilled.

(1) The aspiration disadvantage would become greater than 3 or the strength of the concession maker is at least 2. Then the first concession maker will make a concession of 1 level unit.
(2) The absolute difference between the first demands is very small (<.25% wage increase). Then the other player's first demand is accepted.
(3) A normal concession would go beyond the other player's first demand. Then the first concession maker concedes to the midpoint between the first demands, or by 1% wage increase, whatever is smaller.

From now on the players alternate in making concessions. The size of a counterconcession can be determined by the *accountable preconcession*. This is the amount in wage units by which the opponent's last demand is more favorable than the conflict limit. The *size of the first counterconcession* is the accountable preconcession multiplied by the strength of the first decision maker, unless ... (similar exceptions as in the case of the first concession). The size of the *further concessions*, in wage units, is equal to the size of the opponent's last concession, unless ... (details are omitted).

This theory, which we do not present in full detail, completely determines the bargaining process. It was tested with a sample of 30 periods of 12 games.

Many periods had to be excluded since the theory does not directly apply to bargaining over more than one variable and does not cover the effects of central bank intervention. The planning difference theory predicted the end result correctly in 73% of all cases and the dynamic aspiration balance theory even in 93% of the sample. In 60% of all cases the dynamic aspiration balance theory correctly predicts the complete process up to deviations of .05% (wage increase percentages).

Table 7.1. gives an overview over the empirical evidence for the static aspiration balance principle and the dynamic aspiration balance theory.

Even if not always exactly the same theories were tested the overall result of these studies is quite impressive. Of course, in view of the importance of the subject matter many more experiments are needed before a final judgement can be passed. The static aspiration balance theory is quite simple and seems to be well supported. The dynamic aspiration balance theory is a complex set of rules and it has not always been applied in exactly the same form. One does not find equal support for all details. Some cases covered by special rules only rarely occur in the data. Obviously, more experimental research is necessary.

Table 7.1. Empirical evidence

	Interaction				Bargaining variables	Support		Sample size
	verbal	formal	repeated	one shot		static	dynamic	
Tietz & Weber 1972	✓		✓		wage	+	++	30
Tietz & Weber 1978	✓		✓		wage, hours, period of notice	+	+ [4]	39
Scholz 1980		✓		✓ [3]	price	+	0 [5]	96
Scholz, Fleischer & Bentrup 1983	✓			✓	price quantity	+		55 [7]
Tietz & Bartos 1983	✓		✓ [1]		5 political 0-1-variables	+	+ [8]	80
Crößmann & Tietz 1983	✓		✓ [2]		price quantity		++ [6]	114

[1] 4 repetitions
[2] with partner choice
[3] one repetition with seemingly different opponents
[4] weak predictions of a version adapted to many bargaining variables; equal support for dynamic aspiration security theory (not explained here)
[5] better support for planning difference theory
[6] modified version
[7] 60 expert subjects with practical bargaining experience in business and 50 novices
[8] combined with a proposal search model

The principle of balanced aspiration levels has the potential to be a powerful tool of economic modelling. Unfortunately, up to now the way in which economic behavior is modelled is heavily influenced by empirically unsupported normative presumptions. It would be better to make more use of experimentally based behavioral theories.

8. Cooperation in normal form games

Ostmann (1988) conducted experiments with 3×3×3-games $G=(A_1,A_2,A_3;u)$. In the following the notation will be explained and some basic concepts will be introduced.

A_i is the *pure strategy* set of player i (i=1, 2, 3).

$a=(a_1,a_2,a_3)$ with $a_i \in A_i$ is a *strategy combination* of the three players.

$u(a)=(u_1(a),u_2(a),u_3(a))$ is the *payoff function*. $u_i(a)$ is i's payoff for a.

$a_C=(a_i)_{i\in C}$ is a *coalition strategy*. The notation $a=a_C a_{N\backslash C}$ is used.

$x_C=(x_i)_{i\in C}$ is a *coalition payoff vector*.

A coalition C can *absolutely secure* $y_C=(y_i)_{i\in C}$ iff it has a strategy a_C such that for every $a_{N\backslash C}$ $u_i(a_C a_{N\backslash C}) \geq y_i$ for all $i\in C$.

A coalition C can *conditionally secure* $y_C=(y_i)_{i\in C}$ iff for every $a_{N\backslash C}$ it has a strategy a_C with

$u_i(a_C a_{N\backslash C}) \geq y_i$ for all $i\in C$.

$x=(x_1,x_2,x_3)$ is in the *α-core* iff no coalition C can absolutely secure an $y_C=(y_i)_{i\in C}$ with $y_C \neq x_C$ and $y_i \geq x_i$ for $i\in C$. $x=(x_1,x_2,x_3)$ is in the *β-core* iff no coalition C can conditionally secure an $y_C=(y_i)_{i\in C}$ with $y_C \neq x_C$ and $y_i \geq x_i$ for $i\in C$. The β-core is a subset of the α-core.

$a_{N\backslash C}$ is a *Pareto-best reply* to a_C iff for no $a'_{N\backslash C}$ we have $u_i(a_C a'_{N\backslash C}) \geq u_i(a_C a_{N\backslash C})$ for $i\in N\backslash C$ with $u_i(a_C a'_{N\backslash C}) > u_i(a_C a_{N\backslash C})$ for at least one $i\in N\backslash C$. $B(a_C)$ is the *set of all Pareto-best replies* to a_C.

A coalition C can attain $y_C=(y_i)_{i\in C}$ as a *leader payoff* iff it has a strategy a_C such that for all $a_{N\backslash C}\in B(a_C)$

$u_i(a_C a_{N\backslash C}) \geq y_i$ holds for $i\in C$.

$x=(x_1,x_2,x_3)$ is in the *γ-core* iff no coalition C can attain an $y_C=(y_i)_{i\in C}$ with $y_C \neq x_C$ and $y_i \geq x_i$ for $i\in C$ as a leader payoff. The γ-core is a subset of the α-core.

The *minimal core* is the intersection of all those cores (α, β, and γ), which are non-empty.

An *equilibrium* (in pure strategies) is a strategy combination $a=(a_1,a_2,a_3)$ with $a_i\in B(a_{N\backslash i})$ for i=1, 2, 3.

A *Rawls optimum* is a strategy combination $a=(a_1,a_2,a_3)$ with the property that for no other strategy combination $a'=(a'_1,a'_2,a'_3)$ the inequality

$\min_{i=1,2,3} u_i(a') > \min_{i=1,2,3} u_i(a)$ holds.

The notions of the α- and the β-core are well known game theoretic concepts (Aumann 1961 and Scarf 1967), but the γ-core has first been introduced in the work of Ostmann.

Ostmann conducted an experiment involving various 3×3×3-games. The payoffs in the game were in points. Money payoffs were determined by a non-linear transformation from points to money, a different one for each player only known to the player himself. Therefore, the payoffs in the game were not interpersonally comparable. Nevertheless, the players sometimes acted as if interpersonal comparability was possible. Bargaining was mostly face to face and lasted until an agreement was reached. The agreement could be on the grand coalition, a pair coalition or on the fact that no agreement can be reached. The coalition agreements were not binding. After the end of bargaining the players independently and simultaneously selected their strategies. So, finally this was a non-cooperative game.

In table 8.1. a distinction is made between normal form games with and without a pure strategy equilibrium in the minimal core. It can be seen that the results are quite different for these two categories of games. An agreement is called *stable* if it is not violated by the final choice. The numbers of stable agreements are shown in brackets. The categories in table 8.1. are partially overlapping.

Table 8.1. Experimental results

	Equilibrium in minimal core ?	
	Yes	No
no coalition	3	13
pair coalition	18 (17)	49 (45)
grand coalition	95 (92)	148 (78)
minimal core	78 (76)	78 (33) [1]
Rawls optimum	83 (82)	130 (62)
equilibrium	75 (74)	10 (9) [2]
number of plays	116	210

in brackets: agreements not violated by final choice
[1] out of 177 games with non-empty minimal core
[2] out of 135 games with equilibria outside the minimal core

The results of the experiment can be summarized as follows:
- only in few cases no agreement is reached
- most agreements are grand coalitions
- pair coalitions are very stable
- grand coalitions are very stable if there is an equilibrium in the minimal core but rather unstable otherwise

- equilibrium occurs often if there is an equilibrium in the minimal core but rarely otherwise

Cooperation in normal form games is a very interesting subject matter. Ostmann's results indicate that the minimal core is a useful predictive concept. In their bargaining the players seem to use arguments related to the α-, β-, and γ-core. It is understandable that an argument related to one of these cores does not have much force if the players find out that the concerning core is empty. Therefore, it is reasonable to define the minimal core as the intersection of the non-empty cores of this kind.

A more detailed analysis of Ostmann's data will probably be published in the near future.

9. Duopoly strategies programmed by experienced players

It is maybe not an exaggeration to say that in the many years after the work of Cournot (1838) economics did not yet produce an empirically well supported duopoly theory. Selten, Mitzkewitz, and Uhlich (1988) applied the strategy method to an asymmetric Cournot duopoly. The *strategy method* is an experimental procedure in which the players first play the game interactively, so they experience the strategic structure of the game. Afterwards, each player has to develop a computer program specifying a strategy for the game. The strategies are matched in a computer tournament. A strategy study may involve several tournaments between which the subjects have the opportunity to change their strategies.

The strategy method was proposed by Selten (1967) and found further applications, inter alia, in Axelrod (1984), Fader and Hauser (1988), Keser (1992), and Kuon (1994).

The study of Selten, Mitzkewitz, and Uhlich concerned a 20-period supergame of an asymmetric quantity variation model with linear cost and linear demand. The parameters were as follows:

costs: $C_1 = 9820+9x_1$ $\qquad$ $C_2 = 1260+51x_2$

demand: $p = \max[0,300-x_1-x_2]$.

Here, x_1 and x_2 are the production quantities, C_1 and C_2 are the costs of duopolist 1 and 2, resp., and p is the price.

Figure 9.1. shows a graphical representation of the model.

In this strategy experiment 23 subjects participated in the framework of a student's seminar. The seminar started with three game playing rounds where the players interacted anonymously by computer terminals. After this experience phase three strategy programming rounds with computer tournaments followed. Each subject had to write a strategy for both sides. The motivation of the students was by grades.

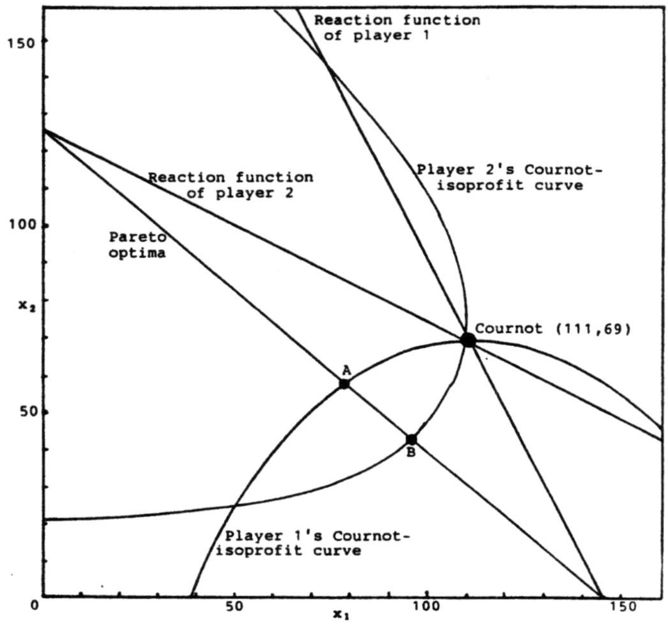

Figure 9.1. Graphical representation of the model

Figures 9.2. to 9.4. show the results of the game playing rounds. In these diagrams each point represents the profits of player 1 and 2 in a play. In the first game playing round (figure 9.2.) many players had payoffs below the Cournot profits. In the second game playing round (figure 9.3.) this already improved, but there are still many profits below the Cournot profit. In the third game playing round (figure 9.4.) cooperation was learned to a large extent.

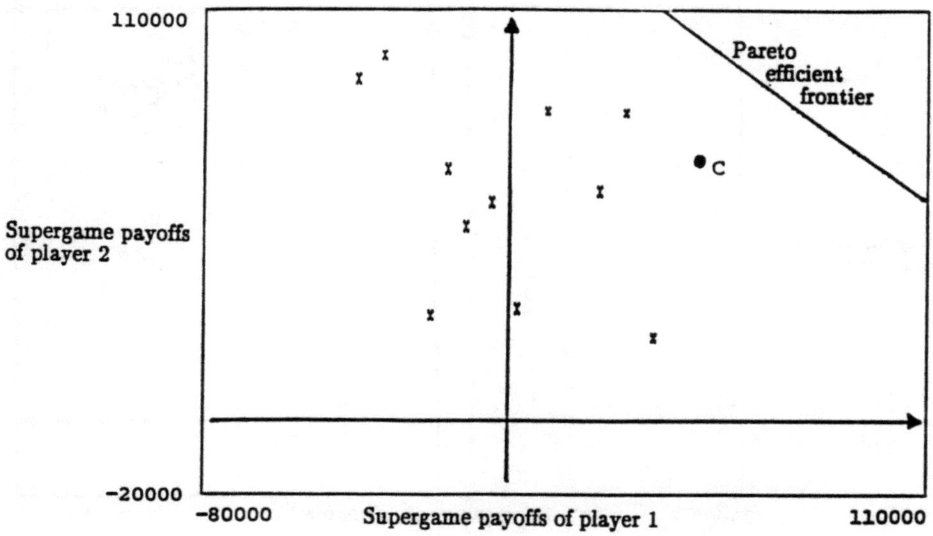

Figure 9.2. Result of the first game playing round

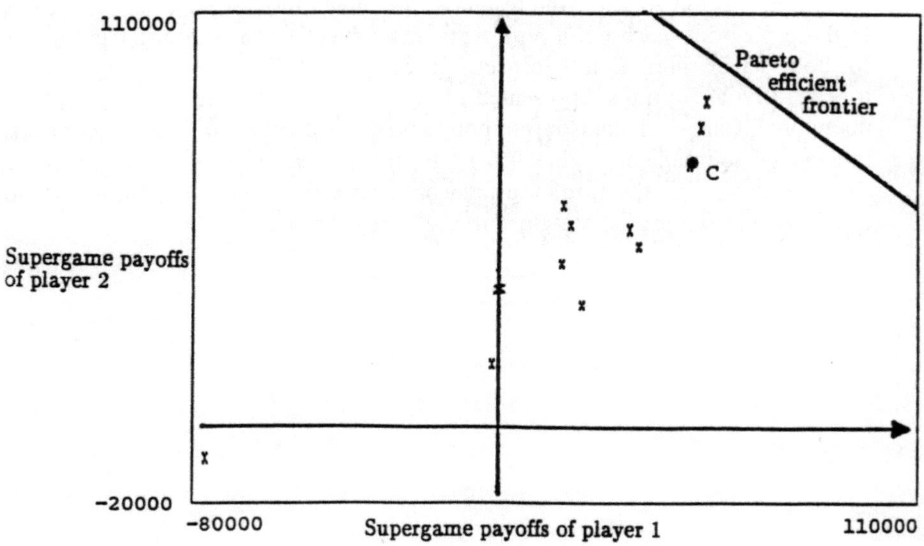

Figure 9.3. Result of the second game playing round

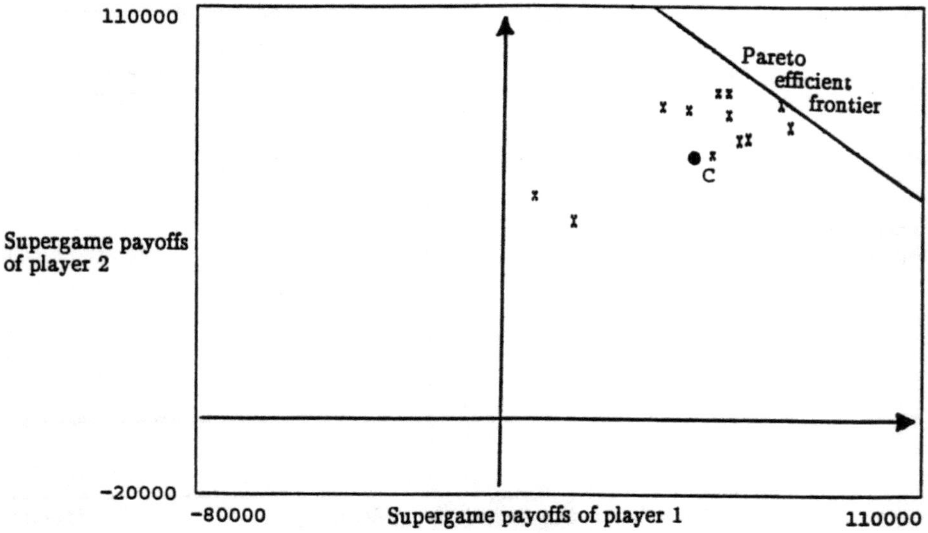

Figure 9.4. Result of the third game playing round

Figures 9.5. to 9.7. show the results of the three tournaments of the strategies. In these diagrams each point represents one subject's pair of tournament profits for both player roles. In this respect figures 9.5. to 9.7. differ from figures 9.2. to 9.4., in which points represented plays. In the first tournament (figure 9.5.) many profits are near the Cournot point and some profits are below Cournot. In the second tournament (figure 9.6.) the bulk of the observations is above the Cournot profit. In the third tournament (figure 9.7.) full cooperation is not achieved, but the profits are close to the Pareto frontier.

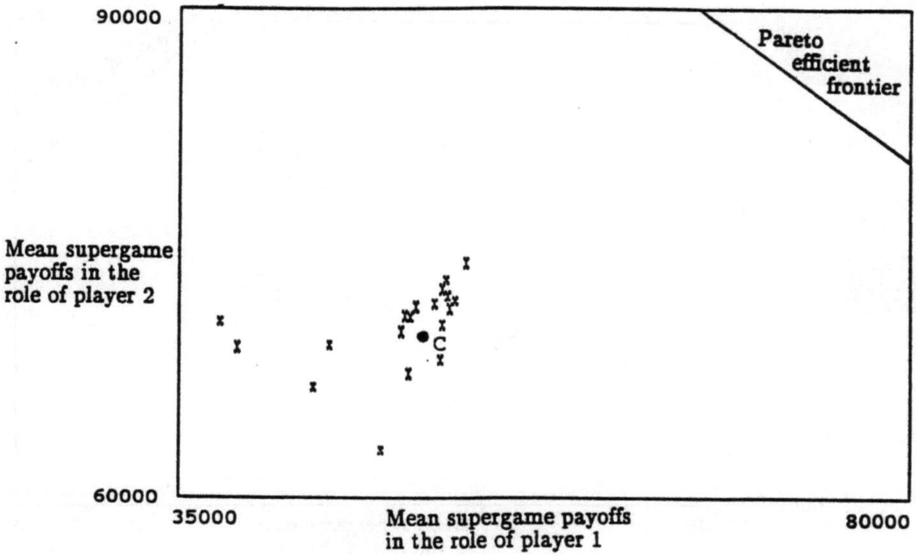

Figure 9.5. Result of the first tournament of the strategies

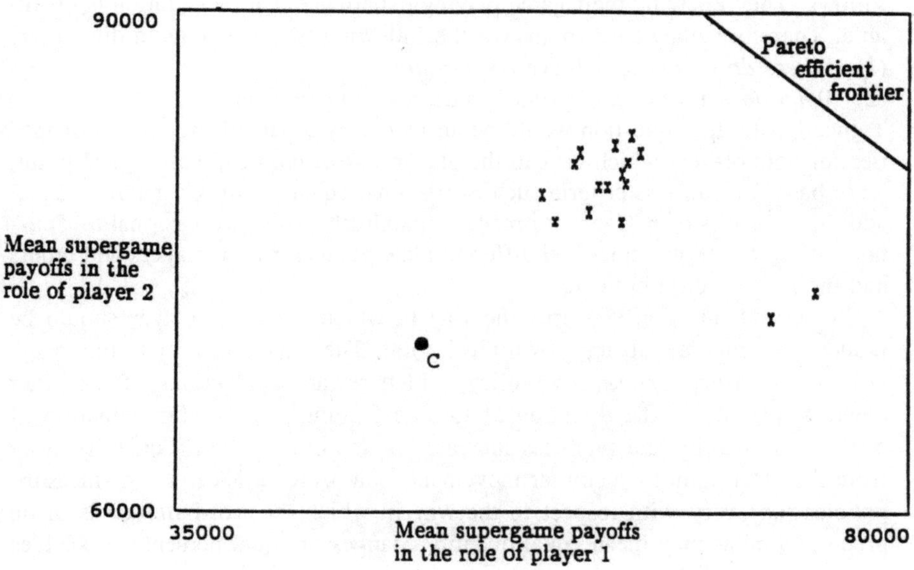

Figure 9.6. Result of the second tournament of the strategies

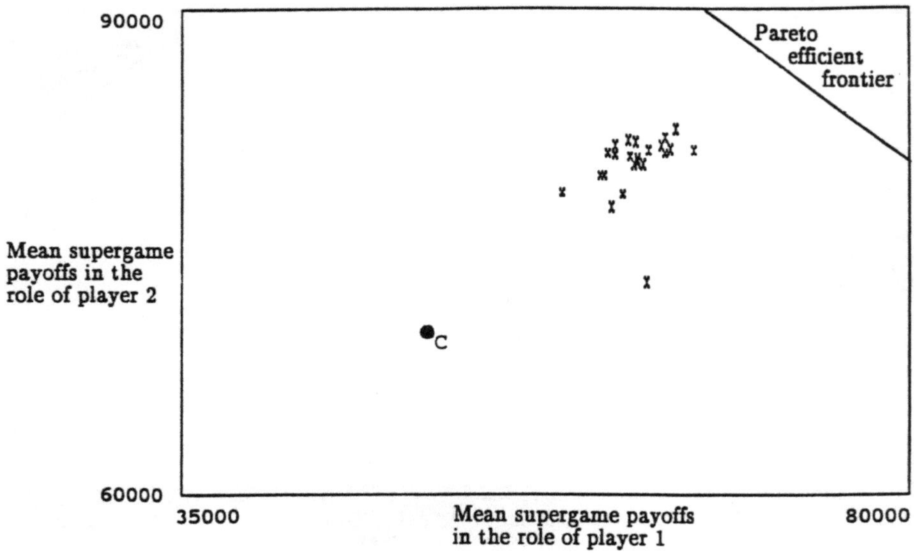

Figure 9.7. Result of the third tournament of the strategies

Typically, the participants approached the task of strategy construction as follows. They knew by their game playing experience that cooperation is profitable. Therefore, they tried to answer the following two questions in this order:
(1) *Where do I want to achieve cooperation?*
(2) *What do I have to do in order to achieve cooperation?*
Typically, the first question would be answered by a pair of production quantities for both players which we call the player's *ideal point*. Usually, ideal points were based on fairness criteria such as maximal equal profits or maximal equal additional profits over Cournot profits or maximal profits proportional to Cournot profits. Some strategies had different ideal points for both players and others had the same one on both sides.

The second question concerns the way in which the other player should be induced to cooperate at one's own ideal point. The typical answer to this question is a *measure-for-measure* policy, which rewards reductions of the other player's quantity in the direction of the ideal point by a similar reduction of one's own quantity and punishes increases of the other player's quantity away from the ideal point by a similar movement away from it. Measure-for-measure policies may vary with respect to the way in which the word *similar* is made precise. Similar may mean equal quantity changes or equal percentage changes in terms of the difference between Cournot quantity and ideal point quantity, or something more complicated, like equal change in profits.

A measure-for-measure policy can be considered to be a generalization of the *tit-for-tat* rule, which has turned out to be most successful in Axelrod's (1984) tournaments. However, in the case of the prisoner's dilemma there is no question about where cooperation should take place. There is only one answer to this question which respects the symmetry of the game. Moreover, there are only two actions which can be chosen. Therefore, the question does not arise what is a similar response to a deviation. Tit-for-tat is a special case of a measure-for-measure policy with a very restricted domain of application.

Typically, a strategy distinguished three phases of playing the game: an *initial phase*, a *main phase*, and an *end phase*. The initial phase served the purpose to signal cooperativeness by a fixed sequence of decreasing quantities. In the main phase a measure-for-measure policy was applied. In the end phase cooperation was abolished in favor of some kind of non-cooperative behavior. Of course, this is only a rough picture of typical behavior, which, however, is supported by an analysis of the final tournament strategies.

The evaluation of the final tournament strategies identified certain structural properties, referred to as *characteristics*. Each characteristic was present in the majority of strategies to which this characteristic is applicable. The list of these characteristics follows.

Typical characteristics of the strategies of the last tournament

General principles

1. No prediction
2. No random decision
3. Not only integer outputs

Initial phase

4. fixed outputs for the first two to four periods
5. last fixed output at least 8% below Cournot

Main phase (*Measure-for-measure-principle)*

6. decisions are guided by one or two ideal points
7. response with own ideal output if opponent's output below own ideal point
8. Cournot-output if opponent's output above Cournot
9. response to Cournot-output at most 5% below Cournot
10. reduction as response to reduction (between ideal point and Cournot)
11. increase as response to increase (between ideal point and Cournot)

End phase

12. end phase in the last two to four periods
13. Cournot output in all periods of end phase

The first characteristic, *no prediction*, means that no attempt is made to predict the opponent's quantity in the next period. The oligopoly theories in the literature, at least as far as they are known to me, all involve some prediction about the opponent's quantity of the next period. It is therefore surprising that only five of the 23 strategies of the last tournament make such predictions. This is connected to the fact that typically no attempt was made to optimize against the expected behavior of the other player.

The second characteristic has the meaning that the strategy involved no random elements. The third characteristic means that typically quantities were determined by smooth formula rather than by regions with fixed integer outputs based on case distinctions. Casuistic strategies of the latter type were also sometimes observed.

Characteristic 4 has the meaning that in the initial phase outputs do not depend on past history. Characteristic 5 means that the initial phase provides a sufficiently strong signal of cooperativeness. Of course, the reduction by at least 8% is to some degree arbitrary, but it has been chosen in such a way that the characteristic is shared by a majority of strategies with an initial phase.

The characteristics 6 to 11 for the main phase concern aspects of the measure-for-measure principle. The presence of one or two ideal points is covered by characteristic 6. Characteristics 7 and 8 restrict response quantities to the interval between Cournot quantity and ideal point quantity. Some times strategies signal cooperativeness by an output below Cournot production. Characteristic 9 restricts the percentage by which such signalling outputs remain below the Cournot quantity. The typical response to reductions and increases of the other player's quantity is covered by characteristic 10 and 11.

Characteristic 12 describes the presence of an end phase. Characteristic 13 means that the Cournot output is offered in all periods of the end phase. Some strategies behave differently in the end phase, e.g. they gradually increase outputs.

The distribution of the characteristics among the strategies is shown in the incidence matrix displayed in figure 9.8. The strategies are ranked according to success. Visual inspection of figure 9.8. shows that less successful strategies tend to have less of the characteristics. This suggests a connection between typicalness and success. However, it would be wrong to measure the typicalness of a strategy by the number of its characteristics, since the characteristics should be weighted according to their typicalness which in turn should depend on the typicalness of the strategies in which they occur. In order to justice to these considerations the typicalness of strategies and characteristics is measured by numbers between zero and one, called *typicities*, defined by the following principles.

1. The typicity of a strategy is the sum of the typicities of its characteristics
2. The typicity of a characteristic is proportional to the sum of the typicities of the strategies with it
3. The typicities of the characteristics sum up to 1
4. The typicities are Eigenvectors connected to the greatest Eigenvalue of the Eigenvalue problem resulting from 1., 2., and 3.

A definition of the typicities as a least-squares approximation and questions of existence and uniqueness are discussed in Kuon (1993).

CHARACTERISTICS	STRATEGIES																							Typicity
1	▬	▬	▬	▬	▬	▬		▬	▬	▬	▬		▬	▬	▬		▬	▬	▬		▬		▬	.0917
2	▬	▬	▬	▬	▬	▬	▬	▬	▬	▬	▬	▬	▬	▬	▬	▬	▬	▬	▬	▬		▬	▬	.1062
3	▬	▬	▬	▬	▬	▬	▬	▬	▬	▬	▬	▬	▬	▬	▬	▬	▬				▬	▬		.1005
4	▬	▬		▬			▬				▬	▬	▬	▬	▬		▬	▬			▬			.0609
5	▬	▬	▬	▬		▬	▬			▬	▬	▬		▬	▬	▬	▬							.0710
6	▬	▬	▬	▬	▬	▬	▬	▬	▬	▬				▬	▬	▬	▬	▬	▬		▬		▬	.0927
7	▬	▬	▬	▬	▬				▬						▬	▬					▬		▬	.0545
8	▬		▬		▬	▬		▬	▬			▬	▬	▬		▬			▬		▬			.0653
9	▬	▬	▬		▬	▬		▬	▬	▬		▬	▬	▬	▬	▬		▬			▬	▬		.0851
10	▬		▬		▬		▬	▬	▬	▬	▬	▬	▬			▬					▬	▬	▬	.0729
11	▬	▬	▬			▬	▬	▬	▬	▬		▬	▬			▬	▬					▬	▬	.0749
12	▬	▬	▬	▬		▬	▬				▬		▬			▬	▬	▬						.0591
13	▬		▬					▬	▬	▬	▬	▬		▬		▬	▬		▬		▬			.0652
Ranking of success	1	2	3	4	5	6	7	8	9	10	11	12	13	14	15	16	17	18	19	20	21	22	23	
Ranking of typicity	1	5	2	17	14	8	16	7	4	6	18	12	11	9	15	3	10	19	22	23	13	21	20	
Typicity	1.0000	.7966	.9391	.6366	.6689	.7465	.6382	.7545	.8090	.7602	.6275	.7020	.7166	.7386	.6626	.8474	.7222	.4957	.4211	.1062	.6888	.4396	.4929	

Figure 9.8. Typicity of characteristics and strategies

The Spearman rank correlation between the success and the typicity of a strategy is .619, which is significant at the 1% level (two-sided). This means that *the more typical the strategies are the more successful they tend to be.*

The typical approach to the strategic problem is fundamentally different from that of traditional oligopoly theories including applications of non-cooperative normative game theory. These theories assume that a player has quantitatively specified expectations about the other player's behavior and tries to optimize against them. Contrary to this, a typical strategy does not involve any quantitatively specified expectations and does not try to optimize. Of course, some vague qualitative expectations are connected to an ideal point and a measure-for-measure policy. The strategy designer expects that his strategy will be successful in achieving his cooperative goal. However, the cooperative goal is not determined by a procedure which maximizes the player's short run or long run expected profit, but rather as the result of the application of fairness criteria. Similarly, a measure-for-measure policy is not derived as a solution of an optimization problem.

Typically, a strategy designer takes an active approach to the strategic problem. It is seen as the task of the strategy to exert an active influence on the opponent's behavior. A more passive optimization approach would need quantitatively specified expectations, but the active approach completely avoids them. The strategy designer first fixes a cooperative goal and then looks for a way to induce the other player to conform to it. The usual answer to this inducement problem is a measure-for-measure policy which administers the right kind of rewards and punishments in order to guide the opponent's behavior towards one's own cooperative goal.

The structure of a typical strategy shows that experienced subjects approach the strategic problem in a rational way. However, the rationality underlying their behavior is not the full rationality of Bayesian decision theory and game theory, but rather a non-optimizing bounded rationality of goal formation and goal pursuit.

10. Conclusion

In this lecture I have tried to give an impression of the nature of descriptive theories applicable to problems of cooperation. The typical behavior of experimental subjects is by no means irrational. It is based on its own rationality, which is quite different from that of normative decision and game theory. Of course, up to now we have only a limited insight into the ways in which strategic problems are approached by human decision makers. However, it can be hoped that in the next decades our knowledge in this area will grow rapidly.

I hope that it has become clear that the few behavioral theories already available have a great potential to be fruitfully applied in economic modelling. However, it is necessary to develop many more such theories until finally a coherent picture of boundedly rational human strategic behavior will emerge. Of course, this will require much painstaking experimental and other empirical work.

Reality is still full of undiscovered regularities. We must discover these regularities.

References

Albers, W. and G. Albers (1983) *On the Prominence Structure of the Decimal System*, in: R.W. Scholz (ed.), Decision Making under Uncertainty, Elsevier (North Holland), 271-287.

Aumann, R.J. (1961) *The core of a cooperative game without side payments*, Trans. Amer. Math. Soc., 98, 539-552.

Aumann, R.J. and M. Maschler (1964) *The bargaining set for cooperative games*, in: M. Dresher, L.S. Shapley, and A.W. Tucker (eds.), Advances in game theory, Princeton University Press, Princeton, N.J., 443-476.

Axelrod, R. (1984) *The evolution of cooperation*, Basic Books, New York.

Cournot, A. (1838) *Recherches sur les principes mathématiques de la théorie des richesses*, Hachette, Paris.

Crößmann, H.J. and R. Tietz (1983) *Market behavior based on aspiration levels*, in: R.Tietz (ed.) Aspiration levels in bargaining and economic decision making, Lecture Notes in Economics and Mathematical Systems No. 213, Springer, Berlin, 170-185.

Fader, P.S. and J.R. Hauser (1988) *Implicit coalition in a generalized prisoner's dilemma*, Journal of Conflict Resolution, 32, 553-582.

Homans, G.C. (1961) *Social behavior: its elementary forms*, Hartcourt, Brace and World, New York.

Keser, C. (1992) *Experimental duopoly markets with demand inertia*, Springer Lecture Notes in Economics and Mathematical Systems No. 391, Springer, Berlin.

Kuon, B. (1993) *Measuring the typicalness of behavior*, Mathematical Social Sciences, 26, 35-49.

Kuon, B. (1994) *Two-Person Bargaining Experiments with Incomplete Informa tion*, Springer Lecture Notes in Economics and Mathematical Systems No. 412, Springer, Berlin.

Kuon, B., and G.R. Uhlich (1993) *The Negotiation Agreement Area: An Experi mental Analysis of Two-Person Characteristic Function Games*, Group Decision and Negotiation, 2, 323-345.

Maschler, M. (1978) *Playing an n-Person Game - An Experiment*, in: H. Sauermann (ed.), Coalition Forming Behavior, Contributions to Experimental Economics, Vol. VIII, J.C.B. Mohr (Paul Siebeck), Tübingen, 231-328.

Medlin, S.M. (1976) *Effects of grand coalition payoffs on coalition formation in three-person games*, Behavioral Science, 21, 48-61.

Murnighan, J.K. and A.E. Roth (1977) *The effects of communication and infor mation availability in an experimental study of a three-person game*, Management Science, 23, 1136-1348.

Ostmann, A. (1988) *Limits of rational behavior in cooperatively played normal form games*, in: R. Tietz, W. Albers and R. Selten (eds.), Bounded Rational Behavior in Experimental Games and Markets, Lecture Notes in Economics and Mathematical Systems, No. 314, Springer, Berlin, 317-332.

Rapoport, Amn. and J.P. Kahan (1976) *When three is not always two against one: coalitions in experimental three-person cooperative games*, Journal of Experimental Social Psychology, 12, 253-273.

Scarf, H. (1967) *The core of an n-person game*, Econometrica, 35, 50-69.

Schmeidler, D. (1969) *The nucleolus of a characteristic function game*, SIAM, Journal of Applied Mathematics, 17, 1163-1170.

Shapley, L. (1953) *A value for n-person games*, in: H.W. Kuhn and H.W. Tucker (eds.), Contributions to the theory of games, Vol. II, Ann. Math. Stud., 28, Princeton University Press, Princeton, N.J., 307-317.

Selten, R. (1967) *Die Strategiemethode zur Erforschung des eingeschränkt rationalen Verhaltens im Rahmen eines Oligopolexperiments*, in: H. Sauermann (ed.), Beiträge zur experimentellen Wirtschaftsforschung, J.C.B. Mohr, Tübingen, 136-168.

Selten, R. (1987) *Equity and Coalition Bargaining in Experimental Three-Person Games*, in: A.E. Roth (ed.) Laboratory Experimentation in Economics, Cambridge University Press, 42-98.

Selten, R. (1988) *The Equity Principle in Economic Behavior*, in: Models of Strategic Rationality, Kluwer, 269-281.

Selten, R. (1991) *Properties of a measure of predictive success*, Mathematical Social Sciences, 21, 153-167.

Selten, R. and S. Krischker (1983) *Comparison of two theories for characteristic function experiments*, in: R. Tietz (ed.), Aspiration levels in Bargaining and Economic Decision Making, Springer, Berlin, 259-264.

Selten, R., M. Mitzkewitz, and G.R. Uhlich (1988) *Duopoly Strategies Pro grammed by Experienced Players*, Discussion Paper No. B-106, Sonderforschungsbereich 303, University of Bonn.

Scholz, R.W. (1980) *Dyadische Verhandlungen*, Verlag Anton Hain, Königstein.

Scholz, R.W., Fleischer, A., and A. Bentrup (1983) *Aspiration forming and predictions based on aspiration levels compared between professional and non-professional bargainers*, in: R. Tietz (ed.) Aspiration levels in bargaining and economic decision making, Lecture Notes in Economics and Mathematical Systems No. 213, Springer, Berlin, 104-121.

Tietz, R. (1973), *Ein anspruchsanpassungsorientiertes Wachstums- und Konjunkturmodell (KRESKO)*, in: H. Sauermann (ed.), Contributions to experimental economics, Vol IV, J.C.B. Mohr (Paul Siebeck), Tübingen.

Tietz, R. (1975) *An Experimental Analysis of Wage Bargaining Behavior*, Zeitschrift für die Gesamte Staatswissenschaft, 131, 1, 44-91.

Tietz, R. (1976) *Anspruchsausgleich in experimentellen Zwei-Personen-Verhandlungen mit verbaler Kommunikation*, Zeitschrift für Sozialpsychologie, Beiheft 2, Entscheidungsprozesse in Gruppen, 123-141.

Tietz, R. and O.J. Bartos (1983) *Balancing of aspiration levels as fairness principle in negotiations*, in: R. Tietz (ed.) Aspiration levels in bargaining and economic decision making, Lecture Notes in Economics and Mathematical Systems No. 213, Springer, Berlin, 52-66.

Tietz, R., Daus, W., Lautsch, J. and P. Lotz (1988) *Semi-normative properties of bounded rational bargaining theories*, in: Tietz, R., Albers, W. and R. Selten (eds.) Bounded rational behavior in experimental games and markets, Lecture Notes in Economics and Mathematical Systems No. 314, Springer Berlin, 142-159.

Tietz, R. and H.J. Weber (1972), *On the nature of the bargaining process in the KRESKO-game*, in: H. Sauermann (ed.), Contributions to experimental economics, Vol III, J.C.B. Mohr (Paul Siebeck), Tübingen, 305-334.

Tietz, R. and H.J. Weber (1978), *Decision behavior in multi-variable negotia tions*, in: H. Sauermann (ed.), Bargaining behavior, Contributions to experimental economics, Vol VII, J.C.B. Mohr (Paul Siebeck), Tübingen, 60-87.

Uhlich, G.R. (1989) *Descriptive Theories of Bargaining*, Lecture Notes in Economics and Mathematical Systems No. 341, Springer, Berlin.

NATO ASI Series F

Including Special Programmes on Sensory Systems for Robotic Control (ROB) and on Advanced Educational Technology (AET)

Vol. 1: Issues in Acoustic Signal - Image Processing and Recognition. Edited by C. H. Chen. VIII, 333 pages. 1983. *(out of print)*

Vol. 2: Image Sequence Processing and Dynamic Scene Analysis. Edited by T. S. Huang. IX, 749 pages. 1983.

Vol. 3: Electronic Systems Effectiveness and Life Cycle Costing. Edited by J. K. Skwirzynski. XVII, 732 pages. 1983. *(out of print)*

Vol. 4: Pictorial Data Analysis. Edited by R. M. Haralick. VIII, 468 pages. 1983.

Vol. 5: International Calibration Study of Traffic Conflict Techniques. Edited by E. Asmussen. VII, 229 pages. 1984.

Vol. 6: Information Technology and the Computer Network. Edited by K. G. Beauchamp. VIII, 271 pages. 1984.

Vol. 7: High-Speed Computation. Edited by J. S. Kowalik. IX, 441 pages. 1984.

Vol. 8: Program Transformation and Programming Environments. Report on a Workshop directed by F. L. Bauer and H. Remus. Edited by P. Pepper. XIV, 378 pages. 1984.

Vol. 9: Computer Aided Analysis and Optimization of Mechanical System Dynamics. Edited by E. J. Haug. XXII, 700 pages. 1984. *(out of print)*

Vol. 10: Simulation and Model-Based Methodologies: An Integrative View. Edited by T. I. Ören, B. P. Zeigler, M. S. Elzas. XIII, 651 pages. 1984.

Vol. 11: Robotics and Artificial Intelligence. Edited by M. Brady, L. A. Gerhardt, H. F. Davidson. XVII, 693 pages. 1984.

Vol. 12: Combinatorial Algorithms on Words. Edited by A. Apostolico, Z. Galil. VIII, 361 pages. 1985.

Vol. 13: Logics and Models of Concurrent Systems. Edited by K. R. Apt. VIII, 498 pages. 1985.

Vol. 14: Control Flow and Data Flow: Concepts of Distributed Programming. Edited by M. Broy. VIII, 525 pages. 1985. *Reprinted as Springer Study Edition 1986.*

Vol. 15: Computational Mathematical Programming. Edited by K. Schittkowski. VIII, 451 pages. 1985.

Vol. 16: New Systems and Architectures for Automatic Speech Recognition and Synthesis. Edited by R. De Mori, C.Y. Suen. XIII, 630 pages. 1985. *(out of print)*

Vol. 17: Fundamental Algorithms for Computer Graphics. Edited by R. A. Earnshaw. XVI, 1042 pages. 1985. *Reprinted as Springer Study Edition 1991.*

Vol. 18: Computer Architectures for Spatially Distributed Data. Edited by H. Freeman and G. G. Pieroni. VIII, 391 pages. 1985. *(out of print)*

Vol. 19: Pictorial Information Systems in Medicine. Edited by K. H. Höhne. XII, 525 pages. 1986.

Vol. 20: Disordered Systems and Biological Organization. Edited by E. Bienenstock, F. Fogelman Soulié, G. Weisbuch. XXI, 405 pages.1986

Vol. 21: Intelligent Decision Support in Process Environments. Edited by E. Hollnagel, G. Mancini, D. D. Woods. XV, 524 pages. 1986.

Vol. 22: Software System Design Methods. The Challenge of Advanced Computing Technology. Edited by J. K. Skwirzynski. XIII, 747 pages. 1986. *(out of print)*

Vol. 23: Designing Computer-Based Learning Materials. Edited by H. Weinstock and A. Bork. IX, 285 pages. 1986. *(out of print)*

Vol. 24: Database Machines. Modern Trends and Applications. Edited by A. K. Sood and A. H. Qureshi. VIII, 570 pages. 1986.

NATO ASI Series F

Including Special Programmes on Sensory Systems for Robotic Control (ROB) and on Advanced Educational Technology (AET)

Vol. 25: Pyramidal Systems for Computer Vision. Edited by V. Cantoni and S. Levialdi. VIII, 392 pages. 1986. *(ROB)*

Vol. 26: Modelling and Analysis in Arms Control. Edited by R. Avenhaus, R. K. Huber and J. D. Kettelle. VIII, 488 pages. 1986. *(out of print)*

Vol. 27: Computer Aided Optimal Design: Structural and Mechanical Systems. Edited by C. A. Mota Soares. XIII, 1029 pages. 1987.

Vol. 28: Distributed Operating Systems. Theory und Practice. Edited by Y. Paker, J.-P. Banatre and M. Bozyiğit. X, 379 pages. 1987.

Vol. 29: Languages for Sensor-Based Control in Robotics. Edited by U. Rembold and K. Hörmann. IX, 625 pages. 1987. *(ROB)*

Vol. 30: Pattern Recognition Theory and Applications. Edited by P. A. Devijver and J. Kittler. XI, 543 pages. 1987.

Vol. 31: Decision Support Systems: Theory and Application. Edited by C. W. Holsapple and A. B. Whinston. X, 500 pages. 1987.

Vol. 32: Information Systems: Failure Analysis. Edited by J. A. Wise and A. Debons. XV, 338 pages. 1987.

Vol. 33: Machine Intelligence and Knowledge Engineering for Robotic Applications. Edited by A. K. C. Wong and A. Pugh. XIV, 486 pages. 1987. *(ROB)*

Vol. 34: Modelling, Robustness and Sensitivity Reduction in Control Systems. Edited by R.F. Curtain. IX, 492 pages. 1987.

Vol. 35: Expert Judgment and Expert Systems. Edited by J. L. Mumpower, L. D. Phillips, O. Renn and V. R. R. Uppuluri. VIII, 361 pages. 1987.

Vol. 36: Logic of Programming and Calculi of Discrete Design. Edited by M. Broy. VII, 415 pages. 1987.

Vol. 37: Dynamics of Infinite Dimensional Systems. Edited by S.-N. Chow and J. K. Hale. IX. 514 pages. 1987.

Vol. 38: Flow Control of Congested Networks. Edited by A. R. Odoni, L. Bianco and G. Szegö. XII, 355 pages. 1987.

Vol. 39: Mathematics and Computer Science in Medical Imaging. Edited by M. A. Viergever and A. Todd-Pokropek. VIII, 546 pages. 1988.

Vol. 40: Theoretical Foundations of Computer Graphics and CAD. Edited by R. A. Earnshaw. XX, 1246 pages. 1988. *(out of print)*

Vol. 41: Neural Computers. Edited by R. Eckmiller and Ch. v. d. Malsburg. XIII, 566 pages. 1988. *Reprinted as Springer Study Edition 1989, 1990.*

Vol. 42: Real-Time Object Measurement and Classification. Edited by A. K. Jain. VIII, 407 pages. 1988. *(ROB)*

Vol. 43: Sensors and Sensory Systems for Advanced Robots. Edited by P. Dario. XI, 597 pages. 1988. *(ROB)*

Vol. 44: Signal Processing and Pattern Recognition in Nondestructive Evaluation of Materials. Edited by C. H. Chen. VIII, 344 pages. 1988. *(ROB)*

Vol. 45: Syntactic and Structural Pattern Recognition. Edited by G. Ferraté, T. Pavlidis, A. Sanfeliu and H. Bunke. XVI, 467 pages. 1988. *(ROB)*

NATO ASI Series F

Including Special Programmes on Sensory Systems for Robotic Control (ROB) and on Advanced Educational Technology (AET)

Vol. 46: Recent Advances in Speech Understanding and Dialog Systems. Edited by H. Niemann, M. Lang and G. Sagerer. X, 521 pages. 1988.

Vol. 47: Advanced Computing Concepts and Techniques in Control Engineering. Edited by M. J. Denham and A. J. Laub. XI, 518 pages. 1988. *(out of print)*

Vol. 48: Mathematical Models for Decision Support. Edited by G. Mitra. IX, 762 pages. 1988.

Vol. 49: Computer Integrated Manufacturing. Edited by I. B. Turksen. VIII, 568 pages. 1988.

Vol. 50: CAD Based Programming for Sensory Robots. Edited by B. Ravani. IX, 565 pages. 1988. *(ROB)*

Vol. 51: Algorithms and Model Formulations in Mathematical Programming. Edited by S. W. Wallace. IX, 190 pages. 1989.

Vol. 52: Sensor Devices and Systems for Robotics. Edited by A. Casals. IX, 362 pages. 1989. *(ROB)*

Vol. 53: Advanced Information Technologies for Industrial Material Flow Systems. Edited by S. Y. Nof and C. L. Moodie. IX, 710 pages. 1989.

Vol. 54: A Reappraisal of the Efficiency of Financial Markets. Edited by R. M. C. Guimarães, B. G. Kingsman and S. J. Taylor. X, 804 pages. 1989.

Vol. 55: Constructive Methods in Computing Science. Edited by M. Broy. VII, 478 pages. 1989.

Vol. 56: Multiple Criteria Decision Making and Risk Analysis Using Microcomputers. Edited by B. Karpak and S. Zionts. VII, 399 pages. 1989.

Vol. 57: Kinematics and Dynamic Issues in Sensor Based Control. Edited by G. E. Taylor. XI, 456 pages. 1990. *(ROB)*

Vol. 58: Highly Redundant Sensing in Robotic Systems. Edited by J. T. Tou and J. G. Balchen. X, 322 pages. 1990. *(ROB)*

Vol. 59: Superconducting Electronics. Edited by H. Weinstock and M. Nisenoff. X, 441 pages. 1989.

Vol. 60: 3D Imaging in Medicine. Algorithms, Systems, Applications. Edited by K. H. Höhne, H. Fuchs and S. M. Pizer. IX, 460 pages. 1990. *(out of print)*

Vol. 61: Knowledge, Data and Computer-Assisted Decisions. Edited by M. Schader and W. Gaul. VIII, 421 pages. 1990.

Vol. 62: Supercomputing. Edited by J. S. Kowalik. X, 425 pages. 1990.

Vol. 63: Traditional and Non-Traditional Robotic Sensors. Edited by T. C. Henderson. VIII, 468 pages. 1990. *(ROB)*

Vol. 64: Sensory Robotics for the Handling of Limp Materials. Edited by P. M. Taylor. IX, 343 pages. 1990. *(ROB)*

Vol. 65: Mapping and Spatial Modelling for Navigation. Edited by L. F. Pau. VIII, 357 pages. 1990. *(ROB)*

Vol. 66: Sensor-Based Robots: Algorithms and Architectures. Edited by C. S. G. Lee. X, 285 pages. 1991. *(ROB)*

Vol. 67: Designing Hypermedia for Learning. Edited by D. H. Jonassen and H. Mandl. XXV, 457 pages. 1990. *(AET)*

Vol. 68: Neurocomputing. Algorithms, Architectures and Applications. Edited by F. Fogelman Soulié and J. Hérault. XI, 455 pages. 1990.

Vol. 69: Real-Time Integration Methods for Mechanical System Simulation. Edited by E. J. Haug and R. C. Deyo. VIII, 352 pages. 1991.

NATO ASI Series F

Including Special Programmes on Sensory Systems for Robotic Control (ROB) and on Advanced Educational Technology (AET)

Vol. 70: Numerical Linear Algebra, Digital Signal Processing and Parallel Algorithms. Edited by G. H. Golub and P. Van Dooren. XIII, 729 pages. 1991.

Vol. 71: Expert Systems and Robotics. Edited by T. Jordanides and B.Torby. XII, 744 pages. 1991.

Vol. 72: High-Capacity Local and Metropolitan Area Networks. Architecture and Performance Issues. Edited by G. Pujolle. X, 536 pages. 1991.

Vol. 73: Automation and Systems Issues in Air Traffic Control. Edited by J. A. Wise, V. D. Hopkin and M. L. Smith. XIX, 594 pages. 1991.

Vol. 74: Picture Archiving and Communication Systems (PACS) in Medicine. Edited by H. K. Huang, O. Ratib, A. R. Bakker and G. Witte. XI, 438 pages. 1991.

Vol. 75: Speech Recognition and Understanding. Recent Advances, Trends and Applications. Edited by P. Laface and Renato De Mori. XI, 559 pages. 1991.

Vol. 76: Multimedia Interface Design in Education. Edited by A. D. N. Edwards and S. Holland. XIV, 216 pages. 1992. *(AET)*

Vol. 77: Computer Algorithms for Solving Linear Algebraic Equations. The State of the Art. Edited by E. Spedicato. VIII, 352 pages. 1991.

Vol. 78: Integrating Advanced Technology into Technology Education. Edited by M. Hacker, A. Gordon and M. de Vries. VIII, 185 pages. 1991. *(AET)*

Vol. 79: Logic, Algebra, and Computation. Edited by F. L. Bauer. VII, 485 pages. 1991.

Vol. 80: Intelligent Tutoring Systems for Foreign Language Learning. Edited by M. L. Swartz and M. Yazdani. IX, 347 pages. 1992. *(AET)*

Vol. 81: Cognitive Tools for Learning. Edited by P. A. M. Kommers, D. H. Jonassen, and J. T. Mayes. X, 278 pages. 1992. *(AET)*

Vol. 82: Combinatorial Optimization. New Frontiers in Theory and Practice. Edited by M. Akgül, H. W. Hamacher, and S. Tüfekçi. XI, 334 pages. 1992.

Vol. 83: Active Perception and Robot Vision. Edited by A. K. Sood and H. Wechsler. IX, 756 pages. 1992.

Vol. 84: Computer-Based Learning Environments and Problem Solving. Edited by E. De Corte, M. C. Linn, H. Mandl, and L. Verschaffel. XVI, 488 pages. 1992. *(AET)*

Vol. 85: Adaptive Learning Environments. Foundations and Frontiers. Edited by M. Jones and P. H. Winne. VIII, 408 pages. 1992. *(AET)*

Vol. 86: Intelligent Learning Environments and Knowledge Acquisition in Physics. Edited by A. Tiberghien and H. Mandl. VIII, 285 pages. 1992. *(AET)*

Vol. 87: Cognitive Modelling and Interactive Environments. With demo diskettes (Apple and IBM compatible). Edited by F. L. Engel, D. G. Bouwhuis, T. Bösser, and G. d'Ydewalle. IX, 311 pages. 1992. *(AET)*

Vol. 88: Programming and Mathematical Method. Edited by M. Broy. VIII, 428 pages. 1992.

Vol. 89: Mathematical Problem Solving and New Information Technologies. Edited by J. P. Ponte, J. F. Matos, J. M. Matos, and D. Fernandes. XV, 346 pages. 1992. *(AET)*

Vol. 90: Collaborative Learning Through Computer Conferencing. Edited by A. R. Kaye. X, 260 pages. 1992. *(AET)*

Vol. 91: New Directions for Intelligent Tutoring Systems. Edited by E. Costa. X, 296 pages. 1992. *(AET)*

NATO ASI Series F

Including Special Programmes on Sensory Systems for Robotic Control (ROB) and on Advanced Educational Technology (AET)

Vol. 92: Hypermedia Courseware: Structures of Communication and Intelligent Help. Edited by A. Oliveira. X, 241 pages. 1992. *(AET)*

Vol. 93: Interactive Multimedia Learning Environments. Human Factors and Technical Considerations on Design Issues. Edited by M. Giardina. VIII, 254 pages. 1992. *(AET)*

Vol. 94: Logic and Algebra of Specification. Edited by F. L. Bauer, W. Brauer, and H. Schwichtenberg. VII, 442 pages. 1993.

Vol. 95: Comprehensive Systems Design: A New Educational Technology. Edited by C. M. Reigeluth, B. H. Banathy, and J. R. Olson. IX, 437 pages. 1993. *(AET)*

Vol. 96: New Directions in Educational Technology. Edited by E. Scanlon and T. O'Shea. VIII, 251 pages. 1992. *(AET)*

Vol. 97: Advanced Models of Cognition for Medical Training and Practice. Edited by D. A. Evans and V. L. Patel. XI, 372 pages. 1992. *(AET)*

Vol. 98: Medical Images: Formation, Handling and Evaluation. Edited by A. E. Todd-Pokropek and M. A. Viergever. IX, 700 pages. 1992.

Vol. 99: Multisensor Fusion for Computer Vision. Edited by J. K. Aggarwal. XI, 456 pages. 1993. *(ROB)*

Vol. 100: Communication from an Artificial Intelligence Perspective. Theoretical and Applied Issues. Edited by A. Ortony, J. Slack and O. Stock. XII, 260 pages. 1992.

Vol. 101: Recent Developments in Decision Support Systems. Edited by C. W. Holsapple and A. B. Whinston. XI, 618 pages. 1993.

Vol. 102: Robots and Biological Systems: Towards a New Bionics? Edited by P. Dario, G. Sandini and P. Aebischer. XII, 786 pages. 1993.

Vol. 103: Parallel Computing on Distributed Memory Multiprocessors. Edited by F. Özgüner and F. Erçal. VIII, 332 pages. 1993.

Vol. 104: Instructional Models in Computer-Based Learning Environments. Edited by S. Dijkstra, H. P. M. Krammer and J. J. G. van Merriënboer. X, 510 pages. 1993. *(AET)*

Vol. 105: Designing Environments for Constructive Learning. Edited by T. M. Duffy, J. Lowyck and D. H. Jonassen. VIII, 374 pages. 1993. *(AET)*

Vol. 106: Software for Parallel Computation. Edited by J. S. Kowalik and L. Grandinetti. IX, 363 pages. 1993.

Vol. 107: Advanced Educational Technologies for Mathematics and Science. Edited by D. L. Ferguson. XII, 749 pages. 1993. *(AET)*

Vol. 108: Concurrent Engineering: Tools and Technologies for Mechanical System Design. Edited by E. J. Haug. XIII, 998 pages. 1993.

Vol. 109: Advanced Educational Technology in Technology Education. Edited by A. Gordon, M. Hacker and M. de Vries. VIII, 253 pages. 1993. *(AET)*

Vol. 110: Verification and Validation of Complex Systems: Human Factors Issues. Edited by J. A. Wise, V. D. Hopkin and P. Stager. XIII, 704 pages. 1993.

Vol. 111: Cognitive Models and Intelligent Environments for Learning Programming. Edited by E. Lemut, B. du Boulay and G. Dettori. VIII, 305 pages. 1993. *(AET)*

Vol. 112: Item Banking: Interactive Testing and Self-Assessment. Edited by D. A. Leclercq and J. E. Bruno. VIII, 261 pages. 1993. *(AET)*

Vol. 113: Interactive Learning Technology for the Deaf. Edited by B. A. G. Elsendoorn and F. Coninx. XIII, 285 pages. 1993. *(AET)*

NATO ASI Series F

Including Special Programmes on Sensory Systems for Robotic Control (ROB) and on Advanced Educational Technology (AET)

Vol. 114: Intelligent Systems: Safety, Reliability and Maintainability Issues. Edited by O. Kaynak, G. Honderd and E. Grant. XI, 340 pages. 1993.

Vol. 115: Learning Electricity and Electronics with Advanced Educational Technology. Edited by M. Caillot. VII, 329 pages. 1993. *(AET)*

Vol. 116: Control Technology in Elementary Education. Edited by B. Denis. IX, 311 pages. 1993. *(AET)*

Vol. 117: Intelligent Learning Environments: The Case of Geometry. Edited by J.-M. Laborde. VIII, 267 pages. 1996. *(AET)*

Vol. 118: Program Design Calculi. Edited by M. Broy. VIII, 409 pages. 1993.

Vol. 119: Automating Instructional Design, Development, and Delivery. Edited by. R. D. Tennyson. VIII, 266 pages. 1994. *(AET)*

Vol. 120: Reliability and Safety Assessment of Dynamic Process Systems. Edited by T. Aldemir, N. O. Siu, A. Mosleh, P. C. Cacciabue and B. G. Göktepe. X, 242 pages. 1994.

Vol. 121: Learning from Computers: Mathematics Education and Technology. Edited by C. Keitel and K. Ruthven. XIII, 332 pages. 1993. *(AET)*

Vol. 122: Simulation-Based Experiential Learning. Edited by D. M. Towne, T. de Jong and H. Spada. XIV, 274 pages. 1993. *(AET)*

Vol. 123: User-Centred Requirements for Software Engineering Environments. Edited by D. J. Gilmore, R. L. Winder and F. Détienne. VII, 377 pages. 1994.

Vol. 124: Fundamentals in Handwriting Recognition. Edited by S. Impedovo. IX, 496 pages. 1994.

Vol. 125: Student Modelling: The Key to Individualized Knowledge-Based Instruction. Edited by J. E. Greer and G. I. McCalla. X, 383 pages. 1994. *(AET)*

Vol. 126: Shape in Picture. Mathematical Description of Shape in Grey-level Images. Edited by Y.-L. O, A. Toet, D. Foster, H. J. A. M. Heijmans and P. Meer. XI, 676 pages. 1994.

Vol. 127: Real Time Computing. Edited by W. A. Halang and A. D. Stoyenko. XXII, 762 pages. 1994.

Vol. 128: Computer Supported Collaborative Learning. Edited by C. O'Malley. X, 303 pages. 1994. *(AET)*

Vol. 129: Human-Machine Communication for Educational Systems Design. Edited by M. D. Brouwer-Janse and T. L. Harrington. X, 342 pages. 1994. *(AET)*

Vol. 130: Advances in Object-Oriented Database Systems. Edited by A. Dogac, M. T. Özsu, A. Biliris and T. Sellis. XI, 515 pages. 1994.

Vol. 131: Constraint Programming. Edited by B. Mayoh, E. Tyugu and J. Penjam. VII, 452 pages. 1994.

Vol. 132: Mathematical Modelling Courses for Engineering Education. Edited by Y. Ersoy and A. O. Moscardini. X, 246 pages. 1994. *(AET)*

Vol. 133: Collaborative Dialogue Technologies in Distance Learning. Edited by M. F. Verdejo and S. A. Cerri. XIV, 296 pages. 1994. *(AET)*

Vol. 134: Computer Integrated Production Systems and Organizations. The Human-Centred Approach. Edited by F. Schmid, S. Evans, A. W. S. Ainger and R. J. Grieve. X, 347 pages. 1994.

Vol. 135: Technology Education in School and Industry. Emerging Didactics for Human Resource Development. Edited by D. Blandow and M. J. Dyrenfurth. XI, 367 pages. 1994. *(AET)*

Vol. 136: From Statistics to Neural Networks. Theory and Pattern Recognition Applications. Edited by V. Cherkassky, J. H. Friedman and H. Wechsler. XII, 394 pages. 1994.

NATO ASI Series F

Including Special Programmes on Sensory Systems for Robotic Control (ROB) and on Advanced Educational Technology (AET)

Vol. 137: Technology-Based Learning Environments. Psychological and Educational Foundations. Edited by S. Vosniadou, E. De Corte and H. Mandl. X, 302 pages. 1994. *(AET)*

Vol. 138: Exploiting Mental Imagery with Computers in Mathematics Education. Edited by R. Sutherland and J. Mason. VIII, 326 pages. 1995. *(AET)*

Vol. 139: Proof and Computation. Edited by H. Schwichtenberg. VII, 470 pages. 1995.

Vol. 140: Automating Instructional Design: Computer-Based Development and Delivery Tools. Edited by R. D. Tennyson and A. E. Barron. IX, 618 pages. 1995. *(AET)*

Vol. 141: Organizational Learning and Technological Change. Edited by C. Zucchermaglio, S. Bagnara and S. U. Stucky. X, 368 pages. 1995. *(AET)*

Vol. 142: Dialogue and Instruction. Modeling Interaction in Intelligent Tutoring Systems. Edited by R.-J. Beun, M. Baker and M. Reiner. IX, 368 pages. 1995. *(AET)*

Vol. 143: Batch Processing Systems Engineering. Fundamentals of Chemical Engineering. Edited by G. V. Reklaitis, A. K. Sunol, D. W. T. Rippin, and Ö. Hortaçsu. XIV, 868 pages. 1996.

Vol. 144: The Biology and Technology of Intelligent Autonomous Agents. Edited by Luc Steels. VIII, 517 pages. 1995.

Vol. 145: Advanced Educational Technology: Research Issues and Future Potential. Edited by T. T. Liao. VIII, 219 pages. 1996. *(AET)*

Vol. 146: Computers and Exploratory Learning. Edited by A. A. diSessa, C. Hoyles and R. Noss. VIII, 482 pages. 1995. *(AET)*

Vol. 147: Speech Recognition and Coding. New Advances and Trends. Edited by A. J. Rubio Ayuso and J. M. López Soler. XI, 505 pages. 1995.

Vol. 148: Knowledge Acquisition, Organization, and Use in Biology. Edited by K. M. Fisher and M. R. Kibby. X, 246 pages. 1996. *(AET)*

Vol. 149: Emergent Computing Methods in Engineering Design. Applications of Genetic Algorithms and Neural Networks. Edited by D.E. Grierson and P. Hajela. VIII, 350 pages. 1996.

Vol. 150: Speechreading by Humans and Machines. Edited by D. G. Stork and M. E. Hennecke. XV, 686 pages. 1996.

Vol. 151: Computational and Conversational Discourse. Burning Issues – An Interdisciplinary Account. Edited by E. H. Hovy and D. R. Scott. XII, 202 pages. 1996.

Vol. 152: Deductive Program Design. Edited by M. Broy. IX, 467 pages. 1996.

Vol. 153: Identification, Adaptation, Learning. Edited by S. Bittanti and G. Picci. XIV, 553 pages. 1996.

Vol. 154: Reliability and Maintenance of Complex Systems. Edited by S. Özekici. XI, 589 pages. 1996.

Vol. 155: Cooperation: Game-Theoretic Approaches. Edited by S. Hart and A. Mas-Colell. VIII, 328 pages. 1997.

Vol. 156: Microcomputer-Based Labs: Educational Research and Standards. Edited by R.F. Tinker. XIV, 398 pages. 1996.

MIX
Papier aus verantwortungsvollen Quellen
Paper from responsible sources
FSC® C105338

If you have any concerns about our products,
you can contact us on
ProductSafety@springernature.com

In case Publisher is established outside the EU,
the EU authorized representative is:
Springer Nature Customer Service Center GmbH
Europaplatz 3, 69115 Heidelberg, Germany

Printed by Libri Plureos GmbH
in Hamburg, Germany